AF522076

MOTION OF CYLINDERS

By

A.K. Sharma

DISCOVERY PUBLISHING HOUSE PVT. LTD.
NEW DELHI-110 002

First Published-2008

ISBN 978-81-8356-344-4

Published by:

DISCOVERY PUBLISHING HOUSE PVT. LTD.
4831/24, Ansari Road, Prahlad Street,
Darya Ganj, New Delhi-110002 (India)
Phone: 23279245 • Fax: 91-11-23253475
E-mail: dphbooks@rediffmail.com
dphtemp@indiatimes.com
Website: www.discoverypublishinghouse.com

Printed at:
Arora Enterprises
Laxmi Nagar, Delhi–110 092

Preface

This book "Motion of Cylinders" has been design to cover the syllabus of Mathematics required for the B.Sc. (Hon.), M.Sc. and Engineering students of the Indian Universities. The subject of Motion of Cylinders deals with the laws Governing the pressure and flow of fluids and the application of these laws to engineering practice. The term fluid includes liquid and gases. The term fluid is applied to all the substance which offer no resistance to change of shape. The subject matter has been arranged as to provide a clear and integrated approach to the subject-care has been taken to make the treatment of the subject simple and accessible to the average students.

Suggestion for the improvement of the book will always be most welcome.

Author

CONTENTS

1

Waves

INTRODUCTION

The continuous transference of a particular state or form from one part of a medium to another is known a wave. This does not mean the transference of the medium itself from one place to another but simply the propagation trough it of a particular form or state. We observe that when a stone is thrown into a well, some disturbance occurs which travels radially over the surface of water; this disturbance is known as *water waves*. Other examples of the wave motion in liquids are high and low tides of the sea, standing waves in lakes etc. The concept of wave motion is of great importance in physical investigations. It constitutes one of the principal modes of transmission of energy. The energy received from the sun is transmitted by waves in Ether, the energy of sound by airways etc.

WAVE MOTION

Wave motion is due to the action of gravity which acts in the direction of restoring the undisturbed state of rest. Wave motion of a liquid acted upon by gravity and having a free surface is a motion in which the elevation of the free surface above a fixed horizontal plane varies.

Consider an arbitrary disturbance y which moves along the X-axis with velocity c, so that it is a function of the variables x and t, let $f(x, t)$. At $t = 0$, the curve $y = f(x)$ is generally known a *wave profile*. If the disturbance travels without any change of shape, the wave profile will move a distance ct in the positive direction of X- axis. Thus the equation to the wave profile is given by $y = f(x - ct)$. If we increase t by t' and x by ct', the profile remains unaltered *i.e.*

$$y = f\{x + ct' - c(t + t')\} = f(x - ct).$$

This shows that the profile $y = f(x)$ moves with velocity c in the positive direction of X-axis.

MATHEMATICAL REPRESENTATION OF WAVE MOTION

A simple harmonic progressive wave represented by a sine curve moving with definite velocity in the direction of its length is of the form

$$y = a \sin (mx - nt) = a \sin m \{x - (n/m)\, t\},$$

where a, m and n are constant. The profile $y = a \sin mx$ at $t = 0$ moves with velocity c (= n/m) along the positive direction of the X-axis, c is called the velocity of propagation of the wave.

The maximum value of the disturbance y, viz., a is called the amplitude of the wave. The points P and P′ of maximum elevation are called the *crests* of the wave and the points Q, Q′ ... of maximum depression are known as troughs. The distance between two successive crests is called the *wave length* and is denoted by λ *i.e.*, $\lambda = 2\pi/m$.

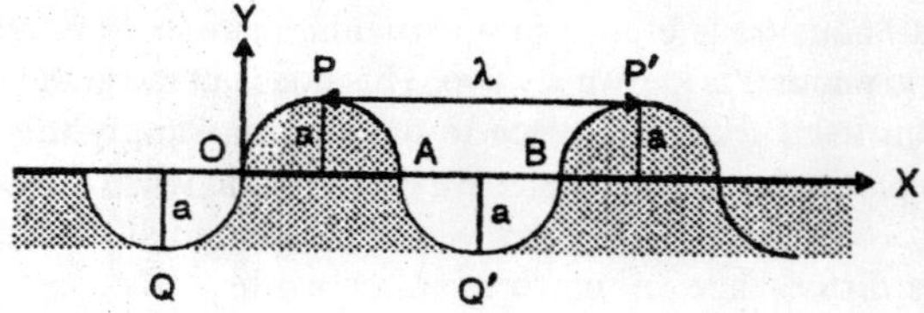

Fig. 1.1

The aspect of the free surface is same at time t and $t + 2\pi/n$. Thus the *period*, T of a wave is $2\pi/n$. The reciprocal of a period is called the *frequency*. It denotes the number of oscillations per-second.

Phase : Let the equation of the wave be taken as

$$y = a \sin (mx - nt + \epsilon).$$

where ∈ represents the phase of the wave at the instant from which t is measured. We notice that wave motion have the same amplitude, wave length and period but they differ in phase. The angle (mx – nt) is called the phase angle and n is called the *phase rate*.

STANDING OR STATIONARY WAVES

Let the two simple harmonic progressive waves of the same amplitude, wave length and period travel in opposite directions are given by the equations

$$y_1 (x, t) = A \sin (mx - nt)$$

and $y_2 (x, t) = A \sin (mx + nt)$ ...(1)

Let y (x, t) represents the free surface profile then by superimposing these two waves, we get

$$y\ (x, t) = y1 + y2 = A\ \{\sin\ (mx - nt) + \sin\ (mx + nt)\},$$

$$\text{or } y\ (x, t) = 2A \sin mx \cos nt. \quad ...(2)$$

A wave if this type is known a *standing or stationary wave* which is not propagated. At any instant t the form of the surface (2) is a sine curve of amplitude 2a cos nt which varies between 0 to 2a. The points of intersection with X-axis are given by

$$\sin mx = 0 \Rightarrow x = (p\pi/m);\ p = 0, \pm 1, \pm 2.$$

These fixed points are called nodes and the intermediate points where the amplitude is greatest are called *anti-nodes*. The points for which mx = (2p + 1) π/2 are points of maximum displacement for a given value of t and are called *loops*. When cos nt = ± 1 the surface is in the form of the sine curve y = ± 2a sin mx which shows the maximum departure from the mean level. When cos nt = 0 the free surface coincide with the mean level.

Similarly a progressive wave system is taken as a combination of two systems of stationary waves of the same amplitude, wave length and period, the crests and troughs of one system coincide with nodes of the other.

Let y_1 = a sin mx cos nt be one of the stationary waves, then the other wave must be y_2 = a cos mx sin nt.

or $y = y_1 \pm y_2 = a\ \{\sin mx \text{ cosn } t \pm \cos mx \sin nt\}$

represents a progressive wave.

Classification of waves

We shall consider the theory of wave motion in an imcompressible liquid (water) in the presence of gravity. The waves are formed by disturbing forces such as wind pressure or the relative motion of a body in the water. The flows associated with be assumed to the two-dimensional and potential in nature. The wave motion in a liquid is generally classified into two parts : (i) *Surface waves,* (ii) *Tidal waves or Long waves.*

Surface waves arise where the vertical acceleration of the fluid is no longer neglected and the wave length is small in comparison with the depth of the water. It moves along the surface of the liquid. The disturbance does not extend far below the surface. Surface waves occur in deep and bounded (in horizontal directions) liquids like ocean and lakes.

Tidal waves describe the alternative limit where the wave length of oscillations is much larger compared to the depth of the water. The vertical accelerations can be assumed negligible compared with the horizontal accelerations. The disturbance effects the motion of the whole of fluid. Tidal waves are also known as long wave in shallow water.

SURFACE WAVES

Consider waves due to small oscillatory motions where the depth of the water may be comparable to the wave length. The motion is supposed to be two dimensional; the axis of X is taken in the undisturbed surface and the axis of Y, vertically upwards. The motion generated from rest by the action of ordinary forces is irrotational, so the velocity vector may be expressed as the gradient of a velocity potential, which in turn, satisfy Laplace's equation:

$$\frac{\partial^2 \phi}{\partial x^2} + \frac{\partial^2 \phi}{\partial y^2} = 0 \text{ throughout the liquid,} \qquad ...(1)$$

with the condition $\frac{\partial \phi}{\partial \eta} = 0$ at a fixed boundary. ...(2)

The surface condition which must be satisfied at the free surface = (Const.). is given by Bernoulli's equation

$$\frac{p}{\rho} = \frac{\partial \phi}{\partial t} - \Omega - \frac{1}{2}q^2 + F\,(t). \qquad ...(3)$$

Substituting $\Omega = gy$ and neglecting the square of the velocity, we have

$$\frac{p}{\rho} = \frac{\partial \phi}{\partial t} - gy + F(t). \qquad ...(4)$$

Let η denote the elevation of the surface at time t above the point (x, 0) and p_0 be the pressure on the surface $y = \eta\,(x, t)$ then from (4), we have

$$\frac{p_0}{\rho} = \left(\frac{\partial \phi}{\partial t}\right)_{y=\eta} - g\eta\,.$$

Since the pressure is uniform, so

$$\eta = \frac{1}{g}\left(\frac{\partial \phi}{\partial t}\right)_{y=\eta}, \qquad ...(5)$$

provided the function F (t) and the constant be absorbed in the value of $\partial f/\partial t$.

The normal to the free surface makes an infinitely small angle $\partial\eta/\partial x$ with the vertical. The kinematic boundary condition that the normal component of the fluid velocity at the free surface must be equal to the normal velocity of the surface itself, provides

$$\frac{\partial \eta}{\partial t}(x,t) = \left\{\frac{\partial \phi}{\partial y}(x, \eta t)\right\}_{y=\eta} \qquad ...(6)$$

From (5) and (6), we have

$$\frac{1}{g}\left(\frac{\partial^2 \phi}{\partial t^2}\right)_{y=\eta} = -\left(\frac{\partial \phi}{\partial y}\right)_{y=\eta}$$

$$\frac{\partial^2 \phi}{\partial t} + g\frac{\partial \phi}{\partial y} = 0, \text{ satisfied at the free surface } y = 0. \qquad ...(7)$$

In the case of simple harmonic motion, the time factor being exp. $\{i(\sigma t + \varepsilon)\}$then the condition (7) becomes

$$\sigma^2\phi = g\frac{\partial \phi}{\partial y} \qquad ...(8)$$

Since the free surface is a surface of equi-pressure p = const., hence on the free surface

$$\frac{dp}{Dt} = \frac{\partial p}{\partial t} + u\frac{\partial p}{\partial x} + v\frac{\partial p}{\partial y} = 0$$

Using the relation (7), the equation (4) reduces to

$$\frac{\partial p}{\partial t} - \frac{\partial \phi}{\partial x} - \frac{\partial p}{\partial x} - \frac{\partial \phi}{\partial y}\frac{\partial p}{\partial y} = 0$$

$$\text{or } \rho\frac{\partial^2 \phi}{\partial t^2} - \rho\frac{\partial \phi}{\partial x}.\frac{\partial^2 \phi}{\partial x \partial t} - \rho\frac{\partial \phi}{\partial y}\left(\frac{\partial^2 \phi}{\partial y \partial t} - g\right) = 0$$

$$\text{or } \frac{\partial^2 \phi}{\partial t^2} - g\frac{\partial \phi}{\partial y} = 0, \text{ neglecting the other terms.}$$

PROGRESSIVE WAVES ON THE SURFACE OF A CANAL

We shall determine the propagation of simple harmonic waves of the form

$$\eta = a \sin(mx - nt), \qquad ...(1)$$

at the surface of a canal of uniform depth h with parallel vertical walls at right angles of the ridges and hollows.

Since the motion produced by natural forces is irrotational the velocity potential ϕ exists

$$\frac{\partial^2 \phi}{\partial x^2} - g\frac{\partial^2 \phi}{\partial y^2} = 0. \qquad ...(2)$$

with the conditions $\left(\frac{\partial \phi}{\partial y}\right)_{y=-h} = 0$

and $$\left(g\frac{\partial\phi}{\partial y}+\frac{\partial^2\phi}{\partial t^2}\right)_{y=-0} = 0 \qquad ...(3, 4)$$

Again $$\left(\frac{\partial\eta}{\partial t}\right)_{y=0} = -\left(\frac{\partial\phi}{\partial y}\right)_{y=0}$$

$= na \cos (mx - nt)$. ...(5)

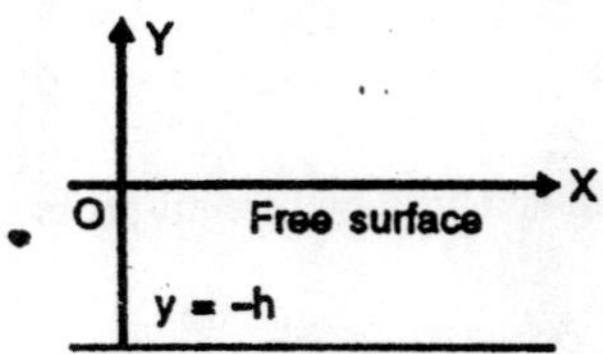

Fig. 1.2

Let the solution be of the form

$\phi = f(y) \cos (mx - nt)$. ...(6)

By substituting in (2), we have

$$\frac{d^2f}{dy^2} - m^2f = 0 \Rightarrow f(y) = Ae^{my} + Be^{-my},$$

where A and B are arbitrary constants.

From the relation (6), we have

$f = (Ae^{my} + Be^{-my}) \cos (mx - nt)$ ...(7)

Using the condition (3), we get

$Ae^{-mh} - Be^{mh} = 0$

$\Rightarrow Ae^{-mh} = Be^{mh} = 1/2\ \mu$ (say)

Therefore $\phi = 1/2\ \mu\ [e^{m(y+h)} + e^{-m(y+h)}] \cos (mx - nt)$

or $\phi = \mu \cosh m (y + h) \cos (mx - nt)$. ...(8)

From the surface condition (4), we obtain

$[gm\mu \sin m (y + h) - n^2 m \cosh m (y + h)] \cos (mx - nt) = 0$

or $n^2 = gm\ [\tanh m (y + h)]_{y=0} = gm$ tah mh. ...(9)

If c (= n/m) denotes the velocity of propagation and λ (= $2\pi/m$) denotes the wave length then

$$c^2 = \frac{g\lambda}{2\pi}\tanh\frac{2\pi n}{\lambda}. \qquad ...(10)$$

The constant m in (8) may be expressed in terms of the emplitude a of the wave. Therefore from (5) and (8), we have

$m\ \mu \sinh mh = na \Rightarrow \mu = na/\ (m\sinh mh).$

Thus $\phi = \dfrac{na}{m}\dfrac{\cosh m(y+h)}{\sinh mh}\cos(mx-nt)$

From the relation (9), we have

$$\text{or } \phi = \frac{ga}{n}\frac{\cosh m(y+h)}{\cosh ma}\cos(mx-nt). \qquad \text{...(11)}$$

Let (X, Y) be the coordinates of a particle relative to the mean position (x,y) then

$$u = \dot{X} = -\frac{\partial\phi}{\partial x} = na\frac{\cosh m(y+h)}{\sinh mh}\sin(mx-nt),$$

$$\text{and } v = \dot{Y} = -\frac{\partial\phi}{\partial y} = -na\frac{\sinh m(y+h)}{\sinh mh}\cos(mx-nt), \qquad \text{...(12, 13)}$$

By integrating the above relations with regard to the time t, we have

$$X = \frac{a\cosh m(y+h)}{\sinh mh}\cos(mx-nt),$$

$$Y = \frac{a\sinh m(y+h)}{\sinh mh}\sin(mx-nt), \qquad \text{...(14)}$$

$$\text{or } \frac{X^2}{\cosh^2 m(y+h)} + \frac{Y^2}{\sinh^2 m(y+h)} = \frac{a^2}{\sinh^2 mn}, \qquad \text{...(15)}$$

which shows that the motion of each particle is elliptic harmonic about its mean position. The distance between the focii is 2a cosech mh which is the same for all such ellipses being independent of y. The major axes of the ellipses are horizontal. The major axes and the minor axes decrease as the depth of the particle increases and the minor axes vanishes when y = – h. *Thus the ellipse degenerates a straight line at the bottom. We notice that a surface particle is moving in the direction of wave propagation when it is at a crest and in the opposite direction when it is a trough.*

Waves on a deep canal

Consider the depth h of the canal be sufficiently great (let h → ∞) in comparison with the wave length λ, e^{-my} is very small, then we must set B = 0. The equation (7) becomes

$$\phi = Ae^{my}\cos(mx - nt). \qquad \text{...(16)}$$

$$\text{or } \phi = \frac{na}{m}e^{my}\cos(mx-nt), A = \frac{na}{m}. \qquad \text{...(17)}$$

Let (X, Y) be the coordinates of a particle relative to its mean position (x, y) then

$$u = \dot{X} = -\frac{\partial \phi}{\partial x} = na\, e^{my} \sin(mx - nt).$$

$$v = \dot{Y} = -\frac{\partial \phi}{\partial y} = -na\, e^{my} \cos(mx - nt).$$

By integrating with regard to the time t, we have

$X = ae^{my} \cos(mx - nt)$, $Y = aemy \sin(mx - nt)$,

or $X^2 + Y^2 = a^2e^{2my} = (ae^{my})^2$

Thus each particle describes a circle with uniform angular velocity a, where

$n = (gm)^{1/2} = (2\pi g/\lambda)^{1/2}$.

ENERGY OF PROGRESSIVE WAVE

Let the wave profile of the progressive waves at the surface of water of depth h, be of the form

$\eta = a \sin(mx - nt)$

$$\text{and } \phi = \frac{ga}{n}\frac{\cosh m(y+h)}{\cosh mh}\cosh(mx - nt). \qquad ...(2)$$

Let V be the energy of the water between two vertical planes, parallel to the direction of propagation at unit distance apart, then the potential energy for a single wavelength is given by

$$V = \frac{1}{2} g\rho \int_0^{\lambda} \eta^2 dx$$

$$\text{or } V = \frac{1}{2} g\rho\, a^2 \int_0^{\lambda} \sin^2(mx - nt)dx$$

$$\text{or } V = \frac{1}{4} g\rho a^2 \lambda; \text{ since } \lambda = 2\pi/m. \qquad ...(3)$$

Since the motion is irrotational, the kinetic energy is given by

$$T = -\frac{1}{2}\rho \int \phi \frac{\partial \phi}{\partial n} dS, \qquad ...(4)$$

integrated along the profile of a wave length and ∂n is measured along the normal into the water. To the order of small quantities (4) can be written as

$$T = \frac{1}{2}\rho \int_0^{\lambda} \left(\phi \frac{\partial \phi}{\partial y}\right)_{y=0} dx$$

or $T = \frac{1}{2} g\rho a^2 \int_0^\lambda \cos^2(mx - nt)dx$

or $T = \frac{1}{4} g\rho a^2\lambda$; since $n^2 = gm \tanh mh$. ...(5)

Thus the total energy E per unit area of the water surface is given by

$E = V + T = \frac{1}{4} g\rho a^2\lambda + \frac{1}{4} g\rho a^2\lambda = \frac{1}{4} g\rho a^2\lambda,$ ...(6)

which is half kinetic and half potential at any instant.

PROGRESSIVE WAVES REDUCED TO A STEADY MOTION

Let c be the velocity of propagation of the wave moving towards the positive direction of X-axis. The wave profile becomes fixed in space by super-imposing on the whole mass a velocity equal and opposite to the velocity of propagation. Since the motion is irrotational so an additional term in ϕ will be given by

$-(\partial\phi/\partial x) = -c, \phi = cx.$

The corresponding complex potential becomes

$w = c(x + iy) = cz$ ...(1)

Let the complex potential be of the form

$w = cz + A \cos mz - iB \sin mz$

or $\phi + i\psi = c(x + iy) + A \cos m(x + iy) - iB \sin m(x + iy)$

Equating real and imaginary parts, we have

$\phi = cx + (A \cosh my + B \sinh my) \cos mx$

and $\psi = cy + (A \sinh my + B \cosh my) \sin mx$ ...(2, 3)

They velocity potential and stream function satisfy the Laplace equation. For the bottom to be a stream line, we must have

$\psi = \text{const.}$, when $y = -h$.

So that $-ch - (A \sinh mh + B \cosh mh) \sin mx = \text{const.}$

or $-A \sinh mh + B \cosh mh = 0$

or $\frac{A}{\cosh mh} = \frac{B}{\sinh mh} = \mu$ (let)

Substituting the values of A and B into (2, 3), we have

$\phi = cx + \mu \cosh m(y + h) \cos mx,$

and $cy + \mu \sinh m(y + h) \sin mx.$ ...(4, 5)

If the free surface be a simple sin curve η = a sin mx the equation (5) willmake this a stream line $\psi = 0$, provided

$$ca - \mu \sin h\, mh = 0, \qquad \text{...(6)}$$

neglecting squares of small quantities.

The pressure is given by Bernoulli's equation

$$\frac{p}{\rho} + gy + \frac{1}{2}q^2 = \text{Const.}$$

At the free surface it reduces

$$\frac{p}{\rho} + ga \sin mx + \frac{1}{2}[\{c - m\mu \cosh m(y+h) \sin mx\}^2$$

$$+ \{m\mu \sinh m(y+h) \cosh mx\}^2] = \text{Const.}$$

$$\text{or } \frac{p}{\rho} + ga \sin mx + \frac{1}{2}[c^2 - 2m\mu c \cosh mh \sin mx] = \text{const,}$$

neglecting the terms of small quantities (*i.e.*, μ^2).

Since p = Const. at the free surface so the coefficients of sin mx must vanish.

$$ga - m\mu c \cosh mh = 0$$

$$\text{or } ga - mc^2 a \coth mh = 0$$

$$\text{or } c^2 = \frac{g\lambda}{2\pi}\tanh\frac{2\pi h}{\lambda};\; m = 2\pi/\lambda. \qquad \text{...(7)}$$

On Deep Water

Now we shall discuss the progressive waves the ratio of the depth to the wave length is sufficiently great. The conditions being that the uper surface is free and total depth infinite.

Consider velocity potential and stream function be of the form

$$\phi = cx + Ae^{my} \cos mx,$$

$$\psi = cy - Ae^{my} \sin mx. \qquad \text{...(8, 9)}$$

The free surface η = a sin mx will be a stream line if ψ = const. then

$$[ca - Ae^{ma \sin mx}] \sin mx = \text{Const.}$$

$$\text{or } [ca - A(1 + ma \sin mx + ...)] \sin mx = \text{Const.}$$

This will hold if coefficients of sin mx vanish

i.e., $ca - A = 0 \Rightarrow ca = A$

then $\phi = cx + ca\, e^{my} \cos mx$,

and $\psi = cy - ca\ e^{my} \sin mx$. ...(10)

At free surface pressure (p = const.), given by Bernoulli's equation, becomes

$$gy + \frac{1}{2}\left\{\left(\frac{\partial\phi}{\partial x}\right)^2 + \left(\frac{\partial\phi}{\partial y}^2\right)\right. = \text{Const.,}$$

$$\text{or } gy + \frac{1}{2}c^2\{1 - 2ma e^{my} \sin mx\} = \text{Const.}$$

neglecting the terms of a^2.

$$\text{or } ga \sin mx + \frac{1}{2}c^2\{1 - 2\ ma\ e^{ma \sin mx} \sin mx\} = \text{Const.}$$

Equating the coefficients of sin mx to zero, we have

$$ga = \frac{1}{2}c^2\ (2ma), \text{ where } c^2 = g/m,$$

Which gives the velocity of propagation of two dimensional waves of given length propagated over infinitely deep liquid and represents that the pressure is constant along each stream line.

STANDING OR STATIONARY WAVES

Now we consider the waves which remain stationary, the surface moves vertically only. The equation for a stationary wave is given by

$$\eta = a \sin mx \cos nt. \qquad ...(1)$$

Being an irrotational motion, the velocity potential ϕ satisfies the Laplace equation such that

$$\frac{\partial^2\phi}{\partial x^2} + \frac{\partial^2\phi}{\partial y^2} = 0. \qquad ...(2)$$

Assuming the solution be of the form

$$\phi = f(y) \sin mx \sin nt \qquad ...(3)$$

$$\text{or } f''(y) - m^2 f(y) = 0$$

$$\text{or } \phi(y) = Ae^{my} + Be^{-my}$$

$$\text{or } \phi = (Ae^{my} + Be^{-my}) \sin mx \sin nt, \qquad ...(4)$$

where A and B are arbitrary constants.

Using the condition $\partial\phi/\partial y = 0$ at $y = -h$, we have

$Ae^{-mh} = Be^{mh} = 1/2\ \mu$ (say),

so that the equation (4) becomes

$\phi = 1/2\ \mu\ [e^{m\ (y+h)} + e^{-m(y+h)}]$ sin mx sin nt

or $\phi = \mu \cosh m\ (y + h) \sin mx \sin nt.$ ...(5)

Again from the condition $g\ \dfrac{\partial\phi}{\partial y} + \dfrac{\partial^2\phi}{\partial t^2} = 0$ at $y = 0$, we have

$[g\mu m \sin h\ m\ (y + h) - \mu n^2 \cosh m\ (y + h)] \sin mx \sin nt = 0.$

or $n^2 = gm \tanh m\ (y + h).$

The phase velocity c is given by

$$c^2 = \frac{n^2}{m^2} = \frac{g}{m}\ [\tan h\ m\ (y + h)]_{y=0} = \frac{g}{m} \tanh mh$$

or $c^2 = \dfrac{g\lambda}{2\pi} \tanh \dfrac{2\pi h}{\lambda}$; as $m = 2p/l.$ ...(6)

The constant μ can be determined in terms of the wave amplitude by using the condition

$(\partial\eta/\partial t) = -(\partial f/\partial y)_{y=0} \Rightarrow na = m\mu \sin h\ mh.$ ...(7)

Substituting the value of μ in (5), we get

$$\phi = \frac{na}{m} \frac{\cosh m(y+h)}{\sinh mh} \sin mx \sin nt. \quad ...(8)$$

Path of the Particles

Consider (X, Y) be the coordinates of a fluid particle relative to its mean position (x, y) then

$$u = X = -\frac{\partial\phi}{\partial x} = -na\ \frac{\cosh m(y+h)}{\sinh mh} \cos mx \sin nt.$$

$$v = Y = -\frac{\partial\phi}{\partial y} = -na\ \frac{\sinh m(y+h)}{\sinh mh} \sin mx \sin nt.$$

By integrating with regard to t, we have

$$X = a\ \frac{\cosh m(y+h)}{\sinh mh} \cos mx \sin nt,$$

and $Y = a \dfrac{\sinh m(y+h)}{\sinh mh} \sin mx \sin nt.$

$\Rightarrow \dfrac{Y}{X} = \tan h\ m\ (y + h) \tan mx,$ ...(8)

which is independent of the time t, thus the motion of each particle rectilinear. The direction varies from vertical below the crests and trough to horizontal below the nodes.

On Deep Water

In the case of standing waves on deep water the depth is considered sufficiently great ($h \to \infty$) in comparison with the wave length, thus we must set the constant B = 0. The equation (4) becomes

$$\phi = Ae^{my} \sin mx \sin nt. \qquad ...(9)$$

Using the condition $g\dfrac{\partial \phi}{\partial y} + \dfrac{\partial^2 \phi}{\partial t^2} = 0$ at the free surface y = 0, we have

$$A\,(gm - n^2)\, e^{my} \sin mx \sin nt = 0,$$

which gives $gm - n^2 = 0$;

$$m = \frac{n^2}{g} \text{ and } c^2 = \frac{n^2}{m^2} = \frac{g}{m} = \frac{g\lambda}{2\pi}. \qquad ...(10)$$

The constant A can be determined in terms of the wave amplitude by using the condition (6), such that

$$\phi = \frac{na}{m} e^{my} \sin mx \sin nt,\ A = \frac{na}{m}. \qquad ...(11)$$

Path of the Particles

$$u = \dot{X} = -\frac{\partial \phi}{\partial x} = -\,na\, e^{my} \cos mx \sin nt,$$

$$\text{and } v = \dot{Y} = -\frac{\partial \phi}{\partial y} = -\,na\, e^{my} \sin mx \sin nt.$$

By integrating with regard to t, we have

$$X = ae^{my} \cos mx \cos nt,\ Y = ae^{my} \sin mx \cos nt$$

$$\Rightarrow Y = X \tan mx,$$

which is independent of the time t. Thus the trajectory of the particle is a straight line with slope tan mx. The particle oscillates in the vertical direction at the antinodes and in the horizontal direction at the nodes. The amplitude of the oscillation decreases as depth increases.

ENERGY OF STATIONARY WAVES

Consider the wave profile be of the form

$$\eta = a \sin mx \cos nt \qquad ...(1)$$

$$\text{and } \phi = \frac{ga}{n}\frac{\cosh m(y+h)}{\cosh mh} \sin mx \sin nt. \qquad ...(2)$$

Let η be the elevation. The mass of liquid standing above a base dx in the -XY plane is $\rho\eta$ dx its centre of mass is at a height $\frac{1}{4}\eta$. Let V be the potential energy relative to the undisturbed system such that

$$V = \int_0^{\lambda} \frac{1}{2}\eta g(\rho\eta dx) = \frac{1}{2}\lambda g\rho a^2 \int_0^{\lambda} \sin^2 mx \cos^2 nt dx$$

$$\text{or } V = \frac{1}{4}\lambda g\rho a^2 \cos^2 nt. \qquad ...(4)$$

If cos nt =1 the potential energy becomes $\frac{1}{4}\lambda g\rho a^2$.

Let T be the kinetic energy then

$$T = \frac{1}{2}\rho \int_0^{\lambda} q^2 dv = \frac{1}{2}\rho \int_0^{\lambda} \phi \frac{\partial\phi}{\partial n} dS = \frac{1}{2}\rho \int_0^{\lambda} \left(\phi \frac{\partial\phi}{\partial y}\right)_{y=0} dx$$

$$\text{or } T = \frac{1}{2}\frac{\rho g^2 a^2}{n^2} m \tanh \sin^2 nt \int_0^{\lambda} \sin^2 mx dx$$

$$\text{or } T = \frac{1}{4}\lambda\rho g a^2 \left(\frac{m^2}{n^2}\right)\left(\frac{g}{m}\tanh mh\right) \sin^2 nt = \frac{1}{4} g\rho a^2 \lambda \sin^2 nt. \qquad ...(4)$$

Therefore the total energy E per wave length at any time t is given by

$E = V + T = 1/4\ g\rho a^2\lambda\ (\cos^2 nt + \sin^2 nt) = 1/4\ g\rho a^2\lambda$.

The kinetic energy and potential energy change continuously with the time

WAVES AT THE COMMON SURFACE OF TWO LIQUIDS

Consider a liquid of density ρ' and depth h' moves with velocity V' over another liquid of density ρ and depth h with velocity V in the same direction, the liquids being bounded above and below by two fixed horizontal planes.

Consider X-axis in the direction of undisturbed interface and Y-axis to be vertically upwards. To make the motiòn steady and the wave form to rest in space, super-imposing on the whole mass a negative velocity equal and opposite to the velocity of propagation of the wave. The velocity of the streams reduce to $V' - C$ and $V - C$.

Let the velocity potential and stream function related to lower liquid be

$$\phi = -(V - C)\, x + A \cosh m\, (y + h) \cos mx,$$

$$\psi = -(V - C)\, y - A \sinh m\, (y + h) \sin mx, \qquad ...(1)$$

and for upper liquid be

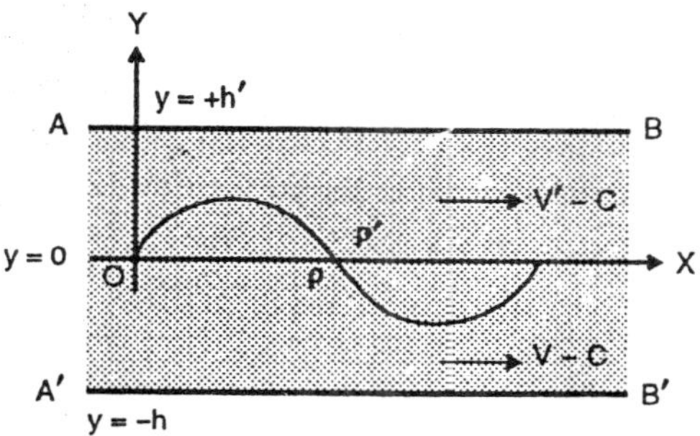

Fig. 1.3

$\phi' = -(V' - C)\,x + A' \cosh m\,(y - h') \cos mx,$

$\psi' = -(V' - C)\,y - A' \sinh m\,(y - h') \sin mx,$ ...(2)

The corresponding complex potentials w and w′ for the lower and upper liquid becomes

$$w = \phi + i\psi = -(V - C)z - \frac{a(V - V)}{\sinh mh} \cos m\,(z + ih)$$

$$w' = \phi' + i\psi' = -(V' - C)z + \frac{(V' - C)}{\sinh mh} \cos m\,(z - ih') \quad ...(3, 4)$$

The lower liquid is moving with – (V – C) in the negative direction of X-axis. Also $\mu = (V - C)\,a/\sin h\,mh$.

Let q be the speed in the lower liquid such that

$$q^2 = (dw/dz).\,(d\overline{w}/d\overline{z})$$

$$\text{or } q^2 = \left\{-(V - C) + \frac{am(V - C)}{\sinh mh} \sin m(z - ih)\right\} \times \left\{-(V - C) + \frac{am(V - C)}{\sinh mh} \sin m(z - ih)\right\}$$

$$\text{or } q^2 = (V - C)^2 - \frac{2ma(V - C)^2}{\sinh mh} \cosh m\,(y + h) \sin mx,$$

neglecting the terms of small quantities of a^2.

Let q_0 be the speed at the interface y = 0, so that

$q_0^2 = (V - C)^2\,[1 - 2ma \coth mh \sin mx],$

Since equation to the free surface is $\eta = a \sin mx$.

So $q_0^2 = (V - C)^2 \, [1 - 2m\eta \coth mh]$ ...(5)

Similarly if q′0 be the speed at the ionterface due to upper liquid then

$q'^2_0 = (V' - C)^2 \, [1 + 2m\eta \cot mh']$. ...(6)

Since the pressure is continuous (p = p′) across the interface due to lower and upper liquid then by Bernoulli's theorem, we have

$$\frac{1}{2}\rho' q'^2_0 - \frac{1}{2}\rho q_0^2 + g\eta(\rho' - \rho) = \text{const.}$$

or $\frac{1}{2}\rho\,'(V' - C)^2 \, [1 + 2m\eta \coth mh']$

$$- \frac{1}{2}\rho \, (V - C)^2 \, [1 - 2m\eta \text{ costh } mh] + g\eta \, (\rho' - \rho) = \text{const.}$$

The R. H.S. of the equation is constant so the coefficients of the variable h must vanish *i.e.*,

$$m\rho' \, (V' - C)^2 \coth mh' + m\rho \, (V - C)^2 \coth mh = g \, (\rho - \rho'), \quad ...(7)$$

which determines the velocity of propagation C of waves of length which determines the velocity of propagation C of waves of length $2\pi/m$ at the common surface of two streams whose velocity are V and V′. Since the tangential velocities on opposite sides of the interface are different, even if V = 0 = V′, the interface must be a vortex sheet.

Case I : If the liquids are at rest i.e., V = 0 = V′ then

$$m\rho' \, C^2 \coth mh' + m\rho C^2 \coth mh = g \, (\rho - \rho')$$

or $$C^2 = \frac{g}{m} \frac{\rho - \rho'}{\rho \coth mh + \rho' \coth mh'}. \quad ...(8)$$

There is no real value for C where $\rho' > \rho$ which shows that the equilibrium position is unstable.

Case II : Consider the upper liquid to be air of specific gravity S $(=\rho'/\rho)$ at an infinite depth. Substituting V = 0 = V′, we have

$$C^2 = \frac{g}{m} \frac{1 - (\rho'/\rho)}{\coth mh + (\rho'/\rho)\coth mh'} = \frac{g}{m} \frac{1 - S}{\coth mh + S}$$

$$C^2 = \frac{g}{m} \tanh mh \, (1 - S) \, (1 - S \tanh mh)$$

$$C^2 = \frac{g}{m} \tanh mh \, [1 - S\{1 + \tanh mh\}],$$

neglecting terms of S^2. Thus the presence of atmosphere tends to decrease wave velocity.

Case III : If the depths of the liquid are so large compared to the wave length so that we may take

$$\coth mh = \text{cotmh } mh' = 1, \text{ then } C^2 = \frac{g}{m}\frac{\rho-\rho'}{\rho+\rho'}$$

WAVES AT AN INTERFACE WITH UPPER SURFACE FREE

Here we shall consider the wave motion when the surface of the upper liquid is free *e.g.*, a layer of oil upon water or a layer of fresh water upon salt water.

Let a liquid of density r′ and depth h′ lie over another liquid of density ρ and depth h. Assuming the liquids tobe at rest for the wave motion and C be the common velocity of wave propagation at the free surface of the upper liquid and at the common surface. The motion becomes steady by superposing a velocity equal and opposite to the common velocity of propagation of waves on the whole mass; the wave profile gets fixed in space and the fluid flows with velocity C in the negative direction of X-axis.

The complex potentials of the lower and the upper liquids are given by

$$w = Cz + \frac{aC}{\sinh mh}\cos m(z + ih), \qquad \text{...(1)}$$

$$w' = Cz + \frac{bC\cos mz}{\sinh mh'} - \frac{aC}{\sinh mh'}\cos m(z - ih) \qquad \text{...(2)}$$

The second term denotes the complex potential of a simple wave $\eta^2 = b \sin mx$ at $t = 0$. Let q be the speed in the lower liquid then

$$q^2 = \frac{dw}{dz}\cdot\frac{\overline{dw}}{\overline{dz}}$$

$$\text{or } q^2 = C^2\left\{1 - \frac{ma}{\sinh mh}\sin m(z + ih)\right\} \times \left\{1 - \frac{ma}{\sinh mh}\sin m(\bar{z} + ih)\right\}$$

$$\text{or } q^2 = C^2\left\{1 - \frac{ma}{\sinh mh}\sin mx \cosh m(y + h)\right\}, \qquad \text{...(3)}$$

neglecting the terms of a^2.

The speed ($= q_0$, say) at the interface due to lower liquid is obtained by substituting $y = 0$ into (3), so

$$q_0^2 = C^2\left\{1 - \frac{2ma}{\sinh mh}\sin mx \cosh mh\right\} \qquad \text{...(4)}$$

Similarly, let q′ be the speed in the upper liquid then

$$q'^2 = \frac{dw'}{dz} \cdot \frac{d\overline{w}}{d\overline{z}}$$

$$\text{or } q'^2 = C^2 \left\{1 - \frac{2bm \sin mx \cosh my}{\sinh mh'} + \frac{2ma \sin mx \cosh m(y-h')}{\sinh mh'}\right\} \quad ...(5)$$

The speed (= q'_0, say) at the interface due to lower liquid is obtained by substituting y = 0 into (5), therefore

$$q'^2_0 = C^2 \left\{1 - \frac{2bm \sin mx}{\sinh mh'} + \frac{2ma \sin mx \cosh mh'}{\sinh mh'}\right\} \quad ...(6)$$

Since the pressure is continuous (p = p′) across the surface due to lower and upper liquid, by using Bernoulli's theorem, we have

$$1/2\ \rho q_0^2 + g\rho\eta_1 + \text{const.} = 1/2\ \rho'\ q'^2_0 + g\rho'\eta_1 + \text{const.},$$

$$\text{or } (\rho'\ q'^2 - \rho q_0^2) + 2g\eta_1\ (\rho' - \rho) = \text{const.}$$

$$\text{or } - C^2\rho' \left\{1 - \frac{2bm \sin mx}{\sinh mh'} + \frac{2am \sin mx \cosh mh'}{\sinh mh'}\right\}$$

$$- C^2\rho \left\{1 - \frac{2am \sin mx \cosh mh}{\sinh mh}\right\} + 2ga\ (\rho' - \rho) \sin mx = \text{const.} \quad ...(7)$$

The R. H. S. of (7) is constant so the coefficients of the variable sin mx must vanish such that

$$g\ (\rho - \rho') = C^2 m\ \{\rho' \coth mh' + \rho \coth mh - \rho'\ (b/a) \operatorname{cosech} mh'\} \quad ...(8)$$

Assuming the pressure p′ is constant at the free surface (y = h′) η_2 = h′ + b sin mx, thus, we have

$$g\rho'\eta_2 + \frac{1}{2}\rho' q'^2 = \text{const.}$$

$$\text{or } g\rho'\ (h' + b \sin mx) + \frac{1}{2}\rho'\ C^2 \left\{1 - \frac{2bm \sin mx \cosh mh'}{\sinh mh'} + \frac{2am \sin mx}{\sinh mh'}\right\} = \text{cons.}$$

Equating the coefficients of sin mx equal to zero, we have

$$g = C^2 m\{\coth mh' - (a/b) \operatorname{cosech} mh'\}$$

$$\text{or } \frac{b}{a} = \frac{C^2 m}{C^2 m \cosh mh' - g \sinh mh'}, \quad ...(9)$$

which determines the ratio of the amplitudes of the waves.

Eliminating the ratio (b/a) from (8) and (9), we have

$$g(\rho - \rho') = C^2 m \left\{\rho' \coth mh' + \rho \coth mh - \frac{\rho' C^2 m \operatorname{cosech} mh'}{C^2 m \cosh mh' - g \sinh mh'}\right\}$$

or C^2m^2 (ρ coth mh coth mh′ + ρ')

$- C^2mg\rho$ (coth mh + coth mh′) + g^2 ($\rho - \rho'$) = 0. ...(10)

Thus we notice that the above equation in C gives two possible velocities of propagation for a given wave length, provided $\rho > \rho'$.

Particular Case : When the lower liquid is deep.

Substituting coth mh = 1 in the equation (10), we have

C^4m^2 (ρ coth mh′ + ρ') – $C^2mg\ \rho$ (1 – coth mh′) + g^2 ($\rho - \rho'$) = 0

or (mC^2– g) {mC^2 (ρ coth mh′ + ρ') – g ($\rho - \rho'$)} = 0

which gives $C^2 = \frac{g}{m}.\frac{\rho - \rho'}{\coth mh' + \rho'}$.

GROUP VELOCITY

When the waves are started by a local disturbance e.g., dropping of a stone into a canal or the motion of a boat through water, the successive waves have different lengths and are propagated with different velocities. Since the velocity of propagation of a simple-harmonic train varies with the wave length so the waves of slightly different wave lengths will be sorted out into groups. The velocity with which an isolated group of waves, of considerably the same length advances over relatively deep water is called the *group velocity.*

Consider the case of two systems of simple harmonic waves of the same amplitude, and of nearly but not quite the same wave lengths such that

η_1 = a sin (mx – nt),

η_2 = a sin [(m + δm) x – (n + δn) t],]

where δm and δn are infinitesimals.

The corresponding equation of the free surface will be of the form

η = a [sin (mx – nt) + sin {(m + δm) x– (n + δn) t}]

or η = 2a sin (mx..nt) cos 1/2 (x δm – tδn) = A sin (mx – nt), ...(1)

where A = 2a cos1/2 (x δm – tδn); $\delta m \ll m$, $\delta n \ll n$.

The cosine in this expression varies very slowely with x; so that the wave profile at any instant has the form of a sine curve in which the amplitude varies between the values 0 to 2a. Thus the surface (1) represents a series of waves and the motion of each group is independent of the presence of the others. The distance between the centre of two successive groups is $2\pi/\delta m$, and the time taken by the system in shifting through this space is $2\pi/\delta n$ so group velocity (C_g say) is $C_g = \delta n/\delta m$.

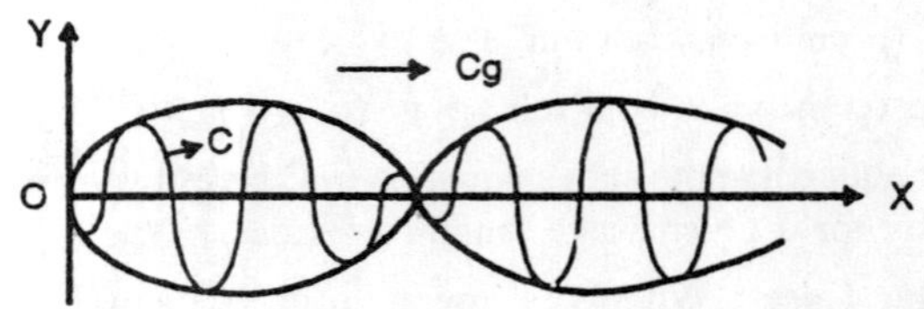

Fig. 1.4

It may be expressed in terms of the wave length C, we get

$$C_g = \frac{dn}{dm} = \frac{d}{dm}(mC);\ C = n/m$$

$$\text{or } C_g = \left\{1 + \frac{m}{C}\frac{dC}{dm}\right\} = C\left\{1 + \frac{m}{2C^2}\frac{dC^2}{dm}\right\}$$

On the surface of water of depth h, we have

$C^2 = (g/m) \tanh mh$

$$\text{then } C_g = C\left\{1 + \frac{m^2}{2g \tanh mh}.\frac{d}{dm}\left(\frac{g}{m}\tanh mh\right)\right\}$$

$$\text{or } C_g = \frac{1}{2}C\left\{1 + \frac{2mh}{\sinh 2mh}\right\} \qquad ...(3)$$

which shows that in the case of waves on the surface of water of depth h, the, ratio of the group velocity to the wave velocity is $1/2 + mh/\sinh(2mh)$.

Case 1: For Shallow water; when h is small compared with the wave length the ratio Lt $\{\sinh(2mh)/2mh\}$ tends to unity, then

$C_g = 1/2\ C\ (1 + 1) = C.$

$\Rightarrow$ *that the group velocity for shallow water equals to the wave velocity.*

Case II : On deep water; when h increases to infinity *i.e.*, the ratio is very large then $C_g = 1/2\ C \Rightarrow$ that the group velocity for deep water waves is half the wave velocity.

RATE OF TRANSMISSION OF ENERGY

(Dynamical significance of group velocity).

The rate of transmission of energy measured by taking a vertical section of the liquid at right angles to the direction of propagation. We shall determine the rate at which the liquid on one side of this section is doing work on the liquid on the other side. Consider the depth of the liquid be h, we know that

$$\phi = \frac{ga}{n}\frac{\cosh m(y+h)}{\cosh mh}\cos(mx - nt). \qquad ...(1)$$

Neglecting squares of small quantities in the Bernoulli's equation, the pressure is given by

$$\partial p = \rho\,\frac{\partial\phi}{\partial t} = \rho ga\frac{\cosh m(y+h)}{\cosh mh}\sin(mx - nt). \qquad ...(2)$$

Thus the work done W in unit time is given by

$$W = -\int_{-h}^{0}\partial p\left(\frac{\partial\phi}{\partial x}\right)dy \text{ as } u = -\frac{\partial\phi}{\partial x} \qquad ...(3)$$

$$\text{or } W = rga\int_{-h}^{0}\left\{\frac{\cosh m(y+h)}{\cosh mh}\sin(mx-nt)\right\}$$

$$\times\left\{\frac{ga}{n}.\frac{m\cosh m(y+h)}{\cosh my}\sin(mx-nt)\right\}dy$$

$$\text{or } W = \frac{\rho g^2a^2m}{n}\frac{\sin^2(mx-nt)}{\cosh^2 mh}\int_{-h}^{0}\cosh^2 m(y+h)dy$$

$$\text{or } W = \frac{\rho g^2a^2m}{2n}\frac{\sin^2(mx-nt)}{\cosh^2 mh}\frac{\sinh 2mh}{2m}\left(1+\frac{2mh}{\sin 2mh}\right)$$

The average value of sin2 (mx – nt) is 1/2 , so we have

$$W = \frac{\rho g^2a^2m}{4n}\frac{\tanh mh}{m}\left(1+\frac{2mh}{\sinh 2mh}\right).$$

$$\text{or} \qquad W = \frac{1}{4}\rho ga^2C\left(1+\frac{2mh}{\sinh 2mh}\right);$$

where $C^2 = (g/m)\tanh mh$ and $C = n/m$

$$\text{or } W = \frac{1}{2}\rho ga^2C_g;\ C_g = \frac{1}{2}C\left(1+\frac{2mh}{\sinh 2mh}\right). \qquad ...(4)$$

Thus the expression for the energy transmitted in unit time is equal to $1/2\ \rho ga^2 \times$ group velocity, where $1/2\ \rho ga^2$ is the whole energy per unit length at any instant. Hence the energy is transmitted at a rate equal to the group velocity which is known is as dynamical significance of group velocity.

LONG WAVES

Long waves occur in canals and rivers which have a free surface and depth of the liquid is much smaller than the length of the canal. The wave length is much large compared to the depth of the water, the vertical acceleration can be neglected in comparison with the horizontal. Let h be the depth of water in a horizontal canal; the ratio η/λ is small where λ is a typical wave length and the quantities η/h and $d\eta/dx$ are small. Let the axis of X be parallel to the length of the canal and that of Y vertically upwards.

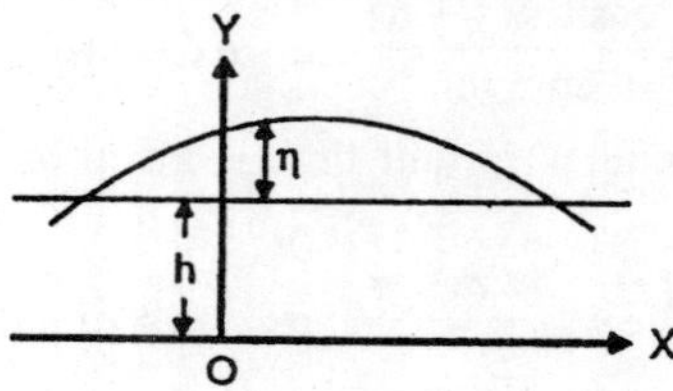

Fig. 1.5

Thus conditions to be satisfied by the waves are

$$\left(\frac{\partial^2\phi}{\partial t^2} - g\frac{\partial\psi}{\partial x}\right)_{y=h} = 0, \ (\psi)_{y=0} = 0. \qquad ...(1,2)$$

Let the complex potential be of the form

$$V = V(z, t) = \phi(x, y, t) + i\psi(x, y, t). \qquad ...(3)$$

From (2), we notice that the complex potential V is real when $y = 0$. By the principle of analytic continuation, V can be expressed in the region – h ″ y < 0.

$$\text{So } V(\bar{z}, t) = \phi(x, y, t) - i\psi(x, y, t). \qquad ...(4)$$

From (3) and (4), we have

$$\phi(x, y, t) = 1/2\,\{V(z, t) + V(\bar{z}, t)\},$$

$$\psi(x, y, t) = -\frac{1}{2}i\,\{V(z, t) - V(\bar{z}, t)\}. \qquad ...(5)$$

Again from (1) and (5), we obtain

$$\frac{\partial^2}{\partial t^2}[V(x + ih, t) + V(x - ih, t)]$$

$$+ ig\frac{\partial}{\partial x}[V(x + ih, t) - V(x - ih, t)] = 0 \qquad ...(6)$$

Since V is an analytic function so the above relation must be true for any point in the domain, putting z for x, we have

$$\frac{\partial^2}{\partial t^2}[V(z+ih, t)+V(z-ih, t)]$$

$$+ ig\frac{\partial}{\partial z}[V(z+ih, t)-V(z-ih, t)] = 0$$

$$\text{or} \frac{\partial^2}{\partial t^2}(V_1+V_2) - ig\frac{\partial}{\partial z}(V_1 - V_2) = 0, \quad \ldots (7)$$

where $V_1 = V(z + ih)$, and $V_2 = V(z - ih, t)$,

which is known as Cissott's equation.

Again $V_1 = V(z + ih, t) = V(z, t) + ih\, \partial/\partial z\, V(z, t) + \ldots$

$V_2 = V(z + ih, t) = V(z, t) + ih\, \partial/\partial z\, V(z, t) + \ldots$

From equation (7) we have

$$\frac{\partial^2}{\partial t^2}[2V(z, t) + \text{terms containing } h^2 \text{ and higher orders}]$$

$$+ ig.\ 2ih\ \frac{\partial}{\partial z}\left[\frac{\partial}{\partial z}V(z,t) + \text{terms containing } h^2 \text{ and higher orders}\right] = 0$$

$$\text{or} \frac{\partial^2 V}{\partial z^2} = \frac{1}{gh}\frac{\partial^2 V}{\partial t^2} = \frac{1}{c^2}\frac{\partial^2 V}{\partial t^2}; gh = c^2, \quad \ldots (8)$$

which represents the wave equation whose general solution is given by

$$V = V_1(z + ct) + V_2(z - ct), \quad \ldots (9)$$

where V_1 and V_2 are arbitrary analytic functions, subject to the condition that V is real when $y = 0$.

Equating real and imaginary parts for the velocity potential and the stream function, we have

$$\phi(x, y, t) = \phi_1(x + ct, y) + \phi_2(x - ct, y),$$

$$\text{and } \psi(x, y, t) = \psi_1(x + ct, y) + \psi_2(x - ct, y), \quad \ldots (10)$$

Since V is real when $y = 0$ so $\psi(x, 0, t) = 0$, expanding by Maclaurin's theorem, we have

$$\phi = \phi(x, 0, t) + y\left[\frac{\partial}{\partial y}\phi(x,y,t)\right]_{y=0} + \ldots$$

The other terms in the expansion is infinitesimal as compared with the first because y lies between 0 and h. Thus, we have

$$\phi = \phi(x, 0, t) = \phi_1(x + ct) + \phi_2(x - ct),\ \psi = 0, \quad \ldots (11)$$

which is the complete solution of the long waves and represents that all particles which are in the same vertical plane have the same horizontal velocity $(-\partial\phi/\partial x)$ and so remain in the same vertical plane.

ENERGY OF A LONG WAVE

For a wave in a canal of rectangular section the potential energy is due to the elevation of depression of the liquid above the mean level. Mass of the liquid in an undisturbed state standing above a base dx is $\rho\eta$ dx, its centre of mass lies at a height $\frac{1}{2}$h. Thus the total potential energy, V for a unit breadth over the whole length of the wave is given by

$$V = \frac{1}{2} g\rho \int \eta^2 dx. \qquad ...(1)$$

Similarly the kinetic energy, T is given by

$$T = \frac{1}{2} g\rho \int u^2 dx. \qquad ...(2)$$

where u is the velocity of liquid due to the wave motion at points where the elevation is η. For a wave motion travelling in one direction, we have

$$\eta = (c/g)u^*$$

or $\eta^2 g = hu^2$; $C^2 = gh$.

From (2), we have

$$T = \frac{1}{2} \rho g \int \eta^2 dx.$$

From (1) and (3), we notice that the kinetic energy is equal to the potential energy and each equals to half the total energy at any instant.

LONG WAVES AT THE COMMON SURFACE OF TWO LIQUIDS BOUNDED ABOVE AND BELOW BY TWO FIXED HORIZONTAL PLANES

Consider ρ and ρ' be the densities of liquids, A and A′ the cross-sections of the two liquid streams and b the breadth of the common surface. Suppose on the whole mass of liquid, a velocity equal and opposite to the velocity of propagation U of the wave to make the wave form stationary. Let η be the elevation of the common surface due to the wave motion and u, u′ the small additional velocities due to the wave motions in the two liquids.

The equations of continuity in the two liquids

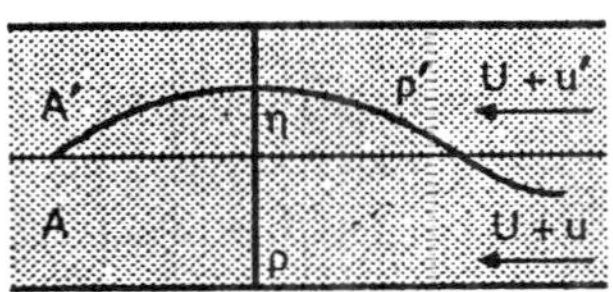

Fig. 1.6

$(A + b\eta)\ (U + u) = AU,$

$(A' + b\eta)\ (U + u') = A'\ U,$

or $Au + bU\eta = 0$ and $A'u' - bU\eta = 0,$ ...(1, 2)

neglecting the quantities of second order.

Let δp and $\delta p'$ denote increments of pressure close to the common surface in the two liquids due to the waves, then by Bernoulli's equation, we have

$$\frac{\delta p}{\rho} + g\eta + \frac{1}{2}(U + u)^2 = \frac{1}{2}U^2,$$

$$\frac{\delta p'}{\rho'} + g\eta + \frac{1}{2}(U + u') = \frac{1}{2}U^2, \qquad \text{...(3, 4)}$$

Neglecting the surface tension so that $\delta p = \delta p'$, we have

$g\ (\rho - \rho')\ \eta = (\rho'\ u' - \rho u)\ U$

Using the relation (1) and (2), we have

$$g\ (\rho - \rho')\ \eta = \left(\rho'\frac{bU\eta}{A'} + \rho\frac{bU\eta}{A}\right)U$$

or $g\ (\rho - \rho') = U^2 b\ (\rho/A + \rho'/A')$

or $$U^2 = \frac{g(\rho - \rho')}{b(\rho / A + \rho' / A')}.$$

SOLVED EXAMPLES

Example 1: *A fixed buoy in deep water is observed to rise and fall twenty times in a minute, prove that the velocity of the wave is about 10.5 miles per hour.*

Solution: Consider c be the velocity of propagation propagated over infinitely deep liquid, then $c^2 = (g\lambda/2\pi)$.

Frequency of the wave = 20 times/minute = $\frac{3g}{2\pi}$ ft/sec; $\lambda = 3c,$

$$\Rightarrow c^2 = \frac{3}{2} \times \frac{32 \times 7}{22} = \frac{168}{11} \text{ ft/sec.}$$

$$= \frac{168}{11} \times \frac{60 \times 60}{1760 \times 3} = 10.5 \text{ miles/hr. } \textbf{Proved.}$$

Example 2: *The crests of rollers which are directly following a ship 220 ft. long aer observed to overtake it at intervals of 16.5 seconds and it takes a crest 6 seconds to run along the ship. Find the length of the wave and the velocity of the ship.*

Solution: Let c represents the velocity of propagation of the wave and v that of ship, so in deep water, we have

$$c^2 = (g\lambda/2\pi). \quad \text{...(1)}$$

$$\text{and } (c - v) \frac{33}{2} = \lambda \Rightarrow (c - v)\, 6 = 220. \quad \text{...(2, 3)}$$

From (2) and (3), we have

$$\lambda = \frac{220}{6} \times \frac{33}{2} = 605 \text{ ft.}$$

Substituting the value of l in (1), we have

$$c = \sqrt{\left(\frac{32 \times 605 \times 7}{2 \times 22}\right)} \text{ ft/sec.} = 55 \text{ft./sec. (app.)} \quad \textbf{Ans.}$$

$$\text{and } v = \left(c - \frac{110}{3}\right) \text{ft./sec.} = \left(55 - \frac{110}{3}\right) \text{ft./sec.}$$

$$= 12\frac{1}{2} \text{ miles/hour.} \quad \textbf{Ans.}$$

Example 3: *Find the type of waves that would travel on deep water at 30 knots. How much is the velocity of wave affected by the presence of atmosphere above the water, its density being .0013?*

Solution: Let the fluid be at rest save for the wave motion, the wave velocity c is given by

$$c^2 = \frac{g}{m} \frac{\rho - \rho'}{\rho \coth mh + \rho' \coth mh'},$$

where ρ and ρ' are the densities of the fluids and h and h′, the depths of the fluids.

Since the wave is affected by the presence of atmosphere, so the wave velocity is given by

$$c^2 = \underset{h,h' \to \infty}{\text{Lt}} \frac{g(\rho - \rho')}{m(\rho \coth mh + \rho' \coth mh')}$$

$$\text{or } c^2 = \frac{g(\rho - \rho')}{m(\rho - \rho')} = \frac{g(1 - \rho'/\rho)}{m(1 + \rho'/\rho)} = \frac{g(1-\sigma)}{m(1+\sigma)}; \ \sigma = \rho'/\rho.$$

Let c_0 be the corresponding value of the wave velocity if the wave is not affected by the presence of atmosphere, then

$$c_0^2 = \frac{\lambda_0 g}{2\pi}, \text{ where c0} = 30 \text{ nm/hr} = \frac{30 \times 6082}{60 \times 60} \text{ ft/sec.}$$

$$\text{or } \lambda_0 = \frac{2\pi c_0^2}{g} = \frac{2 \times 22}{7 \times 33} \cdot \left(\frac{30 \times 6082}{60 \times 60} \right)^2 = 504 \text{ ft. (app.).}$$

If the atmospehere also be present then

$$\left(\frac{c}{c_0} \right)^2 = \frac{\lambda}{\lambda_0} = \frac{1-\sigma}{1+\sigma} = \frac{0.9987}{1.0013} = 0.9974 \text{ (app.). } \textbf{Ans.}$$

Example 4: *When simple harmonic waves of length l are propagated over the surface of deep water, prove that, at a point whose depth below the undisturbed surface is n, the pressure at the instant when the disturbed depth of the point is h + η bears to the undisturbed pressure at the same point in the ratio,*

$$1 + \frac{\eta}{h} e^{-2\pi h/\lambda;} 1,$$

atmospheric pressure and surface tension being neglected.

Solution: For a single harmonic wave profile

$$\eta = a \sin (mx - nt), \quad \text{...(1)}$$

the velocity at any point (x, y) is given by

$$\phi = \frac{an}{m} e^{my} \cos (mx - nt)$$

$$\text{or } \frac{\partial \phi}{\partial t} = \frac{an^2}{m} e^{my} \sin (mx - nt) = g\eta e^{my}; \ c^2 = \frac{\eta^2}{m^2} = \frac{g}{m} \quad \text{...(2)}$$

The pressure is given by Bernoulli's equation as

$$\frac{p}{\rho} + gy - \frac{\partial \phi}{\partial t} = \frac{\Pi}{\rho}, \quad \text{...(3)}$$

where Π is the atmospheric pressure at the surface.

From (2) and (3), we have

$$\frac{p-\Pi}{\rho} = [g\eta e^{-my} - gy]_{y=-h} = g\,(\eta e^{-mh} + h). \qquad ...(4)$$

Let p_0 be the pressure in the undisturbed state at a depth h, then

$$p_0 - \Pi = \rho gh. \qquad ...(5)$$

Neglecting the atmospheric pressure Π, we have

$$\frac{p}{p_0} = \frac{h+\eta e^{-mh}}{h} = \left\{1+\frac{\eta}{h}e^{-mh}\right\}.$$

Thus $p : p_0 : : 1 + \frac{\eta}{h}e^{-2\pi h/\lambda} : 1;\ m = 2\pi/\lambda$. **Proved.**

Example 5: *Two fluids of densities ρ_1, ρ_2 have a horizontal surface of separation but are otherwise unbounded. Shew that when waves of small amplitude are propagated at their common surface, the particles of the two fluids describe circles about their mean positions; and that at any point of the surface of the separation where the elevation is h, the particles on either sides have a relative velocity*

$4\pi c\eta/\lambda$.

Solution: Consider the wave profile at the surface be

$\eta = a \sin(mx - nt)$.

where $m = (2\pi/\lambda)$ and a, 1.

Let ϕ and ϕ' be the velocity potentials in the lower and upper fluids and may be represented as

$$\phi = Ae^{my}\cos(mx + nt),\ \phi' = Be^{-my}\cos(mx - nt). \qquad ...(2, 3)$$

Both the functions ϕ and ϕ' satisfy Laplace equation and are such that

$\frac{\partial\phi}{\partial y} \to 0$ as $y \to -\infty$ (for lower fluid).

$\frac{\partial\phi'}{\partial y} \to 0$ as $y \to +\infty$ (for upper fluid).

At the common surface, we have

$$\frac{\partial\eta}{\partial t} = \left(-\frac{\partial\phi}{\partial y}\right)_{y=0} = \left(-\frac{\partial\phi'}{\partial y}\right)_{y=0};\ \eta \text{ is small.}$$

From (1), (2) and (3), we have

$$-an\cos(mx - nt) = -\{Ame^{my}\cos(mx - nt)\}_{y=0}$$

$$= -\{-Bme^{-my}\cos(mx - nt)_{y=0}$$

or $-$ an cos (mx$-$ nt) $= -$ Am cos (mx $-$ nt) $=$ Bm cos (mx $-$ nt)

or A $=$ an/m and B $= -$ (an/m).

Substituting the values of A and B in (2) and (3), we get

$$\phi = \frac{an}{m} e^{my} \cos (mx - nt) = ace^{my} \cos m (x - ct),$$

and $$\phi' = -\frac{an}{m} e^{my} \cos (mx - nt) = - ace^{-my} \cos m (x - ct).$$

Consider (x_0, y_0) be the co-ordinates in the equilibrium position of any point (x, y) in the lower fluid, such that

$x = x_0 + X$ and $y = y_0 + Y$.

or $$\frac{dX}{dt} = x = -\frac{\partial \phi}{\partial x} = - acme^{my_0} \sin m (x_0 - ct), \qquad ...(4)$$

and $$\frac{\partial Y}{\partial t} = y = -\frac{\partial \phi}{\partial y} = - acme^{my_0} \cos m (x_0 - ct), \qquad ...(5)$$

Integrating (4) and (5) with regard to t, we have

$X = - ae^{my_0} \cos m (x_0 - ct)$; $Y = ae^{my_0} \sin m (x_0 - ct)$,

which gives $X^2 + Y^2 = a^2e^{2my_0}$. ...(6)

It represents that the path is a circle with centre at (x_0, y_0).

Similarly we can prove that the path of a particle in the upper fluid is also a circle.

Again the relative velocity at the interface becomes

$$= \left\{\left(-\frac{\partial \phi}{\partial x}\right) - \left(\frac{\partial \phi'}{\partial x}\right)\right\}_{y=0}$$

$= \{acme^{mj} \sin m (x - ct) + acme^{-my} \sin m (x - ct)\}_{y=0}$

$= 2acm \sin m (x - ct) = 2cm\eta = 4\pi c\eta/\lambda$,

and $$= \left\{\left(-\frac{\partial \phi}{\partial y}\right) - \left(-\frac{\partial \phi'}{\partial y}\right)\right\}_{y=0}$$

$= - amc \cos m (x - ct) \{e^{my} - e^{-my}\} = 0$.

Hence the relative velocity at the surface is equal to $4\pi c\eta/\lambda$. **Proved.**

Example 6: *If a canal of rectangular section contains a depth h of liquid of density ρ on which is superposed a depth h′ of liquid of density ρ', the free surface of the latter being exposed to constant atmospheric pressure, prove that the velocities of propagation of waves of length $(2\pi/m)$ are given by $c^2 = (gu/m)$,* *where* $\rho\ (u \coth mh - 1)\ (u \coth mh' - 1) = \rho'\ (1 - u^2)$.

Solution: We know that

$$c^4m^2\ (\rho \coth mh \coth mh' + \rho') - c^2\ mg\rho\ (\coth mh + \coth mh')$$
$$+ g^2\ (\rho - \rho') = 0 \quad ...(1)$$

Since $c^2 = (gm/m) \Rightarrow c^2m = gu$. ...(2)

From (1) and (2), We have

$$g^2u^2\ (\rho \coth mh \coth mh' + \rho') - g^2\rho u\ (\coth mh + \coth mh')$$
$$+ g^2\ (\rho - \rho') = 0.$$

or $\rho\ [u^2 \coth mh \coth mh' - (u \coth mh + \coth mh') + 1] = \lambda'\ (1 - u^2)$

or $\rho\ (u \coth mh - 1)\ (u \coth mh' - 1) = \rho'\ (1 - u^2)$, **Proved.**

Example 7: *If there be two liquids in a straight canal of uniform section, of densities* σ_1, σ_2 *and depths* l_1, l_2; *shew that the velocity of propagation of long waves is given by the equation*

$$[(c^2/l_1g) - 1]\ [(c^2/l_2g) - 1] = \sigma_2/\sigma_2.$$

where $\sigma_2 > \sigma_2$, and it is assumed that the liquid do not mix.

Solution: Substituting $\rho = \sigma_2$, $\rho' = \sigma_1$, $h = l_2$ and $h' = l_1$, in the equation (10) § 7.12, we have

$$c^4m^2\ (\sigma^2 \coth ml_2 \coth ml_1 + \sigma_1) - c^2\ gm\sigma_2\ (\coth ml_2 + \coth ml_1)$$
$$+ g^2\ (\sigma_2 - \sigma_1) = 0. \ ...(1)$$

From long waves, we have

$$\coth mh \approx (1/mh)$$

Therefore the equation (1) becomes

$$c^4m^2\left(\sigma^2 \frac{1}{ml_2}.\frac{1}{ml_1} + \sigma_1\right) - c^2gm\sigma_2\left(\frac{1}{ml_2} + \frac{1}{ml_1}\right)$$
$$+ g^2\ (\sigma_2 - \sigma_1) = 0.$$

$$\text{or} \quad \frac{c^4\sigma_2}{l_1l_2} + c^4m^2\sigma_1 - c^2g\sigma_2\left(\frac{1}{l_1} + \frac{1}{l_2}\right) + g^2\ (\sigma_2 - \sigma_1) = 0.$$

Again for long waves $m\ (= 2\pi/\lambda)$ is small; neglecting the terms containing m^2, we have

$$\frac{c^4\sigma_2}{l_1l_2} - c^2g\sigma_2\left(\frac{1}{l_1} + \frac{1}{l_2}\right) + g^2\sigma_2\left(1 - \frac{\sigma_1}{\sigma_2}\right) = 0,$$

$$\Rightarrow \quad \frac{c^4}{g^2 l_1 l_2} - \frac{c^2}{g}\left(\frac{1}{l_1} + \frac{1}{l_2}\right) + 1 = \frac{\sigma_1}{\sigma_2},$$

$$\Rightarrow \quad \left(\frac{c^2}{l_1 g} - 1\right)\left(\frac{c^2}{l_2 g} - 1\right) = \frac{\sigma_1}{\sigma_2}.$$

Proved.

Example 8: *Two dimensional waves length $2\pi/m$ are produced at the surface of separation of two liquids which are of densities ρ, ρ' ($\rho > \rho'$) and depths h, h′ confined between two fixed horizontal planes. Prove that, if the potential energy is reckoned zero in the position of equilibrium, the total energy of the lower liquid is to that of the upper in the ratio*

ρ {($2\rho - \rho'$) coth mh + ρ' coth mh′} : ρ'{($\rho - 2\rho'$) coth mh¢ − ρ coth nh}.

Solution: Let the wave profile be of the form

$$\eta = a \sin (mx - nt) = a \sin m (x - ct). \qquad ...(1)$$

Consider ϕ and ϕ' be the velocity potentials due to the lower and upper liquids, then

$$\phi = \frac{ac}{\sinh mh} \cosh m (y + h) \cos m (x - ct), \qquad ...(2)$$

$$\text{and } \phi' = \frac{ac}{\sinh mh'} \cosh m (y - h') \cos m (x - ct), \qquad ...(3)$$

where $c^2 (m\rho \coth mh + m\rho' \coth mh') = g (\rho - \rho')$. ...(4)

Let T be the kinetic energy of the lower liquid per wave length then

$$T = -\frac{1}{2}\rho \int \phi \frac{\partial \phi}{\partial n} ds = \frac{1}{2}\rho \int_0^\lambda \left(\phi \frac{\partial \phi}{\partial y}\right)_{y=0} dx$$

$$\text{or } T = \frac{1}{2}\rho m\, a^2 c^2 \coth mh \int_0^\lambda \cos^2 m(x - ct) dx$$

$$\text{or } T = \frac{1}{4}\rho\, \lambda a^2 c^2 m \coth mh. \qquad ...(5)$$

Also V be the potential energy in the lower liquid per wave length of liquid is given by

$$V = \frac{1}{2} gr \int_0^\lambda \eta^2 dx = \frac{1}{2} g\rho a^2 \int_0^\lambda \sin^2 m(x - ct) dx = \frac{1}{4} g\, \rho a^2 \lambda. \qquad ...(6)$$

Similarly kinetic energy of the upper liquid per wave length

$$= \frac{1}{4}\rho' \lambda a^2 c^2\, m \coth mh' \qquad ...(7)$$

and the potential energy is given by

$$= -\frac{1}{4}g\rho' a^2\lambda \qquad \text{...(8)}$$

Thus the total energy of lower liquid becomes

$$= \frac{1}{4}g\rho\, a^2\lambda \;+\; \frac{1}{4}\rho\lambda a^2c^2m \text{ coth mh}$$

$$= \frac{1}{4}g\rho a^2\lambda \left\{1+\frac{(\rho-\rho')\coth mh}{\rho\coth mh+\rho'\coth mh'}\right\}$$

$$= \frac{1}{4}g\rho a^2\lambda \; \frac{(2\rho-\rho')\coth mh+\rho'\coth mh'}{\rho\coth mh+\rho'\coth mh'}. \qquad \text{...(9)}$$

Also the total energy of upper liquid becomes

$$= \frac{1}{4}\rho'\lambda a^2c^2m \text{ coth } -\frac{1}{4}g\rho' a^2\lambda$$

$$= \frac{1}{4}g\rho' a^2\lambda \left\{\frac{(\rho-\rho')\coth mh'}{\rho\coth mh+\rho'\coth mh'}-1\right\}$$

$$= \frac{1}{4}g\rho' a^2\lambda \left\{\frac{(\rho-2\rho')\coth mh'-\rho\coth mh}{\rho\coth mh+\rho'\coth mh'}\right\} \qquad \text{...(10)}$$

Therefore the ratio of the total energy of lower liquid to that of the upper liquid is given by

ρ $\{2\rho - \rho')$ coth mh + ρ' coth mh + ρ' coth mh'$\}$:

ρ' $\{\rho - 2\rho')$ coth mh' – ρ coth mh$\}$ **Proved.**

Example 9: *An open rectangular box of length b contains two liquids of densities ρ, ρ' and depths h, h' respectively, that of density ρ being at the bottom. Prove that the periods of oscillation when the liquids are slightly disturbed so that there is no motion perpendicular to the sides of the box are determined by the equations of the type*

$$\left(p^2\coth\frac{n\pi h}{b}-\frac{gn\pi}{b}\right)\left(p^2\coth\frac{n\pi h'}{b}-\frac{gn\pi}{b}\right)+\frac{\rho}{\rho'}\left(p^4-\frac{g^2n^2\pi^2}{b^2}\right)=0,$$

where n is an integer.

Solution: Let the stationary wave propagated at the common surface be of the form

$$\eta = a \cos mx \cos pt. \qquad \text{...(1)}$$

Assuming ϕ and ϕ' be the velocity potentials due to the lower and upper liquids respectively. Since the upper surface of the upper liquid bears constant pressure so the velocity potentials are of the form

$\phi = c \cosh m (y + h) \cos mx \sin pt$ ...(2)

and $\phi' = (A \cosh my + B \sinh my) \cos mx \sin pt.$ ...(3)

At the common surface boundary conditions are

$$\frac{\partial \eta}{\partial t} = \left(-\frac{\partial \phi}{\partial y}\right)_{y=0} = \left(-\frac{\partial \phi'}{\partial y}\right)_{y=0} \quad ...(4)$$

$$\left\{\left(-\frac{\partial \phi}{\partial t} + g\eta\right)\rho\right\}_{y=0} = \left\{\left(-\frac{\partial \phi'}{\partial t} + g\eta\right)\rho'\right\}_{y=0} \quad ...(5)$$

$$\left\{\frac{\partial^2 \phi'}{\partial t^2} + g\frac{\partial \phi'}{\partial y}\right\}_{y=0} = 0. \quad ...(6)$$

From the conditions (4) we have

$- ap \cos mx \sin pt = - cm \sinh mh \cos mx \sin pt$

$= - Bm \cos mx \text{ pin } pt$

or $ap = cm \sinh mh = Bm$

or $c = \dfrac{ap}{m \sinh mh}$

and $B = \dfrac{ap}{m}$

Using the condition (5), we get

$\rho \{- cp \cosh mh + ga\} = \rho' \{- AP + ga\},$ ...(8)

Also from the condition (6), we have

$p^2 \{A \coth mh' + B\} = gm \{A + B \coth mh'\}.$ ...(9)

From (8), we have

$$A\rho' = \frac{pa\rho}{p} \coth mh - \frac{na}{p}(\rho - \rho').$$

Substituting the values of A, B and C in (9), we get

$$p^2 \left\{\coth mh'\left(\frac{ap}{m}\rho \coth mh - \frac{ga}{p}(\rho - \rho')\right) + \frac{ap}{m}\rho'\right\}$$

$$= gm \left\{\frac{ap}{m}\rho \coth mh - \frac{ga}{p}(\rho - \rho') + \rho' \coth mh'\right\}$$

or $(p^2 \coth mh - mg)(p^2 \coth mh' - mg) + \frac{\rho'}{\rho}(p^4 - m^2g^2) = 0$...(10)

But normal velocity of the fluid particles must vanish at the end x = b, such that

$(-\partial\phi/\partial x)_{x=b} = 0 \Rightarrow \sin mb = 0 \Rightarrow m = np/b,$

$$\text{or } \left(p^2 \coth\frac{n\pi h}{b} - \frac{n\pi g}{b}\right)\left(p^2 \coth\frac{n\pi h'}{b} - \frac{n\pi g}{b}\right) + \frac{\rho'}{\rho}\left(p^4 - \frac{n^2\pi^2 g^2}{b^2}\right) = 0.$$

Proved.

Example 10: *If a horizontal rectangular canal of great depth has two vertical barriers at a distance l apart, prove that the periods of oscillations of the water are* $2\sqrt{(\pi l/sg)}$, *where s is a positive integer; and that corresponding to any mode; all the particles of fluid oscillate in straight lines of length inversely proportional to exp. (sπz/l) where z is the depth.*

Solution: Let η be the elevation of the stationary waves at the surface, so that

$\eta = a \sin mx \cos nt$ ...(1)

$\phi = (an/m)\, e^{my} \sin nt \cos mx,\ n^2 = mg,$ ...(2)

Now $\partial\phi/\partial x = 0$ for x = 0 and l. ... (3)

From (2) and (3), we have

$\sin ml = 0 \Rightarrow ml = s\pi \Rightarrow m = s\pi/l$, s is a positive integer.

Periods of oscillation are given by

$$= \frac{2\pi}{n} = \frac{2\pi}{\sqrt{(mg)}} = 2\sqrt{\left(\frac{\pi l}{sg}\right)} \quad \text{Proved.}$$

Also $-\frac{\partial\phi}{\partial x} = u = \frac{dx}{dt} = an\, e^{my} \sin nt \sin mx,$ ...(4)

and $-\frac{\partial\phi}{\partial y} = v = \frac{dy}{dt} = -\, an\, e^{my} \sin nt \cos mx.$ ...(5)

Let (x_0, y_0) be the equilibrium points, then substituting

$x = x_0 + X,\ y = y_0 + Y$ in (5), we have

$X = an e^{my_0} \sin mx_0,\ Y = an\, e^{my_0} \sin nt \cos mx_0$ ...(6)

Since the path is a straight line

or $\frac{Y - B}{X - A} = -\cot mx_0 = \text{const.}$

and $\sqrt{\{(X-A)^2} + (Y - B)^2\} = a \cos nt\, e^{my}{}_0$ ...(7)

The maximum value of (7) is aemy0. From (9), we have

$= ae^{-my}{}_0,\ y_0 = -z$

$$= \frac{a}{e^{mz}} = \frac{a}{e^{(\pi sz/l)}}.$$

Thus all the particles of fluid oscillate in straight line of length inversely proportional to exp. (πsz/l). **Proved.**

Example 11: *Show that if the velocity of the wind is just great enough to prevent the propagation of waves of length λ against it, the velocity of propagation of waves with the wind is 2c {σ (1 + σ)}$^{1/2}$, where σ is is the specific gravity of the air and c the wave velocity when no air is present.*

Solution: Let V be the velocity of the wave, U and U′ the velocities of lower and upper fluids of height h and h′ respectively. Let ρ and ρ′ be densities of the fluids. We know that

$$g(\rho - \rho') = m\{(U - V)^2 \rho \coth mh + (U - V)\rho' \coth mh'\}$$

$$\text{or } g\left(1-\frac{\rho}{\rho}\right) = m\left\{(U-V)^2 \coth mh + (U'-V)^2 \frac{\rho'}{\rho}\coth mh'\right\} \quad ...(1)$$

Since the sea is at rest, so U = 0, h = h′ tends to ∞ and σ = ρ′/ρ

$$\text{or } g(1 - \sigma) = m\{V^2 + (V' - V)^2\sigma\}. \quad ...(2)$$

If no wind is present U′ = 0 and V = c, then

$$g(1 - \sigma) = m\{c^2 + c^2\sigma\}$$

$$\text{or } c^2 = \frac{g}{m}\frac{1-\sigma}{1+\sigma}. \quad ...(3)$$

The velocity U′ which prevents the propagation of waves is given by

$$g(1 - \sigma) = mU'^2\sigma. \quad ...(4)$$

Thus the velocity of propagation of the wind is

$$g(1 - \sigma) = m\{V^2 + (U' - V)^2\sigma\}$$

$$\text{or } \frac{g}{m}(1 - \sigma) = V^2(1 + \sigma) - 2U'V\sigma + U'^2\sigma$$

$$\text{or } U'^2\sigma = V^2(1 + \sigma) - 2U'V\sigma + U'^2\sigma$$

$$\text{or } V^2(1 + \sigma) - 2U'V\sigma = 0$$

$$\text{or } V = \frac{2\sigma}{1+\sigma}\sqrt{\left(\frac{g}{m}\frac{1-\sigma}{\sigma}\right)}$$

$$\text{or } V = \frac{2\sigma}{1+\sigma} c \sqrt{\left(\frac{1+\sigma}{1-\sigma}\right)} \sqrt{\left(\frac{1-\sigma}{\sigma}\right)} = 2c \sqrt{\left(\frac{\sigma}{1+\sigma}\right)}.$$ **Proved.**

Example 12: *A canal, of infinite length and rectangular section, is of uniform depth h and breadh b in one part but changes gradually to uniform depth h′ and breadth b′ in another part. An infinite train of simple harmonic waves travelling in one direction only is propagated along the canal. Prove that, if a, a′ are the heights and 2π/m, 2π/m′ the lengths of the waves in the two uniform portions,*

$$m \tanh mh = m' \tanh mh'$$

$$\text{and } a^2 b \operatorname{sech}^2 mh (\sinh 2mh + 2mh)$$

$$= a'^2 b' \operatorname{sech}^2 m'h' (\sinh 2m'h' + 2m'h').$$

Solution: Let the mean profiles of the wave in the two parts be

$$\eta = a \sin (mx - nt) = a \sin m (x - ct), \quad ...(1)$$

$$\text{and } \eta' = a' \sin (m'x - n't) = a' \sin m' (x - c't), \quad ...(2)$$

$$\text{So that } c^2 = \frac{g}{m} \tanh mh,$$

$$c'^2 = \frac{g}{m'} \tanh m' h', \quad ...(3)$$

$$\text{or } \frac{c^2}{c'^2} = \frac{m'}{m} \frac{\tanh mh}{\tanh m'h'}. \quad ...(4)$$

The period of simple harmonic wave must remain the same all along the canal which gives

$$\frac{2\pi}{n} = \frac{2\pi}{n'}$$

$$\text{or } \frac{m}{n} \cdot \frac{1}{m} = \frac{m'}{n'} \cdot \frac{1}{m'} \Rightarrow \frac{m'}{m} = \frac{c}{c'}.$$

$$\text{or } \frac{m'^2}{m^2} = \frac{m' \tanh mh}{m \tan m'h'} \Rightarrow \frac{m'}{m} = \frac{\tan mh}{\tanh m'h'}$$

$$\text{or } m' \tanh m'h' = m \tanh mh.$$ **Proved.**

Also, the energy transmitted shall also be the same in either part of the canal. Thus, we have

$$1/2\, g\rho a^2 bc (1 + 2mh \operatorname{cosech} 2mh)$$

$$= 1/4\, g\rho a'^2 b'c (1 + 2m'h' \operatorname{cosech} 2m'h)$$

or $\frac{a^2 b \sinh mh(2mh + \sinh 2mh)}{\cosh mh \sinh 2mh}$

$= \frac{a'^2 b' \sinh m'h'(2m'h' + \sinh 2m'h')}{\sinh 2m'h' \cosh m'h'}$

or $a^2 b \sec^2 mh(\sinh 2mh + 2mh)$

$= a'^2 b' \sec^2 m'h'(\sinh 2m'h' + 2m'h')$. **Proved.**

EXERCISE

1. Prove that the condition to be satisfied by the velocity potential f of a surface wave at the surface is

   $$\frac{\partial^2 \phi}{\partial t^2} + g\frac{\partial \phi}{\partial y} = 0,$$

   where g is the acceleration due to gravity, and y measures the vertical distance.

2. Prove that the velocity of propagation of a surface wave of wave length l in a sea of depth h, is given by

   $$c = \left[\left(\frac{gl}{2\pi} + \frac{2\pi\sigma}{\rho l}\right)\tanh\frac{2\pi h}{l}\right]^{1/2},$$

   where s is the surface tension and ρ, the density of the sea water.

3. A fluid of density ρ′ and depth h′ moves with velocity U′ over a fluid of density ρ, depth h and moving with velocity U. Prove that a wave η = a sin a (x – Ct) propagates in this media with velocity C given by

   $g(\rho - \rho') = a\rho'(U' - C)^2 \coth ah' + a\rho(U - C)^2 \coth ah.$

4. Find the velocity of ocean rollers, 20 yds long from crest to crest in miles/hour.

5. If in the irrotational motion homogeneous liquid in two dimensions under gravity there be a free surface exposed toan atmosphere of constant pressure. Find the surface of equal pressure.

6. The velocity of propagation of capillary waves of length 2π/m along a uniform canal of depth h is C, and ρ is the density of the liquid. Show that, if the waves are produced by a distribution of external surface pressure of the type P sin m (x – Vt) travelling with a velocity V > C, then the form of the surface is given by

$\eta = a \sin m (x - Vt)$, $a = P \tanh mh/\rho m (V^2 - C^2)$.

Discuss : (i) $V = C$, (ii) $V < C$.

7. If water of depth h be following with velocity proportional to the distance from the bottom. V being the velocity of the stream at the surface prove that the velocity U of propagation of waves in the direction of the stream is given by

   $(U - V)^2 + V (U - V) W^2/gh - W^2 = 0,$

   where W is the velocity propagation in still water.

8. A stream is running with mean velocity U in the plane-XY between a horizontal bottom $y = 0$ and a fixed upper boundary $y = h + a \cos mx$, where a is small. Find the character of the motion by determining its velocity potential or stream function.

2

Motion in Two Dimensions

ELLIPTIC COORDINATES

Now, we introduce the elliptic coordinates connected with x and y by the relation

$z = \cosh \zeta$, where $z = x + iy$, $\zeta = \xi + i\eta$

and c is a real constant. Then

$x = c \cosh \xi \cos \eta$, $y = c \sinh \xi \sin \eta$

or $$\frac{x^2}{c^2 \cosh^2 \xi} + \frac{y^2}{c^2 \sinh^2 \xi} = 1 \qquad ...(1)$$

and $$\frac{x^2}{c^2 \cos^2 \eta} - \frac{y^2}{c^2 \sin^2 \eta} = 1, \qquad ...(2)$$

where ξ varies from 0 to ∞, and η varies from 0 to 2π. Thus ξ = const.,and η = const., represent confocal ellipses and confocal hyperbolas respectively, and the distance between the focii ($\pm$ c, 0) in each case is 2c. The parameters ξ and η are called the *elliptic coordinates*.

When ξ = const = α (say) then, from (1), we have

$a = c \cosh \alpha, b = c \sinh \alpha$

Thus $a + b = ce^{\alpha}$, $a - b = ce^{-a}$; $a^2 - b^2 = c^2$ ...(3)

From (3), we have

$$\frac{a+b}{a-b} = \frac{ce^{a}}{ce^{-a}} = e^{2\alpha},\ \alpha = \frac{1}{2}\log\left(\frac{a+b}{a+b}\right).$$

The distance between the focii is given by

$$= 2ae = 2\sqrt{(a^2 - b^2)}\ ;\ b^2 = a^2\ (1 - e^2)$$

$$= 2c\sqrt{(\cosh^2 \alpha - \sinh^2 \alpha)} = 2c,$$

which shows that x = c cosh a cos η and y = c sinh α sin η are the coordinates of a point of equation (1), η is an ecentric angle of the point.

With the transformation z = c cosh ζ the equation

$$\frac{\partial^2 \psi}{\partial x^2} + \frac{\partial^2 \psi}{\partial y^2} = 0$$

transforms into $\frac{\partial^2 \psi}{\partial \xi^2} + \frac{\partial^2 \psi}{\partial \eta^2} = 0$ whose solution is of the type

$$\left.\begin{matrix} \cosh \\ \sinh \\ \exp. \end{matrix}\right\}(n\xi) \left.\begin{matrix} \cos \\ \\ \sin \end{matrix}\right\}(n\eta),$$

for positive or negative integral values of n, and that e–nx must be used when the property of vanishing at infinity is employed. Also, for confocal ellipses the form (A coshnξ + B sin h nξ) $\left.\begin{matrix} \cos \\ \sin \end{matrix}\right\}$ (nη) may be used.

MOTION OF AN ELLIPTIC CYLINDER

(a) To determine the stream function and velocity potential when an elliptic cylinder moves in an infinite liquid with velocity U parallel to the major axis of the cross-section.

The stream function for the most general type of motion of the cylinder is given by

$$y = Vx - Uy + \frac{1}{2}\omega\,(x^2 + y^2) + A.$$

where U and V are velocities of the cylinder moving parallel to the axes and rotating with an angular velocity $\underline{\omega}$.

Here V = 0, ω = 0 ⇒ ψ = – Uy + A. ...(2)

Let the cross-section of the cylinder be an ellipse, then the elliptic coordinates reduce to

x = c cosh α cos η and y = c sinh α sin η. ...(3)

Since the effect vanishes at infinity and sin η is the only variable factor in the boundary condition then the stream function ψ must be of the form $e^{-\xi}$ sin η. Assuming the complex potential of the form

$$\phi + i\psi = Be^{-(\xi + i\eta)} = Be^{-\xi}(\cos \eta - i \sin \eta)$$

$$\Rightarrow \quad \psi = -Be^{-\xi} \sin \eta. \qquad ...(4)$$

At the boundary $\xi = \alpha$, the relations (2) and (4) are same for all values of η, i.e.,

$-\text{Be}^{-\alpha} \sin \eta = -\text{Uc} \sinh \alpha \sin \eta + A,$

which gives A = 0 and $B = \text{Uc}\, e^{\alpha} \sinh \alpha$

From (4), we have

$$\psi = -\text{Uc}\, e^{\alpha-\xi} \sinh \alpha \sin \eta, \qquad ...(5)$$

determines the stream function which will make the boundary of the ellipse a stream line, when the cylinder moves parallel to its major axis with velocity U.

The relation (5) can be written as

$$\psi = -\text{Ub}\, e^{\alpha}\, e^{-\xi} \sin \eta = -\text{Ub}\sqrt{\left(\frac{a+b}{a-b}\right)}\, e^{-\xi} \sin \eta, \qquad ...(6)$$

Similarly $\phi = \text{Ub}\sqrt{\left(\frac{a+b}{a-b}\right)}\, e^{-\xi} \sin \eta$. (as $b = c \sinh \alpha$)

(b) *To determine the stream function and velocity potential when an elliptic cylinder moves in an infinite liquid with a velocity V parallel to the minor axis of the cross-section.*

The stream function for the most general type of motion of the cylinder is given by

$y = Vx - Uy + 1/2\, w\, (x^2 + y^2) + A.$

Here $U = 0$, $\omega = 0 \Rightarrow \psi = Vx + A.$ ...(7)

Since the effect is to vanish at infinity and cos η is the only variable factor in the boundary condition then ψ must be of the form $e^{-\xi}\text{con}\eta$. Assuming that

$$\psi = Be^{-\xi} \cos \eta. \qquad ....(8)$$

At the boundary $\xi = a$, the relations (7) and (8) are same for all values of η, we must have

$Be^{-\alpha} \cos \eta = Vc \cosh \alpha \cos \eta + A,$

which gives A = 0 and $B = Vce^{\alpha} \cosh \alpha$.

From (8), we have

$$\psi = Vce^{\alpha}\, e^{-\xi} \cosh \alpha \cos \eta = Va\sqrt{\left(\frac{a+b}{a-b}\right)}\, e^{-\xi} \cos \eta. \qquad ...(9)$$

Similarly $\phi = Va\sqrt{\left(\frac{a+b}{a-b}\right)}\, e^{-\xi} \sin \eta$. (as $a = c \cosh \alpha$)

The streamilines defined by (6) and (9) are in each case the same for all confocal elliptic forms of the cylinder. The ellipse reduces to the straight line joining the focii. In this case the relation (9) becomes

$\psi = Vce^{-\xi} \cos \eta$ and $\phi = Vce^{-\xi} \sin \eta$,

which gives the motion produced by an infinitely long lamina of breadth 2c moving in an infinite mass of liquid.

STREAMING PAST A FIXED ELLIPTIC CYLINDER

The complex potential representing the flow in the Z-plane due to an elliptic cylinder held in a stream of uniform velocity at infinity is given by

$$w = Ue^{-i\alpha} \left\{\zeta + \frac{(a+b)^2}{4\zeta}\right\} \qquad ...(1)$$

The complex potential describing flow due to an elliptic cylinder may be obtained by conformal transformation of the region outside the ellipse in the Z-plane into the region outside a circle in the ζ-plane. Let the transformation be

$z = \zeta + (c2/4z)$, where c is real constant

or $4\zeta^2 \dot{-} 4z\zeta - c^2 = 0$

$$\text{or } z = \frac{1}{2}\{z \pm \sqrt{(z^2 - c^2)}\} \qquad ...(2)$$

From (1) and (2), the complex potential takes the form

$$w = \frac{1}{2}U\left[e^{-i\alpha}\{z + \sqrt{(z^2 - c^2)}\} + \frac{e^{ia}}{c^2}(a+b)^2\{z - \sqrt{(z^2 - c^2)}\}\right] \qquad ...(3)$$

$$\text{or } w = \frac{1}{2}U\left[e^{i\alpha}.ce^{\zeta} + \frac{e^{i\alpha}}{c^2}(a+b)^2.ce^{-\zeta}\right]$$

as $z + \sqrt{z^2 - c^2} = c \cosh \zeta + c\sqrt{(\cos^2 \zeta - 1)}\,\zeta$

$= c \cosh \zeta + c \sin h\ z = ce^{\zeta}$

$$\text{or } w = \frac{1}{2}U\left[e^{i\alpha}.ce^{\zeta} + \frac{e^{i\alpha}}{c^2}(a+b)^2.ce^{-\zeta}\right]$$

$$\text{or } w = \frac{1}{2}U\sqrt{(a^2 - b^2)}\left\{e^{\zeta - i\alpha} + \frac{(a+b)^2}{(a^2 - b^2)}e^{-(\zeta - i\alpha)}\right\}$$

$$\text{or } w = \frac{1}{2}U\sqrt{\left(\frac{a+b}{a-b}\right)}\{(a-b)\,e^{\zeta - i\alpha} + (a+b)\,e^{-(\zeta - i\alpha)}\}$$

Since $a - b = ce^{-\alpha_1}$ and $a + b = ce^{\alpha_1}$

$$\text{or } w = \frac{1}{2}Uc\sqrt{\left(\frac{a+b}{a-b}\right)}\{ce^{-a_1}e^{\zeta - i\alpha} + ce^{a_1}e^{-(\zeta - i\alpha)}\}$$

$$\text{or } w = \frac{1}{2}Uc\sqrt{\left(\frac{ce_1^{a}}{ce_1^{-a}}\right)}[e^{\zeta - \alpha_1 - ia} + e^{ia - \zeta + a_1}]$$

$$\text{or } w = \frac{1}{2}Uce^{a_1}[e^{\zeta - \alpha_2 - ia} + e^{-(\zeta - a_1 - i\alpha)}]$$

$$\text{or } w = U(a + b)\cosh(\zeta - \alpha_1 - i\alpha), \quad \text{...(4)}$$

which is the complex potential for the streaming motion past a fixed elliptic cylinder.

Let the stream flows parallel to the real axis, then substituting a = 0 in (4), the complex potential becomes

$$w = U(a + b)\cosh(\zeta - a_1). \quad \text{...(5)}$$

Particular Case : Let the whole system is given a velocity U inclined at an angle α with the real axis, then the stream is reduced to rest and the cylinder moves with velocity U. The complex potential reduces to

$$w = \frac{U(a+b)^2 e^{i\alpha}}{4\zeta} = \frac{U(a+b)^2 e^{i\alpha}}{2\{z + \sqrt{(z^2 - c^2)}\}}$$

$$\text{or } w = \frac{U(a+b)^2}{2c^2}\{z - \sqrt{(z^2 - c^2)}\}e^{i\alpha} = \frac{U(a+b)^2}{2c}.e^{-\zeta}e^{i\alpha}$$

$$\text{or } w = \frac{1}{2}U(a + b)e^{a_1}e^{-(\xi + i\eta)}e^{i\alpha}$$

$$\text{or } w = \frac{1}{2}U(a + b)e^{\alpha_1 - \xi}e^{-i(\eta - \alpha)}$$

$$\text{or } \phi + i\psi = \frac{1}{2}U(a + b)e^{\alpha_1 - \xi}\{\cos(\eta - \alpha) - i\sin(\eta - \alpha)\}$$

Equating real and imaginary parts, we get

$$\phi = \frac{1}{2}U(a + b)e^{\alpha_1 - \xi}\cos(\eta - \alpha),$$

$$\text{and } \psi = \frac{1}{2}U(a + b)e^{\alpha_1 - \xi}\sin(\eta - \alpha), \quad \text{...(6)}$$

which gives the velocity potential and stream function for an elliptic cylinder moving in an infinite with velocity U inclined at an angle a with the real axis.

If the elliptic cylinder moves parallel to X-axis, then putting a = 0 in (6), we have

$$\phi = \frac{1}{2}U(a + b)\, e_1^{\alpha-\xi} \cos\eta\,,$$

$$\phi = \frac{1}{2}U(a + b)\, e_1^{\alpha-\xi} \sin\eta\,.$$

ELLIPTIC CYLINDER ROTATING IN AN INFINITE MASS OF LIQUID AT REST AT INFINITY

We know that the stream function for the most general type motion of the cylinder is

$\psi = Vx - Uy + 1/2\ \omega\ (x^2 + y^2) + A,$

where U and V are velocity components parallel to the axes of X and Y and ω be the angular velocity.

Being a simple rotation U = V = 0, then

$\psi = 1/2\ \omega\ (x^2 + y^2) + A$

or $\psi = 1/2\ \omega\ (c^2\cosh^2 \xi\ \cos^2 h + c^2 \sinh^2 x\ \sin^2 \eta) + A.$

On the boundary of the elliptic cylinder x = a, we have

$\psi = 1/2\ \omega c^2\ (\cosh^2 a\ \cos^2\eta + \sinh^2 \alpha\ \sin^2\eta) + A$

or $y = 1/8\ \omega c^2\ [\ (1 + \cos 2\eta)\ (1 + \cosh 2\alpha)$

$+ (1 - \cos 2\eta)\ (\cosh 2\alpha - 1)] + A$

or $y = 1/2\ wc2\ (\cos 2\eta + \cosh 2\alpha) + A.$ ...(2)

Since the velocity vanishes at infinity and η is the only variable, then the stream function ψ must be of the form $e^{-2\xi} \cos 2\eta$.

Assuming $y = Be^{-2\xi} \cos 2\eta$

The relations (1) and (2) are same on the boundary ξ = α, then

$Be^{-2\alpha} \cos 2h = 1/4\ \omega c^2\ (\cos 2\eta + \cosh 2\alpha) + A.$

$\Rightarrow B = 1/4\ \omega c^2 e^{2\alpha}$ and $A = 1/4\ \omega c^2 \cosh 2\alpha.$

Substituting the value of B in (2), we have

$\psi = 1/4\ \omega c^2 e^{2\alpha} e^{-2\xi} \cos 2\eta$

or $\psi = 1/4\ \omega\ (a + b)^2\ e^{-2\xi} \cos 2\eta$, where $ce^{\alpha} = a + b$

Similarly $\phi = 1/4\ \omega\ (a + b)^2\ e^{-2\xi} \sin 2\eta,$

Thus the complex potential w is given by

$w = \phi + i\psi = 1/4\ \omega\ (a + b)^2\ e^{-2\xi}\ (\sin 2\eta + i \cos 2\eta)$

$w = 1/4\ \omega\ (a + b)^2\ e^{-2\xi}$.

KINETIC ENERGY OF ROTATING ELLIPTIC CYLINDER

To determine the kinetic energy when an cylinder rotates in an infinite mass of liquid at rest at infinity.

When an elliptic cylinder rotates in an infinite mass of liquid at rest at infinity, the stream function ψ and the velocity potential ϕ are given as

$\psi = 1/4\ \omega\ (a + b)^2\ e^{-2a} \cos 2\eta$

and $\phi = 1/4\ \omega\ (a + b)^2\ e^{-2a} \sin 2\eta$

where η varies from 0 to 2π.

Let T be the kinetic energy of the system, then

$$T = -\frac{1}{2}\rho \int \phi d\psi$$

$$\text{or } T = \frac{1}{2}\rho \int_0^{2\pi} \frac{1}{4}\omega(a+b)^2 e^{-2a} \sin \eta$$

$$\left\{\frac{1}{2}\omega(a+b)^2 e^{-2\alpha} \sin 2\eta\right\} d\eta$$

$$\text{or } T = \frac{1}{16}\rho\omega^2\ (a + b)^4\ e^{-4\alpha} \int_0^{2\pi} \sin^2 2\eta d\eta$$

$$\text{or } T = \frac{1}{16}\pi\omega^2\ (a + b)^4\ e^{-4\alpha}$$

$$\text{or } T = \frac{1}{16}\pi\rho\omega^2 c^4 = \frac{1}{16}\pi\rho\omega^2\ (a^2 + b^2)^2.$$

KINETIC ENERGY WHEN THE LIQUID CONTAINED IN A ROTATING ELLIPTIC CYLINDER

Let $w = -iAz^2 = -Ar^2e^{2\theta i}$ ...(1)

or $\phi + i\psi = -iAr^2 (\cos 2\theta + i \sin 2\theta)$

Equating real and imaginary parts, we get

$\phi = Ar^2 \sin 2\theta$ and $\psi = -Ar^2 \cos 2\theta$,

or $\phi = 2A\ xy$ and $\psi = -A\ (x^2 - y^2)$. ...(2)

The stream function for the most general type of motion of the cylinder is given by

$\psi = (Vx - Uy) + 1/2\ \omega\ (x^2 + y^2) + C$

Since V = U = 0, so $\psi = \frac{1}{2}\omega\ (x^2 + y^2) + C$ ...(3)

Equating (2) and (3), we have]

$$-A\ (x^2 - y^2) = \frac{1}{2}\ \omega\ (x^2 + y^2) + C$$

or $\left(\frac{1}{2}\omega + A\right)x^2 + \left(\frac{1}{2}\omega - A\right)y^2 = -C$

or $\dfrac{x^2}{\left(-C/\left(\frac{1}{2}\omega + A\right)\right)} + \dfrac{y^2}{\left\{-C/\left(\frac{1}{2}\omega - A\right)\right\}} = 1$

Comparing (4) with the equation to the ellipse, we get

$$C = -\frac{\omega a^2 b^2}{a^2 + b^2} \text{ and } A = \frac{1}{2}\omega\frac{b^2 - a^2}{b^2 + a^2}. \quad ...(5)$$

From (2), (3) and (5), we have

$$\psi = \frac{1}{2}\omega\ (x^2 + y^2) - \omega\frac{a^2 b^2}{a^2 + b^2}$$

and $\phi = \omega\ \dfrac{b^2 - a^2}{b^2 + a^2}xy,$

which determines the motion of the liquid in the rotating elliptic cylinder

Let T be the kinetic energy, then

$$T = \frac{1}{2}\rho\iint q^2 dxdy\text{ , where } q^2 = \left[-\frac{\partial\phi}{\partial x}\right]^2 + \left[-\frac{\partial\phi}{\partial y}\right]^2$$

or $q^2 = \left[-\dfrac{\omega(b^2 - a^2)}{(b^2 + a^2)}y\right]^2 + \left[-\dfrac{\omega(b^2 - a^2)}{(b^2 + a^2)}x\right]^2$

or $q^2 = \omega\left[\dfrac{b^2 + a^2}{b^2 + a^2}\right]^2 (x^2 + y^2).$

or $T = \dfrac{1}{2}\rho\omega^2\left[\dfrac{b^2 - a^2}{b^2 + a^2}\right]^2 \iint(x^2 + y^2)dxdy$

or $T = \dfrac{1}{2}\rho\omega^2\left(\dfrac{a^2 - b^2}{a^2 + b^2}\right)\left[\iint x^2 dxdy + \iint y^2 dxdy\right]$

$$\text{or } T = \frac{1}{2}\rho\omega\left(\frac{a^2+b^2}{a^2+b^2}\right)^2\left\{\pi ab.\frac{a^2}{4} \pi ab.\frac{b^2}{4}\right\}$$

$$\text{or } T = \frac{1}{8}\pi\rho ab\omega^2\,\frac{(a^2-b^2)^2}{a^2+b^2}\text{ . Ans.}$$

(a) To determine the K. E. when the liquid contained in a rotating prism whose section is an equilateral prism.

Consider the complex potential w be of the form

$w = i\,A\,z^3$.

Proceed as in the above case.

SCHWARZ-CHIRSTOFFEL THEOREM

The transformation from the ζ-plane to the Z-plane is defined by

$$\frac{dz}{d\zeta} = K(\zeta-a_1)^{\frac{\theta_1}{\pi}-1}(\zeta-a_2)^{\frac{\theta_2}{\pi}-1}\ldots.(\zeta-a_n)^{\frac{\theta_n}{\pi}-1}$$

which transforms the real axis (η = 0) in the z (= $\xi + i\eta$) -plane into the boundary of a closed polygon in the z (= x + iy)- plane in such a way that the vertices of the simple closed polygon correspond to the points $a_1, a_2, \ldots, a_n$ and the interior angles of the polygon $\theta_1, \theta_2, \ldots, \theta_n$. K is a constant which may be complex.

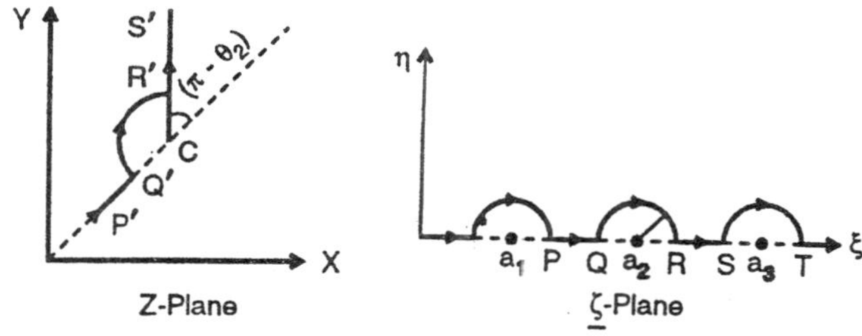

Fig. 2.1

Let $a_1, < a_2, \ldots., a_n$ be n points on the real axis in the ζ-plane such that $a_1 < a_2 < \ldots < a_n$, and $\theta_1, \theta_2, \ldots, \theta_n$ be interior angles of a simple closed polygon of n vertices so that

$$\theta_1 + \theta_2 + \ldots + \theta_n = (n-2)\,\pi.$$

The transformation from the z-plane to the Z-plane is given by

$$\frac{dz}{d\zeta} = K(\zeta - a_1)^{\frac{\theta_1}{\pi}-1}(\zeta - a_2)^{\frac{\theta_2}{\pi}-1}...(\zeta - a_n)^{\frac{\theta_n}{\pi}-1} \qquad ...(1)$$

Since $(\zeta - a_m)$ vanishes at $\zeta = a_m$, it follows that $dz/dz = 0$ if $a_m > \pi$ and $dz/d\zeta = \infty$ if $am < \pi$, for $m = 1, 2, ..., n$. Therefore, we avoid the points $a_1, a_2,...a_m, a_n$ on the real axis by surrounding them by means of semi-circles with these points as centres in the upper half of the ζ-plane.

The semi-circles with centres $a_1, a_2,...,a_n$ cut the real axis $(\eta = 0)$ between the points P, Q, R, S, T, ...Let P′, Q′, R′, S′, T′,... be the points in the Z-plane which correspond to the points P, Q, R, S, T,...in the ζ-plane.

Let $K = \lambda\ e^{i\beta}$, where λ is a real positive constant and b is real. Taking argument of both sides of (1), we have

$$\text{arg. } (dz) - \text{arg. } (d\zeta)$$

$$= 1 + \left(\frac{\theta_1}{\pi} - 1\right) \text{arg. } (\zeta - a_1) + \left(\frac{\theta_2}{\pi} - 1\right) \text{arg. } (\zeta - a_2) + ...+ \left(\frac{\theta_n}{\pi} - 1\right) \text{arg. } (\zeta - a_n).$$

$$= \lambda + \sum_{m=1}^{n} \left(\frac{\theta_m}{\pi} - 1\right) \text{arg. } (\zeta - a_m).$$

arg. $(d\zeta)$ remains zero as ζ moves from P to Q. arg $(\zeta - a_1)$ is zero because $(\zeta - a_1)$ is real and positive where as each of arg. $(\zeta - a_m)$, $(m = 2, 3,...n)$ is equal to π, since $\Sigma\ (\zeta - a_m)$, $m = 2, 3...$are all real and negative. Thus

$$\text{arg. } (dz) = \lambda + \left(\frac{\theta_2}{\pi} - 1\right)\pi + \left(\frac{\theta_3}{\pi} - 1\right)\pi + ... + \left(\frac{\theta_n}{\pi} - 1\right)\pi,$$

$$\Rightarrow \text{arg. } (dz) = \lambda + (\theta_2 + \theta_3 + ...\theta_n) - (n - 1)\pi,$$

$$\Rightarrow \text{arg. } (dz) = \lambda + [(n - 2)\pi - \theta_1] - (n - 1)\pi = \lambda - \theta_1 - \pi.$$

It follows that arg. (dz) is consistant as ζ moves from P to Q therefore z describes a straight line segment P′ Q′.

Likewise z describes the straight line R′ S′ as ζ moves from R to S *i.e.*, arg. $(dz) = 1 - \theta_1 - \theta_2$.

$$\text{arg. } (dz) \text{ on } R'S' - \text{arg. } (dz) \text{ on } P'Q' = \lambda - \theta_1 - \theta_2 - \lambda + \theta_1 + \pi$$

$$= \pi - \theta_2,$$

Thus the direction of motion of z has turned through the angle $(\pi - \theta_2)$ in the positive sense. Now, on the semi circle QR

$\zeta - a_2 = re^{\theta i}$, $d\zeta = ire^{\theta i}\, d\theta$.

Assuming r to be infinitesimal, we have

$$dz/\,(ire^{\theta i}\, d\theta) = \lambda\, e^{i\beta}\, (a_2 - a_1)^{\frac{\theta_1}{\pi}-1} (re^{\theta i})^{\frac{\theta_2}{\pi}-1} (a_2 - a_1)^{\frac{\theta_3}{\pi}-1} \ldots$$

$$\frac{dz}{d\theta} = ir\, e^{\theta i}.\, \lambda e^{i\beta}\, (re^{\theta i})^{\frac{\theta_2}{\pi}-1} (a_2 - a_1)^{\frac{\theta_1}{\pi}-1} (a_2 - a_3)^{\frac{\theta_3}{\pi}-1} \ldots$$

$$\frac{dz}{d\theta} = ir\frac{\theta_2}{\pi}{}^{i}_{e}\left(\beta + \frac{\theta_2}{\pi}\theta\right)F, \qquad \ldots(2)$$

where F is independent of r and θ.

Integrating (2) with regard to θ, we have

$$z = z_1 + \frac{\pi}{\theta_2} r\frac{\theta_2}{\pi_e} i\left(\beta + \frac{\theta_2}{\pi}\theta\right)F, \qquad \ldots(3)$$

where z_1 is an integration constant.

Since θ_2 is positive, we notice that $z \to z_1$ when $r \to 0$, so that z_1 is the point C where the lines P′Q′ and R′ S′ meet.

Thus the transformation makes z to describe a polygon whose vertices correspond to the points a_1, a_2, a_3,...,a_n and whose interior angles are θ_1, θ_2,...θ_n.

Again from (3), we have

$$\arg.\,(z - z_1) = \beta + \frac{\theta_2}{\pi}\theta + \arg.F.$$

As ζ describes the semi-circle, θ decreases form π to 0. arg. $(z - z_1)$ decreases by θ_2, and therefore z describes a circular are with centre C situated inside the polygon when it is a simple polygon. Thus the points in the upper half of the ζ-plane correspond to points with in the polygon.

When a vertex of the polygon in Z-plane corresponds to a point at infinity on the real axis of the ζ-plane (η = 0) *i.e.*, let $a_1 \to -\infty$, the transformation (1) can be written by suitable choice of λ as

$$\frac{dz}{d\zeta} = \lambda\, e^{i\beta}\, (-a_1)^{\frac{\theta_1}{\pi}-1} (\zeta - a_1)^{\frac{\theta_1}{\pi}-1} (\zeta - a_2)^{\frac{\theta_2}{\pi}-1} \ldots$$

$$\frac{dz}{d\zeta} = \lambda e^{i\beta}\left(\frac{\zeta - a_1}{a_1}\right)^{\frac{\theta_1}{\pi}-1} (\zeta - a_2)^{\frac{\theta_2}{\pi}-1} \ldots.$$

When $a_1 \to \infty$, $\left(\dfrac{\zeta - a_1}{a_1}\right)^{\frac{\theta_1}{\pi}-1} \to 1$, and the transformation reduces to

$$\frac{dz}{d\zeta} = \lambda\, e^{i\beta}\, (\zeta - a_2)^{\frac{\theta_2}{\pi}-1}\, (\zeta - a_3)^{\frac{\theta_3}{\pi}-1} \ldots$$

Thus the factor corresponding to $a_1 = -\infty$ vanishes from the equation of transformation.

TRANSFORMATION OF A SEMI-INFINITE STRIPS

(I) Semi-infinite Strip

Consider a semi-infinite strip P_∞, QRS_∞ in Z-plane bounded by the lines $x = 0$, $y = 0$, with two vertices at infinity. By transforming P_∞, Q, R on the points $\zeta = -\infty$, $\zeta = -1$, $\zeta = 1$ of the real axis in the ζ-plane, then in accordance with the Schwarz and Christoffel theorem,

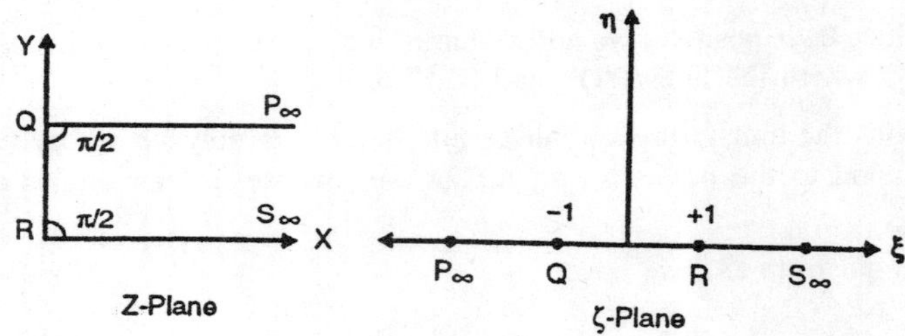

Fig. 2.2

we have

Here $a_1 = -\infty$, $a_2 = -1$, $a_3 = 1$, $a_4 = +\infty$, $\theta_2 = \pi/2 = \theta_3$.

then $\dfrac{dz}{d\zeta} = K\,(\zeta + 1)\,(\zeta+1)^{-\frac{1}{2}}(\zeta-1)^{-\frac{1}{2}} = \dfrac{K}{\sqrt{(\zeta-1)}}$ ...(1)

By integrating (1) with regard to ζ, we have

$z = K \cosh^{-1} \zeta + C.$

When $z = 0$, $\zeta = 1$, equation (2) gives $C = 0$.

When $z = ai$, $\zeta = -1$, equation (2) gives $ai = K(\pi i) \Rightarrow K = a/\pi$.

Thus $z = (a/\pi) \cosh^{-1}\zeta \Rightarrow \zeta = \cosh(\pi z/a)$.

(II) Infinite Strip

Consider an infinite strip P_∞ Q_∞ R_∞ S_∞ in Z-plane bounded by the line $y = 0$ and $y = a$. To transform it into the upper half of the

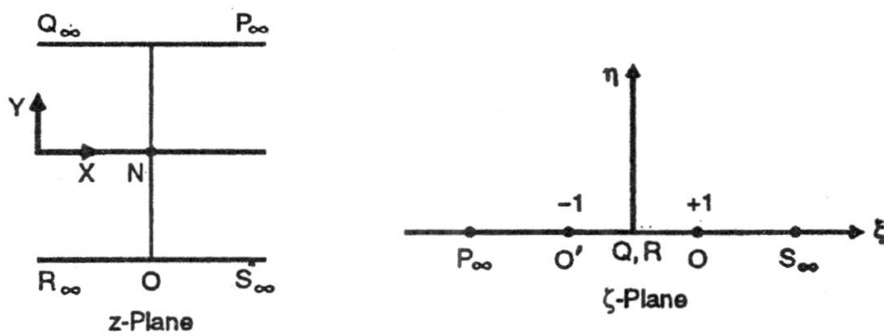

Fig. 2.3

ζ-plane, taking Q_∞, R_∞, as coincident and map the points Q_∞, R_∞ on $\xi = 0$, the origin O on $\zeta = 1$ and O′ (z = ai) on $\xi = -1$. Hence by Schwarz-Christoffel theorem, we have

$$\frac{dz}{d\zeta} = K\,\zeta^{-1} \Rightarrow z = K \log \zeta + D.$$

Here $a_1 = -\infty = a_4$. Also the angle at Q_∞, R_∞ is zero

$\Rightarrow \theta_2 = 0 = \theta_3$. ...(1)

When z = 0 ζ = 1, equation (1) gives 0 = K log 1 + D.

When z = ai, ζ = – 1, equation (1) gives ai = k log (–1) + D.

Thus we must have D = 0 and ai = kπi ⇒ K = (a/π).

Therefore $z = (a/\pi) \log \zeta \Rightarrow \zeta = \exp. (\pi z/a)$. ...(2)

The lines x = const. transform into circles $|\zeta|$ = const. The lines y = const. transform into lines arg. z = const. radiating from the origin in ζ-plane.

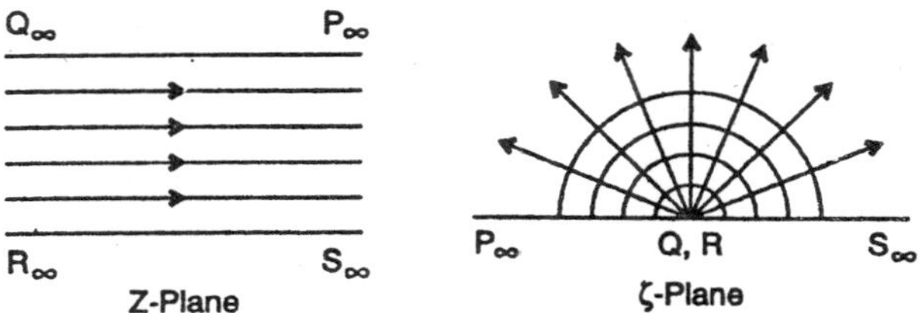

Fig. 2.4

If the origin is considered at the point M midway between the walls in the Z-plane then the corresponding transformation is obtained by substituting (z + ia/2) for z in (2).

Therefore $z = (a/\pi) \log \zeta - ia/2 \Rightarrow \zeta = i \exp. (\pi z/a)$.

FLOW INTO A CHANNEL THROUGH A NARROW SLIT IN A WALL

Let one side of the channel P_∞ Q_∞ R_∞ S_∞ bounded by the lines y = 0 and y = a in Z-plane be taken as a real axis. Let the slit be taken at the origin. The infinite strip in the Z-plane is transformed into the upper half of the real axis is ζ-plane by the transformation $\zeta = \exp. (\pi z/a)$. The origin O in the Z-plane transforms to ζ = 1 and Q_∞, R_∞ coincide at the origin (Q, R) of the ζ-plane.

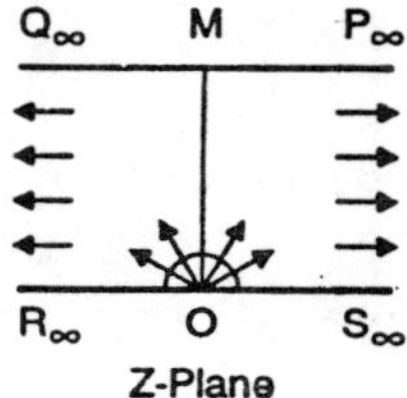

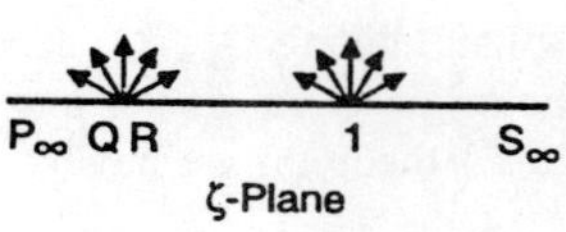

Fig. 2.5

Let (+ m) be the strength of the source then flow at O upwards in the Z-plane per unit time is πm. The flow at O will be due to a source or output 2πm. At infinite distance from O there will be parallel flow, and therefore at P_∞ and Q_∞ there will be sinks of strength m/4. Thus, there will be a sink of strength (m/2) at (Q, R) the origin and a source of strength m at ζ = 1in the ζ-plane. The complex potential is given by

$$w = -m \log (\zeta - 1) + \frac{1}{2} m \log (\zeta)$$

$$w = -m \log (\zeta^{1/2} - \zeta^{-1/2})$$

$$\text{or } w = -m \log (e^{\pi z/2a} - e^{-\pi z/2a}) = -m\log\{2 \sinh (\pi z/2a)\}$$

or $w = -m \log \{\sinh (\pi z/2a)\}$, neglecting the constant term.

Again $\dfrac{dw}{dz} = -(m\pi/2a) \coth (\pi z/2a)$ which vanishes at z = ai.

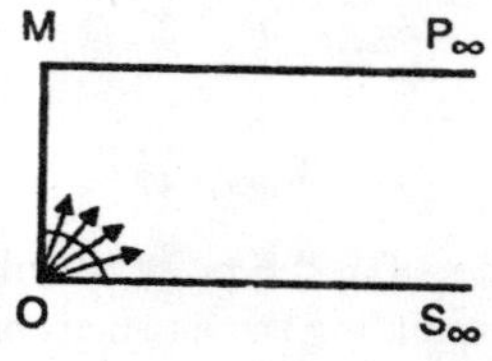

Fig. 2.6

It follows that the dividing stream line moves from O to the opposite wall so there should be a stagnation point at this point M (z = ai). Thus the pressure at points of $P_\infty Q_\infty$ is maximum at M, and is therefore smaller at the remaining points of the wall. The wall be drawn towards the other wall i.e., the channel will collapse. The velocity at an infinite distance from the origin is mπ/2a

Consider the streamline OM to be a rigid wall, we get the motion with in a semi-infinite rectangular channel due to a source at one corner.

FLOW PAST A STEP IN A DEEP STREAM

The step $P_\infty QRS_\infty$ in Z-plane bounded by the straight lines y = a, x = 0 and y = 0 is transformed into ζ-axis of ζ-plane by using schwarz-Christoffel transformation

$$\frac{dz}{d\zeta} = k\ (\zeta - a_2)^{\frac{\theta_2}{\pi}-1}\ (\zeta - a_3)^{\frac{\theta_3}{\rho}-1}\ (\zeta - a_4)^{\frac{\theta_4}{\pi}-1} \qquad ...(1)$$

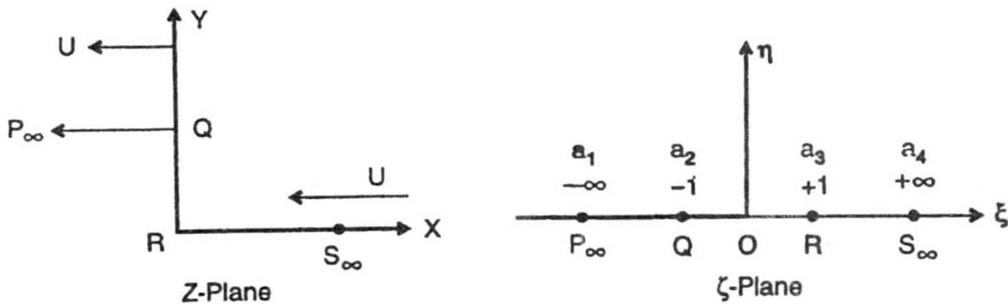

Fig. 2.7

Here $a_1 = -\infty$, $a_2 = -1$, $a_3 = +1$, $a_4 = +\infty$,

$\theta_2 = \frac{3\pi}{2}$ and $\theta_3 = \frac{\pi}{2}$

The terms corresponding to a_1, a_4 will not occur in the transformation (1), hence it reduces to

$$\frac{dz}{d\zeta} = k\ (\zeta + 1)^{-1/2}\ (\zeta - 1)^{-1/2} = k\sqrt{\left(\frac{\zeta+1}{\zeta-1}\right)}$$

$$\Rightarrow \frac{dz}{d\zeta} = k\ \frac{\zeta+1}{\sqrt{(\zeta^2-1)}} \Rightarrow dz = k\left[\frac{\zeta}{\sqrt{(\zeta^2-1)}} + \frac{1}{\sqrt{(\zeta^2-1)}}\right]d\zeta$$

By integrating, we have

$z = k\ [\sqrt{(\zeta^2 - 1) + \cosh^{-1}\zeta}] + A,$

when $z = 0,\ \zeta = 1 \Rightarrow A + k\cosh^{-1}1 = 0 \Rightarrow A + k\log 1 = 0 \Rightarrow A = 0$

Since $\cosh^{-1}\zeta = \log\{\zeta + \sqrt{(\zeta^2 - 1)}\}$.

Here $z = k\ [\sqrt{(\zeta^2 - 1) + \cosh^{-1}\zeta}]$

when $z = ai,\ \zeta = -1 \Rightarrow ai = k\cosh^{-1}(-1) \Rightarrow ai = k\log(-1)$

$\Rightarrow ai = k\log e^{\pi i} = k\pi i \Rightarrow k = \dfrac{a}{\pi}$

$$\Rightarrow z = \frac{a}{\pi}\left[\sqrt{\zeta^2 - 1} + \cosh^{-1}\zeta\right]$$

Let V be the velocity of the stream in ζ-plane, then the complex potential w is given by

$$w = V\zeta$$

$$\frac{dw}{dz} = \frac{dw}{d\zeta}\frac{d\zeta}{dz} = \frac{V}{k}\sqrt{\left(\frac{\zeta - 1}{\zeta + 1}\right)}$$

$$\lim_{\zeta \to \infty}\frac{dw}{dz} = \frac{V}{k}$$

Again $\dfrac{dw}{dz} = U,\ \zeta = \infty$

Thus $U = \dfrac{V}{k} = \dfrac{V\pi}{a} \Rightarrow V = \dfrac{aU}{\Pi} \Rightarrow w = \left(\dfrac{aU}{\Pi}\right)\zeta.$

FLOW PAST A STEP IN A CHANNEL

The breadth of the channel is a at the end $P_\infty S_\infty$ and b at the end $Q_\infty R_\infty$. Let U be the velocity at P_∞ then velocity at Q_∞ is $\dfrac{Ua}{b}$. $Q_\infty R_\infty$ coincide with origin in ζ-plane. A transforms to $\zeta = 1$ and B transforms to $\zeta = c$ then

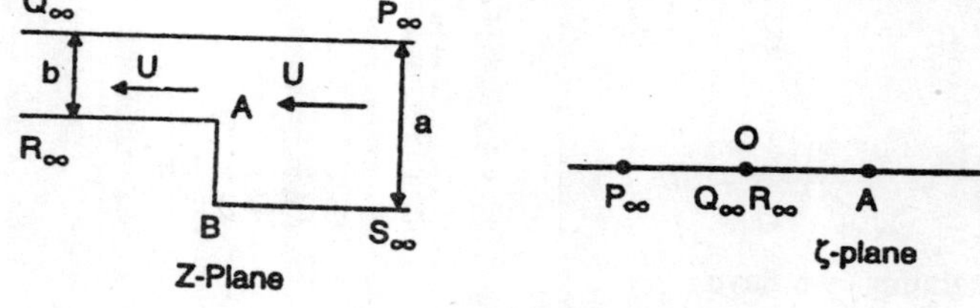

Fig. 2.8

$$\frac{dz}{d\zeta} = k\,\zeta^{-1}\;(\zeta - c)^{\frac{1}{2}}(\zeta - c)^{\frac{1}{2}}.$$

At P_∞ the flow is from a source of output Ua and at Q_∞ the flow is due to a sink of intake Ua. In z- plane, there is a sink at the origin of strength $\frac{Ua}{\Pi}$. The complex potential is given by

$$w = \frac{Ua}{\Pi}\;\log\zeta$$

$$\Rightarrow\;\frac{dw}{dz} = \frac{dw}{d\zeta}.\frac{d\zeta}{dz} = \frac{Ua}{\pi\zeta}.\frac{\zeta(\zeta - c)^{1/2}}{k(\zeta - 1)^{1/2}} = \frac{Ua}{\pi k}\sqrt{\left(\frac{\zeta - c}{\zeta - 1}\right)} \qquad ...(1)$$

As P_∞, $\zeta = \infty$, $\frac{dw}{dz} = U \;\Rightarrow U = \frac{Ua}{\Pi k} \Rightarrow k = \frac{a}{\Pi}$.

As Q_∞ $\zeta = 0$, $\frac{dw}{dz} = \frac{Ua}{b} \Rightarrow \frac{Ua}{b} \Rightarrow = \frac{Ua}{\Pi k}\sqrt{c} \Rightarrow \sqrt{c} = \frac{\Pi k}{n} = \frac{a}{b}$

Substituting $z = \frac{c^2 - t^2}{1 - t^2}$ in (1), we have

$$\frac{\zeta - 1}{\zeta - c^2} = \frac{1}{t^2}\text{ and }\frac{d\zeta}{\zeta} = \left(\frac{2t}{1 - t^2} - \frac{2t}{c^2 - t^2}\right)dt$$

$$\text{Thus } dz = k\left\{\frac{2}{1 - t^2} - \frac{2}{c^2 - t^2}\right\}dt$$

$$\Rightarrow z = \frac{a}{\Pi}\left\{\log\frac{1 + t}{1 - t} - \frac{1}{c}\log\frac{c + t}{c - t}\right\} + A$$

Since z = 0 corresponds to z = c ⇒ t = 0, A = 0

$$\text{Hence } z = \frac{a}{\Pi}\left\{\log\frac{1 + t}{1 - t} - \frac{1}{c}\log\frac{c + t}{c - t}\right\} \qquad ....(2)$$

$$\text{and } w = \frac{Ua}{\Pi}\log\zeta = \frac{Ua}{\Pi}\log\frac{c^2 - t^2}{1 - t^2} \qquad ...(3)$$

By eliminating t from (2) (3), we obtain a relation w = f (z)

MOTION OF A PARABOLIC CYLINDER

Consider a parabolic cylinder moving with velocity U and V parallel to the axes and rotating with an angular velocity $\underline{\omega}$. The stream function is determined as

$$y = Vx - Uy + 1/2\,\omega\,(x^2 + y^2) + \text{const.} \qquad ...(1)$$

Here $V = 0$, $\omega = 0$; the stream function becomes

$\psi = -Uy + \text{const} = -Ur\sin\theta + \text{const.}$

or $\psi = -2Ur\sin\theta/2\cos\theta/2 + \text{const.}$ ...(2)

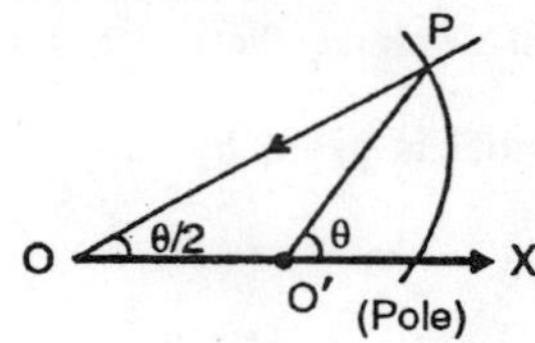

Fig. 2.9

The equation to the curve is

$$r = \frac{2a}{1+\cos\theta} \Rightarrow r(1 + \cos q) = 2a$$

$\Rightarrow r\cos^2\theta/2 = a \Rightarrow r^{1/2}\cos\theta/2 = a^{1/2}.$ ...(3)

From the relations (2) and (3), we have

$\psi = -2Ua^{1/2}r^{1/2}\sin\theta/2 + \text{const.}$ ...(4)

The stream function satisfies Laplace equation $\dfrac{\partial^2\psi}{\partial x^2} + \dfrac{\partial^2\psi}{\partial y^2} = 0$ and the boundary condition as $r^{1/2}\sin\theta/2$ is a plane harmonic.

Also $\phi = -2Ua^{1/2}r^{1/2}\cos\theta/2$

Since both $\dfrac{\partial\phi}{\partial r}$ and $-\dfrac{1}{r}\dfrac{\partial\phi}{\partial\theta}$ vanish at infinity, the liquid is at rest there.

Determine the velocity potential ϕ and the stream function ψ for a liquid streaming past a fixed parabolic cylinder with velocity U parallel to the axis of its section. Also, determine the resultant thrust on the cylinder per unit length.

Superimpose a velocity U on the cylinder and liquid both in the sense opposite to the velocity of the liquid. The stream flows in the direction opposite to that of the initial line, we have

$$-\frac{\partial\phi}{\partial x} = -\frac{\partial\psi}{\partial y} = -U$$

The velocity potential and the stream function is determined as

$\phi = Ux - 2Ua^{1/2}r^{1/2}\cos\theta/2,$

or $\phi = Ur\cos\theta - 2Ua^{1/2}r^{1/2}\cos\theta/2.$ ...(1)

and $\psi = Uy - 2Ua^{1/2}r^{1/2}\sin\theta/2,$

or $\psi = Ur\sin\theta - 2Ua^{1/2}r^{1/2}\sin\theta/2.$ ...(2)

The Bernoulli's equation for steady motion represents that

$$\frac{p}{\rho} = C - \frac{1}{2}q^2$$

Since pressure at infinity is zero so p = 0 when q = U, then

$$C = \frac{1}{2}U^2 \Rightarrow p = \frac{1}{2}\rho\,(U^2 - q^2)$$

At any point on the cylinder, the velocity is given by

$$q^2 = \left(-\frac{\partial\phi}{\partial r}\right)^2 + \left(-\frac{1}{r}\frac{\partial\phi}{\partial\theta}\right)^2 = U^2 \sin^2 \theta/2$$

The pressure P (r, θ) at any point on the boundary of the parabolic section is given by

$$P = \frac{1}{2}\rho\,(U^2 - U^2 \sin^2 \theta/2) = \frac{1}{2}\rho U^2 \cos^2\theta/2. \qquad ...(3)$$

The normal PO makes an angle θ/2 with the initial line X-axis. The resultant thrust on the cylinder, due to symmetry, will be along X-axis.

$$= -\int P\cos\frac{\theta}{2}\,ds \qquad ...(4)$$

Again $(ds)^2 = (dr)^2 + (rd\theta)^2 = a^2 \sec^6\frac{\theta}{2}(d\theta)^2$

$$\Rightarrow ds = a\sec^3\frac{\theta}{2}d\theta$$

From (3) and (4), we have

$$= -\int_{-\pi}^{\pi}\frac{1}{2}\rho U^2\cos^3\frac{\theta}{2}.a\sec^3\frac{\theta}{2}d\theta$$

$$= -\int_{-\pi}^{\pi}\frac{1}{2}\rho U^2 a\,d\theta = -\pi\rho U^2 a.$$

THE AEROFOIL

The *aerofoil* has a profile of fish type. It is used in modern aeroplanes. Such an aerofoil has a blunt leading edge and a sharp trailing edge. The projection of the profile on the double tangent is the chord. The ratio of the span to the chord is the *aspect ratio*. The locus of the point midway between the points in which an ordinate perpendicular to the chord meets the profile is known as the camber line of a profile. The *camber is the ratio of the maximum ordinate of the camber line to the chord.*

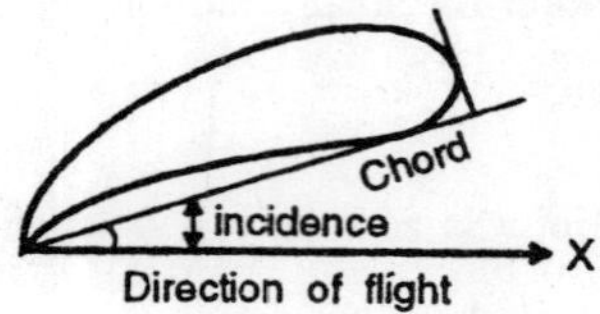

Fig. 2.10

The theory of the flow round such an aerofoil is made on the following assumptions:

(i) The air behaves as an incompressible inviscid fluid.

(ii) The aerofoil is a cylinder whose cross-section is a curve of the fish type.

(iii) The flow is a two-dimensional irrotational cyclic motion.

The assumptions are simply approximation to the actual state of affairs. The profiles obtained by conformal transformation of a circle by the simple Joukowski transformation make good wing shapes, and that the lift can be determined from the known flow with respect to a circular cylinder.

JOUKOWSKI TRANSFORMATION

The transformation

$$\zeta = z + \frac{c^2}{z}, \qquad \text{...(1)}$$

is called the Joukowski transformation, a mapping function from z-plane on the z-plane.

$\zeta = z + z_1$, where $z_1 = c^2/z$.

or $\zeta = re^{\theta i} + (c^2/r)\, e^{-\theta i}$, $z = re^{\theta i}$.

Draw a perpendicular PQ′ to the real axis to meet OQ at Q′ *i.e.*,

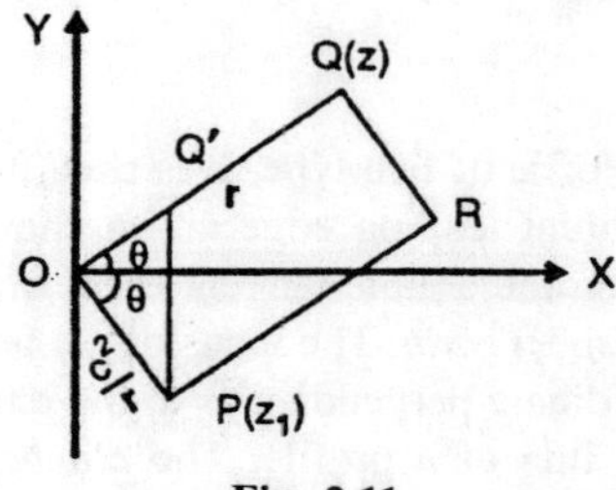

Fig. 2.11

$OQ' = OP = c^2/r$.

Therefore $OQ. OQ' = r(c^2/r) = c^2$

Thus, Q and Q′ are inverse points with regard to O. The position of the point P is obtained by reflecting OQ′ (where Q′ is an inverse point) in the real axis. Again, the position of the point R (z) is obtained by adding the complex numbers represented by the points Q and P by completing the parallelogram OPRQ. Thus the transformation is completed by the fourth vertex R which is represented by ζ. The point O will be made to describe a circle whereas the point Q′ will then describe the inverse of this circle. The point P will therefore describe a circle obtained by reflecting the locus of Q′ in the real axis.

The transformation (1) can be written as

$$\frac{\zeta + 2c}{\zeta - 2c} = \left(\frac{z + c}{z - c}\right)^2. \quad \text{...(2)}$$

From (1), we notice that the singularities $z = 0, \pm c, \infty$ transform to $z = \infty, \pm 2c, \infty$. The inverse transformation becomes $z = \frac{1}{2}\{\zeta \pm \sqrt{(\zeta^2 - 4c^2)}\}$ so that one value of ζ corresponds to two values of z. The positive sign before the radical for large ζ, $z \sim \zeta$ so that infinities in the two planes will correspond, and the negative sign before the radical gives $z \sim 2c^2/\zeta$ *i.e.*, the infinity of the ζ-plane will transform into the neighborhood of $z = 0$.

Assuming $z = -c + re^{\theta i}$ and $\zeta = 2c + Re^{\phi i}$, in the neighbourhood of $z = -c$ and $\zeta = -2c$, where r and R are infinitesimal. Then, approximately, we have

$$\frac{Re^{\phi i}}{-4c} = \frac{r^2 e^{2\theta i}}{4c^2} \Rightarrow \phi + \pi = 2\theta. \quad \text{...(3)}$$

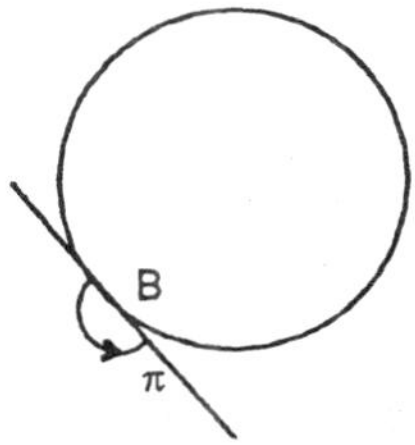

Fig. 2.12

In moving round the point B, θ increases by π, so ϕ increases by 2π. This implies that the two branches of the aerofoil touch one another at the trailing edge, which is therefore a cusp.

Assuming the generalised form of the transformation (2) as

$$\frac{\zeta + nc}{\zeta - nc} = \left(\frac{z+c}{z-c}\right)^n \qquad ...(4)$$

which gives $\phi + \pi = n\theta$.

Let $n = 2 - (\lambda/\pi)$

then $\phi = \pi = \{2 - (\lambda/\pi)\}\ \theta$.

It follows that when θ increases by π, ϕ increases by $2\pi - \lambda$. Thus the two distinct branches of the trailing edge at an angle λ.

The transformation (1) provides

$$\frac{d\zeta}{dz} = 1 - \frac{c^2}{z^2} \Rightarrow \frac{d\zeta}{dz} \text{ vanishes at the point } z = \pm\, c.$$

The representation ceases to be conformal in the immediate neighbourhood of these points.

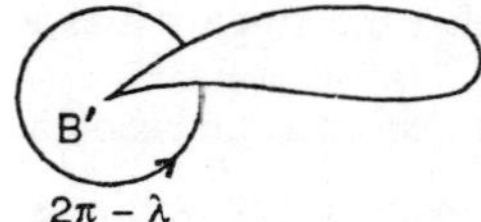

Fig. 2.13

The point z = c, being inside the circle, transforms to a point z = 2c inside the aerofoil and the point z = – c transforms into $\zeta = -2c$, the trailing edge. Thus at the trailing edge $d\zeta/dz$ is zero or $dz/d\zeta$ is infinite. Let **q** be the velocity at the point B of the circle and **q′** be the corresponding velocity at the point B′ (trailing edge) for the aerofoil, then

$$q' = \left|\frac{dw}{d\zeta}\right| = \left|\frac{dw}{dz}\right|\left|\frac{dz}{d\zeta}\right| = q\left|\frac{dz}{d\zeta}\right|.$$

Now the velocity q¢ becomes infinite because $dz/d\zeta$ is infinite at trailing edge. It could be avoided if B were a stagnation point so that **q** = 0.

KUTTA-JOUKOWSKI'S THEOREM

When an aerofoil, at rest, of any shape be placed in a uniform wind of speed U with circulation k around the aerofoil, it experiences then resultant thrust in the form of lift of magnitude ρkU per unit span perpendicular to the wind. The direction to the wind. The direction of the lift is obtained by rotating the wind velocity through a right angle in the sense opposite to that of the circulation.

Since there is a uniform wind, the velocity at a great distance from the aerofoil must tend to the wind velocity. If U be the velocity of the uniform wind at infinity, its direction makes an angle α with the negative direction of the X-axis, k be the circulation around the aerofoil. Since the body is finite, the algebraic sum of the strengths of sources and sinks is zero and thus does not contribute to the residue, but the doublets will produce term at great distance from the cylinder of the type A/z, B/z^2...where A, B,....depend on U and k. The complex potential at infinity from the origin becomes

$$w = Ue^{i\alpha} z + \frac{ik}{2\pi}\log z + \frac{A}{z} + \frac{B}{z^2} + \ ...., \qquad ...(1)$$

where the first term is due to uniform stream velocity U. The direction of the stream makes an angle α with X-axis, the second term is due to circulation k.

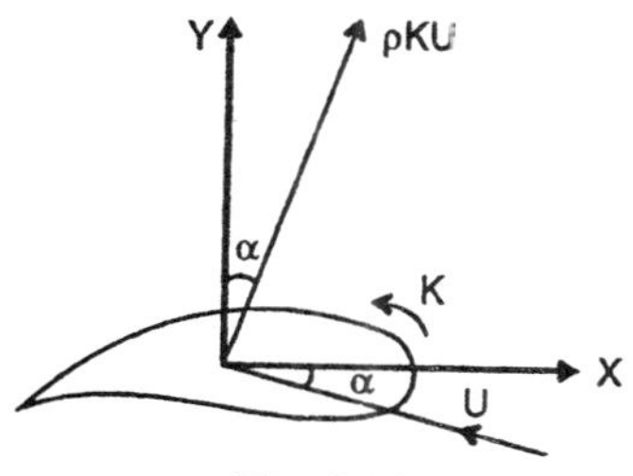

Fig. 2.14

$$\text{or } \frac{dw}{dz} = Ue^{i\alpha} + \frac{ik}{2\pi}\frac{1}{z} - \frac{A}{z^2} - \frac{2B}{z^3} - \ ...$$

$$\text{or } \left(\frac{dw}{dz}\right)^2 = \left(Ue^{i\alpha} + \frac{ik}{2\pi}\frac{1}{z} - \frac{A}{z^2} - \frac{2B}{z^3} \ ...\right)^2 \qquad ...(2)$$

Integrating round a cylinder whose radius is sufficiently large for the expansion (2) to be valid. Using Blasius's theorem, the forces (X,Y) exerted on the cylinder is given by

$$X - iY = \frac{1}{2} i\rho \int_c \left(\frac{dw}{dz}\right)^2 dz \qquad ...(3)$$

$$\text{or } X - iY = \frac{1}{2} i\rho \int_c \left(U^2 e^{2i\alpha} + \frac{ikUe^{i\alpha}}{\pi z} - \frac{k^2 + 8\pi^2 AUe^{i\alpha}}{4\pi^2 z^2} - \ ... \right) dz$$

The sum of the residues at the pole $z = 0$ is

$$X - iY = \frac{1}{2} i\rho \left(\frac{ikUe^{i\alpha}}{\pi}\right).2\pi i = -\, i\, \rho k U e^{\alpha i}.$$

$\Rightarrow X - iY = - i\rho kU (\cos\alpha + i \sin\alpha)$

$\Rightarrow X - iY = - i\rho kU (i \cos\alpha - i \sin\alpha)$

$\Rightarrow X = \rho kU \sin\alpha$ and $Y = \rho\ kU \cos\alpha$.

Thus the lift force L is given by

$$L = \sqrt{(X^2 + Y^2)} = \sqrt{\{\rho^2 k^2 U^2 (\sin^2\alpha + \cos^2\alpha)\}} = \rho kL(4),$$

which is perpendicular to the uniform wind of speed U and is independent of the shape of the profile.

Thus, we observe when a cylinder is inserted in a uniform stream U superposed by a circulation k, the cylinder experience a lift of magnitude ρkU. The lift depends on the uniform velocity of the stream, circulation and density. The resultant force acting on a cylinder can be split into two components-one acting in the direction of the flow, the drag force, and the other acting transversely to it, the lift force.

D' ALEMBERT'S PARADOX

Consider a long straight tube in which an inviscid fluid is flowing with a constant velocity U. Let a small body be placed in the middle of the tube. The fluid in the immediate neighbourhood of the body will be disturbed, but at a great distance from the body the fluid flow will remain undisturbed. Let the body be at rest and experiences a force F in the direction of motion. The external forces such as gravity etc. are not taken into consideration thus the force F is the resultant pressure thrust which the body experiences in the direction of flow.

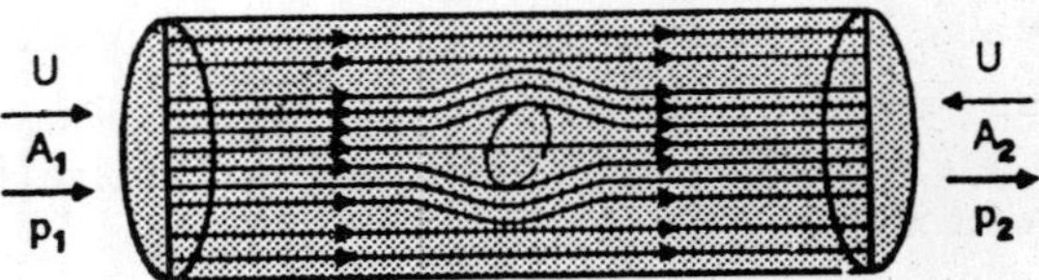

Fig. 2.15

Let A_1 and A_2 be the two cross-sections of the tube at a great distance from the body in the upstream and down stream region. The streamline will be parallel in both the regions. The fluid between these regions can be split up into current flaments. The outer filaments are bounded by the walls of the tube. The resultant pressure thrust components are perpendicular to these currents. The walls of the filaments in contact with the small body are acted on by the solid by a force whose component in the direction of flow is – F. By Euler's theorem, the resultant of the thrusts on the sections A_1 and A_2 of the tube is $(-\rho U_2 A_1 + \rho U^2 A_2)$, which becomes zero since $A_1 = A_2$. By Bernoulli's theorem, the pressure p_1 over A_1 is the same as the pressure p_2 over A_2 *i.e.*,

$$\frac{p_1}{\rho} + \frac{1}{2}\rho q^2 = C = \frac{p_2}{\rho} + \frac{1}{2}\rho q^2,$$

$$\Rightarrow p_1 = C - \frac{1}{2}\rho q^2 = p_2.$$

Thus the forces acting on the fluid in the tube are

$$p_1 A_1 - F - p_2 A_2 = 0 \Rightarrow F = 0.$$

If the walls of the tube reduce then the body immersed in a current will become unbounded in every direction, and the resultant force exerted by the fluid on the small body is zero.

Thus, if we super-impose on the system a uniform velocity U in the direction opposite to that of the current, the inviscid fluid at a great distance remains undisturbed and the body moves with uniform velocity U. The dynamical conditions remains unaltered by superposing a uniform velocity; therefore, the resistance to a body moving with uniform velocity through an unbounded inviscid fluid, otherwise at rest, vanishes. This is in contradiction to the common experience and is called, D′ *Alembert's paradox.*

SOLVED EXAMPLES

Example 1: *Shew that, with proper choice of units, the motion of an infinite liquid produced by the motion of an elliptic cylinder parallel to one of its principle axes is given by the complex function*

$w = e^{-\zeta}$, *where* $z = 2 \cosh \zeta$.

Reduce the formula

$$x = f\left(1 + \frac{1}{\phi^2 + \psi^2}\right),\ y = y\left(1 - \frac{1}{\phi^2 + \psi^2}\right).$$

Solution: The velocity potential and stream function for an elliptic cylinder moving parallel to major axis with a velocity U, is given by

$$f = Ub\sqrt{\left(\frac{a+b}{a-b}\right)} e^{-\xi} \cos \eta\ ;\ \psi = -Ub\sqrt{\left(\frac{a+b}{a-b}\right)} e^{-\xi} \sin \eta.$$

The complex potential becomes

$$w = \phi + i\psi = Ub\sqrt{\left(\frac{a+b}{a-b}\right)} e^{-\zeta}, \text{ where } \zeta = \xi + i\eta, \qquad ...(1)$$

Substituting a = c cosh a, b = c sinh a in (1), we have

$$w = Uc \sin h\ a\ e^{-\zeta}\sqrt{\left(\frac{\cosh\alpha + \sinh\alpha}{\cosh\alpha - \sinh\alpha}\right)} = 2U \sin h\ \alpha.\ e\alpha^{e-\zeta}$$

Thus by proper choice of U and α,we have

$2Ue^{\alpha} \sinh \alpha = 1 \Rightarrow 2U \sinh \alpha = e^{-a}$

Therefore $w = e^{-\zeta} \Rightarrow \phi + i\psi = e(\xi + i\eta),\ z = 2 \cosh \zeta$ **proved.**

or $x + iy = 2 \cosh (\xi + i\eta)$ ...(4)

or $x + iy = e^{(\xi + i\eta)} + e^{-(\xi + i\eta)}$.

From (3) and (4), we have

$$x + iy = \left\{\frac{1}{\phi + i\psi} + (\phi + i\psi)\right\} = \left\{\frac{\phi - i\psi}{\phi^2 + \psi^2} + (\phi + i\psi)\right\}.$$

Separating into real and imaginary parts, we have

$$x = \frac{\phi}{\phi^2 + \psi^2} + \phi = \phi\left(1 + \frac{1}{\phi^2 + \psi^2}\right),$$

and $y = \dfrac{\psi}{\phi^2 + \psi^2} + \psi = \psi\left(1 + \dfrac{1}{\phi^2 + \psi^2}\right)$. **Proved.**

Example 2: *An elliptic cylinder, the semi-axes of whose cross-section are a and b, is moving with velocity U parallel to the major axis of its cross-section, through an infinite liquid of density r which is at rest at infinity, the pressure there being Π. Prove that in order that the pressure may everywhere be positive.*

$$\rho U^2 < \frac{2a^2\Pi}{(2ab + b^2)}.$$

Solution: Let p be the pressure and q be the velocity at any point, then by Bernoulli's theorem for steady motion, we have

$$\frac{p}{\rho} + \frac{1}{2}q^2 = C, \quad \text{...(1)}$$

where C is any constant.

By super-imposing a velocity – U parallel to X-axis both on cylinder and liquid, we have

$$q_\infty = -U,\ p = P;\ C = \frac{\Pi}{\rho} + \frac{1}{2}U^2$$

or $\frac{p}{\rho} = \frac{\Pi}{\rho} + \frac{1}{2}U^2 - \frac{1}{2}q^2$.

The pressure p will be positive every where, if $p \geq 0$

i.e. $\frac{\Pi}{\rho} + \frac{1}{2}U^2 - \frac{1}{2}q^2$ is positive.

or $\frac{\Pi}{\rho} + \frac{1}{2}U^2$ max. value of $\frac{1}{2}q^2$. ...(2)

The complex potential w of the moticn is given by

$w = U(a + b)\cosh(\zeta - a)$

or $\frac{dw}{dz} = U(a + b)\sinh(\zeta - a)\frac{d\zeta}{dz}$

or $\frac{dw}{dz} = U(a + b)\frac{\sinh(\zeta - a)}{c\sinh\zeta}$, $z = c\cosh\zeta$.

The stagnation points can be obtained by putting $\frac{dw}{dz} = 0$

i.e. $\sinh(\zeta - \alpha) = 0$; $\zeta - \alpha = 0$. or $i\pi$

or $(\xi + i\eta - \alpha) = 0\ i\pi \Rightarrow \xi = \alpha$ and $\eta = \pi$

or $q = \left|\frac{dw}{dz}\right| = \left|\frac{U(a+b)}{c}.\frac{\sinh(\zeta - \alpha)}{\sinh\zeta}\right|$

or $q = \left|U\frac{a+b}{\sqrt{(a^2 - b^2)}}.\frac{\sinh\{(\xi - \alpha) + i\eta\}}{\sinh(\xi + i\eta)}\right|$

or $q = U\sqrt{\left(\frac{a+b}{a-b}\right)}\frac{\sqrt{(\sinh^2(\xi - \alpha) + \sin^2\eta\}}}{\sqrt{(\sinh^2\xi + \sin^2\eta)}}$

or $q^2 = U^2\frac{a+b}{a-b}\left[\frac{\sinh^2(\xi - \alpha) + \sin^2\eta}{\sinh^2\xi + \sin^2\eta}\right]$. ...(4)

or $q^2 = U^2\frac{a+b}{a-b}\left[1 - \frac{\sinh^2\xi - \sinh^2(\xi - \alpha)}{\sinh^2\xi + \sin^2\eta}\right]$

But $\sinh\xi > \sin h(\xi - \alpha)$, so q2 will be maximum, when $\sin\eta$ is maximum *i.e.*, $\sin\eta = 1$ or $\eta = \pi/2$.

From (4), we have

$$q^2 = U^2 \frac{a+b}{a-b}\left\{\frac{\sinh^2(\xi-\alpha)+1}{\sinh^2\xi+1}\right\}$$

$$= U^2 \frac{a+b}{a-b}\frac{\cosh^2(\xi-\alpha)}{\cosh^2\xi}.$$

Since it is an elliptic cylinderical boundary surrounded by the liquid, and the minimum value of ξ is α, so q is maximum when $\xi = \alpha$ and $\eta = \pi/2$.

Thus, we have

$$q^2 = U^2\frac{a+b}{a-b}\frac{1}{\cosh^2\alpha} = U^2 \frac{a+b}{a-b}\frac{c^2}{a^2} = U^2\frac{(a+b)^2}{a^2}$$

or $(q)max = U\dfrac{(a+b)}{a}$.

From the relation (2) and (5), we have

$$\frac{\Pi}{\rho}+\frac{1}{2}U^2 > \frac{1}{2}U^2\frac{(a+b)^2}{a^2}$$

or 2Pa2 > rU2(2ab + b2)

$$rU2 < \frac{2a^2\Pi}{2ab+b^2}.$$ **Proved.**

Example 3: *An infinite elliptic cylinder with semi-axes a, b is rotating round its axis with angular velicity ω, in an infinite liquid of density ρ which is at rest at infinity. Shew that if the fluid is under the action of no forces the moment of the fluid pressure on the cylinder round the centre is*

$(1/8)\ \pi\rho c^4\ (d\omega/dt)$, where $c^2 = a^2 - b^2$.

Solution: For an elliptic, cylinder rotating with an angular velocity ω, the velocity potential and stream function are

$$\phi = \frac{1}{4}\omega\ (a+b)^2\ e^{-2\xi}\sin 2\eta,$$

$$\psi = \frac{1}{4}\omega\ (a+b)^2\ e^{-2\xi}\sin 2\eta \qquad ...(1)$$

or $w = \phi + i\psi$

$$w = \frac{1}{4}i\omega\ (a+b)^2\ e^{-2(\xi+i\eta)}$$

$$w = \frac{1}{4}i\omega\ (a+b)^2\ e^{-2\zeta}, \qquad ...(2)$$

where $\zeta = \xi + i\eta$ and $z = c\cosh\xi$.

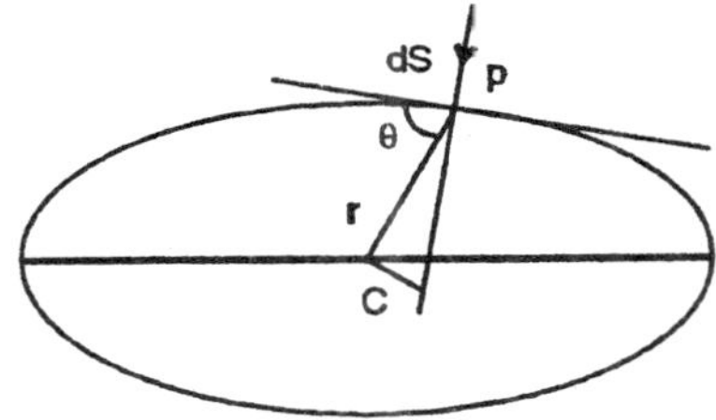

Fig. 2.16

Let q be the velocity at any point, then

$$q = \left|\frac{dw}{dz}\right| = \left|\frac{dw}{d\zeta}\frac{d\zeta}{dz}\right|$$

$$q^2 = \left|\frac{1}{4}i\omega(a+b)^2 e^{-2\zeta}\frac{(-2)}{2\sinh\zeta}\right|^2$$

$$\text{or } q^2 = \frac{\omega^2(a+b)^4}{4c^2}\left|\frac{e^{-2\xi}e^{-2i\eta}}{\sinh(\xi+i\eta)}\right|$$

$$q^2 = \frac{\omega^2(a+b)^4 e^{-4\xi}}{4c^2(\sinh^2\xi+\sin^2\eta)} \qquad \text{....(3)}$$

Also from (1), we have

$$\frac{\partial\phi}{\partial t} = \frac{1}{4}(a+b)^2 e^{-2\xi}\sin 2\eta\frac{d\omega}{dt}.$$

The pressure at any point of the ellipse ($\xi = a$) is given by

$$\frac{p}{\rho} = A - \frac{1}{2}q^2 + \frac{\partial\phi}{\partial t},$$

$$\text{or } \frac{p}{\rho} = A - \frac{1}{2}\frac{\omega^2(a+b)^4}{4c^2}\frac{e^{4a}}{\sinh^2\alpha+\sin^2\eta}$$

$$+ \frac{1}{4}(a+b)^2 e^{-2a}\sin 2\eta\frac{d\omega}{dt}.$$

The moment of the fluid pressure on an element δs of length of the bounding ellipse about the centre C is

$$= -\,pr\cos\phi\,\delta s,\ \cos\phi = dr/ds.$$

Also $r^2 = a^2\cos^2\eta + b^2\sin^2\eta$

$$\text{or } rdr = -\frac{1}{2}(a^2 - b^2)\sin 2\eta\, d\eta$$

Therefore total moment of the fluid pressure on the elliptic cylinder about the centre is given by]

$$= - \int prdr = - \int_0^{2\pi} p\{-\frac{1}{2}(a^2 - b^2)\sin 2\eta d\eta\}$$

$$= \frac{1}{2}(a^2 - b^2)r \int_0^{2\pi} \frac{1}{4}(a + b)^2 e^{-2a} \sin 2\eta \frac{d\omega}{dt} \sin 2\eta d\eta$$

$$= \frac{1}{8}(a^2 - b^2)(a + b)^2 \rho e^{-2a} \int_0^{2\pi} \sin^2 2\eta \frac{d\omega}{dt} d\eta$$

$$= \frac{(a^2 - b^2)(a + b)^2 e^{-2a}}{8} \rho \frac{d\omega}{dt} \pi$$

$$= \frac{1}{8}\pi\rho c^4 \frac{d\omega}{dt}. \text{ Proved.}$$

Example 4: *Prove that when an infinitely long cylinder of density σ whose cross-section is an ellipse of semi-axes a, b is immersed in an infinite liquid of density ρ the square of its radius of gyration about its axis is effectively increased by the equation*

$$\frac{\rho}{8\sigma} \frac{(a^2 - b^2)}{ab}.$$

Solution: The kinetic energy T′ for an elliptic cylinder rotating about its axis with angular velocity $\underline{\omega}$, when the liquid is not present, is given by

$$T' = \frac{1}{2}Mk^2\omega^2,$$

where K is the M.I. of an elliptic plate about a line passing through the centre and perpendicular to its plane and M is the mass of the elliptic plate.

$$T' = \frac{1}{2}\pi ab\sigma\left(\frac{a^2 + b^2}{4}\right)\omega^2.$$

The velocity potential and the stream function for an elliptic cylinder rotating with an angular velocity $\underline{\omega}$ in an infinite liquid becomes

$$\phi = \frac{1}{4}\omega (a + b)^2 e^{-2a} \sin 2\eta$$

$$y = \frac{1}{4}\omega (a + b)^2 e^{-2a} \cos 2\eta$$

The kinetic energy T of the liquid per unit length of the axis is given by

$$T = -\frac{1}{2}\rho \int \phi \frac{\partial \phi}{\partial \eta} dS = -\frac{1}{2}\rho \int \phi d\psi$$

or $T = -\frac{1}{2}\rho\int_0^{2\pi}\frac{1}{4}\omega(a+b)^2 e^{-2a}\sin 2\eta\left\{-\frac{1}{4}\omega(a+b)^2 e^{-2a}\right.$

$\times$ 2 sin $2\eta\}dt$

or $T = (1/16)\ \rho\omega^2\ (a + b)^4\ e{-}4a\int_0^{2\pi}\sin^2 2\eta d\eta$

or $T = (1/16)\ \pi\rho\omega^2\ (a + b)^4\ e^{-4a}$

or $T = (1/16)\ \pi\rho\omega^2\ c^4 = (1/6)\pi\rho\omega^2\ (a^2 - b^2)^2$. ...(2)

Thus the total kinetic energy (T′ + T) becomes

$$= \frac{1}{2}\pi\sigma ab\left(\frac{a^2+b^2}{4}\right)\omega^2 + \frac{1}{16}\pi\rho\,(a^2 - b^2)^2\omega^2. \qquad ...(3)$$

Let μ be the effective increase in radius of gyration then, we have

$$= \frac{1}{2}\pi\sigma ab\frac{a^2+b^2}{4}\omega^2 + \frac{1}{2}\pi\sigma b\mu^2\omega^2$$

$$= \frac{1}{2}\pi ab\sigma\omega^2\left[\frac{a^2+b^2}{4} + \mu^2\right]. \qquad ...(4)$$

From (3) and (4), we have

$$\frac{1}{2}\pi\sigma ab\left(\frac{a^2+b^2}{4}\right)\omega^2 + \frac{1}{16}\pi\rho\omega^2(a^2-b^2)^2$$

$$= \frac{1}{2}\pi ab\sigma\omega^2\left[\frac{a^2+b^2}{4} + \mu^2\right]$$

or $\frac{1}{2}\pi ab\sigma\omega^2\mu^2 = \frac{1}{16}\pi\rho\omega^2(a^2-b^2)^2$

or $\mu^2 = \frac{\rho}{8\sigma}\frac{(a^2-b^2)^2}{ab}$. **Proved.**

Example 5: *A thin shell in the form of an infinitely long elliptic cylinder, semi-axes a and b, is rotating about its axis in infinite liquid otherwise at rest. It is filled with the same liquid. Prove that the ratio of the kinetic energy of the liquid inside to that of the liquid outside is*

$2ab : a^2 + b^2$

Solution: Let the elliptic cylinder is rotating with angular velocity ω in an infinite liquid at rest at infinity then the complex potential is given by

$$w = \frac{1}{4}i\omega\ (a + b)^2\ e^{-2\zeta},\ \text{where}\ \zeta = \xi + i\eta.$$

Equating the real and imaginary parts, we have

$$\phi = \frac{1}{4}\omega\ (a+b)^2\ e^{-2\xi}\ \cos 2\eta,\ \psi = \frac{1}{4}\omega\ (a+b)^2\ e^{-2\xi}\ \cos 2\eta.$$

Let T_1 be the K. E. of the liquid outside the elliptic cylinder, then

$$T_1 = \frac{1}{2}\rho \int \phi\ d\psi$$

$$\text{or } T_1 = \frac{1}{16}\rho\omega^2\ (a+b)^4 e^{-4a} \int_0^{2\pi} \sin^2 2\eta d\eta$$

$$\text{or } T_1 = \frac{1}{16}\pi\rho\omega^2\ (a^2+b^2). \qquad ...(1)$$

The complex potential of the liquid contained in the rotating elliptic cylinder is given by

$$w = iAz^2, \text{ where A is a constant} \qquad ...(2)$$

$$\text{or } \phi + i\psi = iA\ (x+iy)^2 \Rightarrow \psi = -A\ (x^2-y^2). \qquad ...(3)$$

But for rotating cylinder, the stream function is given by

$$\psi = \frac{1}{2}\omega\ (x^2+y^2) - \text{constant}.$$

At the boundary of the ellipse, we must have

$$\frac{1}{2}\ \omega\ (x^2+y^2) - \text{constant} = -A\ (x^2-y^2)$$

$$\text{or } \left(\frac{1}{2}\omega + A\right)x^2 + \left(\frac{1}{2}\omega - A\right)y^2 = \text{constant}.$$

Also, the point (x, y) on the boundary satisfy

$$\frac{x^2}{a^2}+\frac{y^2}{b^2} = 1.$$

From (5) and (6), we have

$$a^2\left(\frac{1}{2}\omega + A\right) = b^2 + \left(\frac{1}{2}\omega - A\right)$$

$$\text{or } A\ (a^2+b^2) = \frac{1}{2}\ \omega\ (b^2-a^2)$$

$$\text{or } A = \frac{1}{2}\ \omega\ \frac{b^2-a^2}{(a^2+b^2)}.$$

Substituting the value of the constant A in (2), we get

$$w = \frac{1}{2}i\ \omega\ \frac{a^2+b^2}{a^2+b^2}z^2,\ z = c\cosh\zeta$$

$$\text{or } w = \frac{1}{2} i\,\omega \frac{a^2 + b^2}{a^2 + b^2} c^2 \cosh^2 (\xi + i\eta)$$

$$\text{or } \phi + i\psi = \frac{1}{2} i\omega \frac{a^2 + b^2}{a^2 + b^2} c^2 (\cosh \xi \cos \eta + i \sinh \xi \sin \eta)^2$$

Equating the real and the imaginary parts, we have

$$\phi = -\,\omega \frac{a^2 + b^2}{a^2 + b^2} c^2 \cosh \xi \sinh \xi \cos hi \sin \eta,$$

$$\text{and } \psi = \frac{1}{2}\omega \frac{a^2 + b^2}{a^2 + b^2} c^2 \{\cosh^2 \xi \cos^2\eta - \sinh^2 \xi \sin^2 \eta\}$$

Let T^2 be the K.E. of the liquid inside the elliptic cylinder $\xi = \alpha$, then

$$T_2 = \frac{1}{2}\rho \int \phi d\psi$$

$$\text{or } T_2 = \frac{1}{2}\rho\omega^2 \left(\frac{a^2 + b^2}{a^2 + b^2}\right)^2 c^4 \cosh \alpha \sinh \alpha$$

$$(\cosh^2 a + \sinh^2\alpha) \int_0^{2\pi} \cos^2 \eta \sin^2 \eta d\eta$$

$$\text{or } T_2 = \frac{1}{8}\pi\rho\omega^2 \left(\frac{a^2 + b^2}{a^2 + b^2}\right)^2 c^4 \cosh \alpha \sinh \alpha (\cosh^2 \alpha + \sinh^2\alpha)$$

$$\text{or } T_2 = \frac{1}{8}\pi\rho\omega^2 \left(\frac{a^2 + b^2}{a^2 + b^2}\right)^2 c \cosh \alpha\, c \sinh \alpha (c^2 \cosh^2 \alpha + c^2 \sinh^2\alpha)$$

$$\text{or } T_2 = \frac{1}{8}\pi\rho\omega^2 \left(\frac{a^2 + b^2}{a^2 + b^2}\right)^2 ab (a^2 + b^2)$$

$$\text{or } T_2 = \frac{1}{8}\pi\rho\omega^2 ab \frac{(a^2 - b^2)^2}{a^2 + b^2}. \qquad ...(7)$$

Dividing (1) and (7), we have

$$\frac{T^2}{T_1} = \frac{16}{8} \frac{\pi\rho\omega^2 (a^2 - b^2)^2 ab}{(a^2 + b^2).\pi\rho\omega^2 (a^2 - b^2)^2}$$

$$= \frac{2ab}{a^2 + b^2}$$

$T_2 : T_1 : : 2ab : a^2 + b^2$. **Proved.**

Example 6: *If an elliptic cylinder of semi-axes a, b filled with a liquid, rotates with a uniform angular velocity about its axis, shew that the kinetic energy of liquid contained is less than if it were moving as a solid in the ratio*

$(a^2 - b^2)^2 : (a^2 + b^2)^2$.

Solution: Let T_1 be the K.E. when the liquid be moving as a solid with angular velocity $\underline{\omega}$, then

$$T_1 = \frac{1}{2}Mk^2\omega^2 = \frac{1}{2}pabr\left(\frac{a^2+b^2}{4}\right)\omega^2. \qquad ...(1)$$

Let T_2 be the K.E. of the liquid contained in the elliptic cylinder rotating with uniform angular velocity $\underline{\omega}$, then

$$T2 = \frac{1}{2}\pi\rho\omega^2 ab\frac{(a^2-b^2)^2}{a^2+b^2} \qquad ...(2)$$

Dividing (1) and (2) we have

$$\frac{T_2}{T_1} = \frac{\pi\rho\omega^2(a^2-b^2)^2 ab}{(a^2+b^2)\pi\rho\omega^2(a^2+b^2)ab} = \frac{(a^2-b^2)^2}{(a^2+b^2)^2}$$

or $T_2 : T_1 :: (a^2 - b^2)^2 : (a^2 + b^2)^2$. **Proved.**

Example 7: *The space between two confocal elliptic cylinders $(a_0 b_0)$ and (a_1, b_1) and two planes perpendicular to their axis is filled with liquid. If both cylinders be made to rotate about their common axis with angular velocity $\underline{\omega}$, the kinetic energy of the motion set up is*

$$\frac{M\omega^2 c^4 (b_1 a_0 - b_0 a_1)}{8(a_1 a_0 - b_1 b_0)(a_1 b_1 - a_0 b_0)},$$

M being the mass of the liquid, and 2c is the distance between the focii.

Solution : Let the elliptic cylinders (a_0, b_0) and (a_1, b_1) be given by $\xi = a_0$ and $\xi = \alpha_1$ respectively, such that

$a_0 = c \cosh a_0,\ b_0 = c \sin h\ a_0,$

and $a_0 = c \cosh a_1,\ b_1 = c \sin h\ a_1,$ ...(1)

Since $z = c \cosh \zeta = c \cosh (\xi + i\eta)$

or $x = c \cosh \xi \cos \eta$ and $y = c \sin h\ \xi \sin \eta$ ...(2)

The stream function in a pure rotation becomes

$$\psi = \frac{1}{2}\omega(x^2 + y^2) + \text{const.}$$

or $\psi = \frac{1}{2}\omega c^2 (\cosh^2 \xi \cos^2 \eta + \sin h^2 \xi \sin^2 \eta) + \text{const.}$

or $\psi = \frac{1}{4}\omega c^2 (\cosh 2\xi + \cos 2\eta) + \text{const.}$ ...(3)

At the boundary on the elliptic cylinders $\xi = a$. we have

$\psi = \frac{1}{4}\omega c^2 (\cosh 2\alpha + \cos 2\eta) + \text{Const.}$

or $\psi = \frac{1}{4}\omega c^2 (\cosh 2\alpha_0 + \cos 2\eta) + \text{Const.},$ ...(4)

At $\alpha = \alpha_0$, $\psi = \psi_0$ (let) on inner elliptic boundary.

and $\psi_1 = \frac{1}{4}\omega c^2 (\cosh 2\alpha_1 + \cos 2\eta) + \text{Const.},$ ...(5)

At $\alpha = \alpha_1$, $\psi = \psi_1$ (let) on outer elliptic boundary.

The suitable form of the stream function ψ, which satisfies Laplace's equation, be chosen in the form as

$\psi = (A \cosh 3\xi + B \sinh 2\xi) \cosh 3\eta.$ ...(6)

Now at $\xi = \alpha = \alpha_0$ (4) and (6) should give the same value and at $\xi = \alpha = \alpha_1$ (5) and (6) should also give the same value.

i.e. $\frac{1}{4}\omega c^2 (\cosh 2\alpha_0 + \cosh 2\eta) + \text{Const.}$

$= (A \cosh 2\alpha_0 + B \sinh 2\alpha_0) \cos 2\eta$

and $\frac{1}{4}\omega c^2 (\cosh \alpha_1 + \cos 2\eta) + \text{Const.}$

$= (A \cosh 2\alpha_1 + B \sinh 2\alpha_1) \cos 2\eta,$

$\Rightarrow A \cosh 2\alpha_0 + B \sinh 2\alpha_0 = \frac{1}{4}\omega c^2$

and $A \cosh 2\alpha_1 + B \sinh 2\alpha_1 = \frac{1}{4}\omega c^2$

or $A (\cosh 2\alpha_0 \sinh 2\alpha_1 - \cosh 2\alpha_1 \sinh 2\alpha_0)$

$= \frac{1}{4}\omega c^2 (\sinh 2\alpha_1 - \sinh 2\alpha_0)$

or $A \sin h (2\alpha_1 - 2\alpha_0) = \frac{1}{4}\omega c^2 \; 2 \cosh (a_1 + a_0) \sinh (a_1 - a_0)$

or $A = \frac{1}{4}\omega c^2 \frac{2\cosh(a_1 + a_0)\sinh(a_1 - a_0)}{2\sinh(a_1 - a_0)\cosh(a_1 - a_0)}$

or $A = \frac{1}{4}\omega c^2 \frac{\cosh(a_1 + a_0)}{\cosh(a_1 - a_0)}$ and $B = \frac{1}{4}\omega c^2 \frac{\sinh(a_1 + a_0)}{\cosh(a_1 - a_0)}$

Substituting the value of A and B into (6), we get

$$\psi = \frac{1}{4}\omega c^2 \left\{\frac{\cosh(a_1 + a_0)\cosh 2\xi + \sinh(a_1 + a_0)\sinh 2\xi}{\cosh(a_1 - a_0)}\right\}\cos 2\eta$$

or $\psi = \frac{1}{4}\omega c^2 \frac{\cosh(2\xi - a_1 - a_0)}{\cosh(a_1 - a_0)}\cos 2\eta$,

and $\phi = \frac{1}{4}\omega c^2 \frac{\sinh(2\xi - a_1 - a_0)}{\cosh(a_1 - a_0)}\sin 2\eta$.

Let T be the K.E. of the liquid, then

$$T = -\frac{1}{2}\rho\int\phi d\psi = \left[-\frac{1}{2}\rho\int_{\xi = a_0}\phi d\psi - \left(-\frac{1}{2}\rho\int_{\xi = a1}\phi d\psi\right)\right]$$

or $T = \frac{1}{16}\rho\frac{\omega^2 c^4 \sinh(a_1 - a_0)}{\cosh(a_1 - a_0)}\left(\int_0^{2\pi}\sin^2 2\eta d\eta + \int_0^{2\pi}\sin^2 2\eta d\eta\right)$

or $T = \frac{1}{8}\pi\rho\omega^2 c^4 \frac{\sinh(\alpha_1 - \alpha_0)}{\cosh(\alpha_1 - \alpha_0)}$.

or $T = \frac{1}{8}\pi\rho\omega^2 c^4 \frac{\sinh\alpha_1\cosh\alpha_0 - \cosh\alpha_1\sinh\alpha_0}{\cosh\alpha_1\cosh\alpha_0 - \sinh\alpha_1\sinh\alpha_0}$,

or $T = \frac{1}{8}\pi\rho\omega^2 c^4 \frac{c\sinh\alpha_1 . c\cosh\alpha_0 - c\cosh\alpha_1 . c\sinh\alpha_0}{c\cosh\alpha_1 . c\cosh\alpha_0 - c\sinh\alpha_1 . c\sinh\alpha_0}$

From (1) and (2), we have

$$T = \frac{1}{8}\pi\rho\omega^2 c^4 \frac{b_1 a_0 - b_0 a_1}{a_1 a_0 - b_1 b_0}.$$

Let M be the mass of liquid occupying the space between two confocal elliptic cylinders (a_0, b_0) and (a_1, b_1) then

$$T = \frac{1}{8}\pi\rho\omega^2 c^4 \frac{(b_1 a_0 - b_0 a_1)}{(a_1 a_0 - b_1 b_0)(a_1 b_1 - a_0 b_0)}; \; M = \pi\rho\,(a_1 b_1 - a_0 b_0).$$ **Proved.**

Example 8: *Liquid is contained in a rotating elliptic cylinder. By making use of the elliptic transformation $z = c\cosh\zeta$ show directly that the stream function of the motion is*

$$\psi = \frac{1}{2}\omega\frac{a^2 - b^2}{a^2 + b^2}(x^2 - y^2).$$

Hence, prove that the paths of the particles are similar ellipses described in time $\frac{\pi(a^2 + b^2)}{\omega ab}$.

Solution: The complex potential of the liquid contained in a rotating elliptic cylinder is given by

$$w = -iAz^2 = -iA(x^2 - y^2 + 2ixy).$$

So $\psi = -A(x^2 - y^2)$, and $\phi = 2Axy$. ...(1)

The stream function for rotating cylinders is given by

$$\psi = \frac{1}{2}\omega(x^2 + y^2) - \text{Const.} \quad ...(2)$$

Equating (1) and (2), we have

$$\frac{1}{2}\omega(x^2 + y^2) - \text{Const.} = -A(x^2 - y^2)$$

$$\text{or } \left(\frac{1}{2}\omega + A\right)x^2 + \left(\frac{1}{2}\omega - A\right)y^2 = \text{Const.} \quad ...(3)$$

Equation to the the elliptic cylinder is

$$\frac{x^2}{a^2} + \frac{y^2}{b^2} = 1. \quad ...(4)$$

Comparing (3) and (4) at the surfaces, we have

$$a^2\left(\frac{1}{2}\omega + A\right) = b^2\left(\frac{1}{2}\omega - A\right)$$

$$\text{or } A(a^2 + b^2) = -\frac{1}{2}\omega(a^2 - b^2)$$

$$\Rightarrow A = -\frac{1}{2}\omega\frac{a^2 + b^2}{a^2 + b^2}.$$

Substituting the value of the constant A in (1), we get

$$\psi = \frac{1}{2}\omega\frac{a^2 + b^2}{a^2 + b^2}(x^2 - y^2),\ \phi = -\omega\frac{a^2 + b^2}{a^2 + b^2}xy.$$ **Proved.**

Consider (x, y) be the coordinates of a point, then velocity of the fluid particle parallel to the axesis

$$-\frac{\partial\phi}{\partial x} = \dot{x} - \omega y,\ -\frac{\partial\phi}{\partial y} = \dot{y} + \omega x. \quad ...(5)$$

$$\text{or } \dot{x} - \omega y = -\frac{\partial\phi}{\partial x} = \frac{a^2 - b^2}{a^2 + b^2}\omega y,$$

$$\text{or } \dot{x} = \omega y\left[\frac{a^2-b^2}{a^2+b^2}\right] = \frac{2a^2}{a^2+b^2}\omega y. \qquad ...(6)$$

$$\text{and } \dot{y} + \omega x = -\frac{\partial\phi}{\partial y} = \frac{a^2-b^2}{a^2+b^2}\omega x$$

$$\text{or } \dot{y} = \omega x\left[\frac{a^2-b^2}{a^2+b^2}\right] = -\frac{2b^2}{a^2+b^2}\omega x. \qquad ...(7)$$

Differentiating (6) and (7) with regard to t, we have

$$\ddot{x} = \frac{2a^2\omega}{a^2+b^2}\dot{y} = \frac{2a^2\omega}{a^2+b^2}\left\{-\frac{2a^2\omega}{a^2+b^2}\dot{x}\right\}$$

$$\ddot{y} = -\frac{2a^2\omega}{a^2+b^2}\cdot\frac{2a^2\omega}{a^2+b^2}y,$$

$$\text{Thus } \frac{d^2x}{dt^2} = -\frac{4a^2b^2\omega^2}{(a^2+b^2)^2}x, \text{ and } \frac{d^2y}{dt^2} = -\frac{4a^2b^2\omega^2}{(a^2+b^2)^2}y, \qquad ...(8)$$

which is a standard equation of S. H. M. whose time period T is given by

$$T = \frac{2\pi}{\sqrt{\mu}} \Rightarrow T = \frac{\pi(a^2+b^2)}{ab\omega}. \qquad \textbf{Proved.}$$

The solution of the differential equation (8) is given by

$$x = A\cos\left(\frac{2ab\omega}{a^2+b^2}t + C\right) \text{ and } y = -\frac{b}{a}A\sin\left(\frac{2ab\omega}{a^2+b^2}t + C\right)$$

Eliminating t, we have

$$\frac{x^2}{a^2} + \frac{y^2}{b^2} = \frac{A^2}{a^2} = 1,$$

which is a similar ellipse. Hence path of the particles are similar ellipses.

Example 9: *An elliptic cylinder of semi-axes a, b is filled with incompressible fluid and rotates about its axis with constant angular velocity* $\underline{\omega}$. *Prove that the velocity component (u, v) parallel to OX, OY (the axis of the ellipse) are given by*

$$u = \omega\frac{a^2-b^2}{a^2+b^2}y,\ v = \omega\frac{a^2-b^2}{a^2+b^2}x.$$

Shew that the coordinates, X, Y (relative to the axes through O fixed in space) of a given particle at time t can be written as

$$X = \lambda\left\{(a+b)\cos\left\{\frac{(a-b)^2\omega t}{a^2+b^2}\right\}+(a-b)\cos\left\{\frac{(a-b)^2\omega t}{a^2+b^2}\right\}\right\},$$

$$Y = \lambda\left\{(a+b)\sin\left\{\frac{(a-b)^2\omega t}{a^2+b^2}\right\}+(a-b)\sin\left\{\frac{(a-b)^2\omega t}{a^2+b^2}\right\}\right\},$$

where λ is a constant depending on the particle and $t = 0$, when the particle crosses the OX.

Solution: The velocity potential ϕ is given by

$$\phi = -\omega\frac{a^2-b^2}{a^2+b^2}xy. \text{ (Ref. Ex. 47).} \qquad ...(1)$$

Let (u, v) are the velocity components parallel to the axes OX and OY, so that from (1), we have

$$u = -\frac{\partial\phi}{\partial x} = \omega\frac{a^2-b^2}{a^2+b^2}y;\; v = -\frac{\partial\phi}{\partial y} = \omega\frac{a^2-b^2}{a^2+b^2}x. \qquad ...(2)$$

Also, we know that

$$x = A\cos\left(\frac{2ab\omega t}{a^2+b^2}+C\right),\; y = -\frac{b}{a}A\sin\left(\frac{2ab\omega t}{a^2+b^2}+C\right) \qquad ...(3, 4)$$

Since $t = 0$; $x = \lambda$, $y = 0$, equation (3) and (4) reduces to,

$$\lambda = A\cos C \text{ and } 0 = -\frac{b}{a}A\sin C \Rightarrow A = \lambda,\; C = 0.$$

$$\Rightarrow x = \lambda\cos\frac{2ab\omega t}{a^2+b^2},\; y = -\frac{b\lambda}{a}\sin\frac{2ab\omega t}{a^2+b^2}, \qquad ...(5)$$

which represent the coordinates with regard to the rotating axes fixed in the elliptic cylinder.

Now we shall determine the coordinates (X, Y) with respect to fixed axis in space.

$$X + iY = e^{i\omega t}(x+iy) = (\cos\omega t + i\sin\omega t)(x+iy)$$

Equating real and imaginary parts, we have

$$X = x\cos\omega t - y\sin\omega t,\; Y = x\sin\omega t + y\cos\omega t.$$

Substituting the value of x and y from the equation (5), we have

$$X = \lambda\cos\frac{2ab\omega t}{a^2+b^2}\cos\omega t - \frac{b}{a}\lambda\sin\frac{2ab\omega t}{a^2+b^2}\sin\omega t$$

$$\text{or } X = \frac{\lambda}{2a}\left\{2a\cos\omega t\cos\frac{2ab}{a^2+b^2}\omega t + 2b\sin\omega t\sin\frac{2ab}{a^2+b^2}\omega t\right\}$$

$$\text{or } X = \lambda'\left[a\left\{\cos\left(\frac{a^2+b^2+2ab}{a^2+b^2}\right)\omega t + \cos\left(\frac{a^2+b^2-2ab}{a^2+b^2}\right)\omega t\right\}\right]$$

$$+ b\left\{\cos\left(\frac{a^2+b^2-2ab}{a^2+b^2}\right)\omega t - \cos\left(\frac{a^2+b^2+2ab}{a^2+b^2}\right)\omega t\right\},\ \lambda' = \frac{\lambda}{2a}.$$

$$\text{or } X = \lambda'\left[(a+b)\cos\left\{\frac{(a-b)^2\omega t}{a^2+b^2}\right\} + (a-b)\cos\left\{\frac{(a+b)^2\omega t}{a^2+b^2}\right\}\right].$$

Similarly

$$Y = \lambda'\left[(a+b)\sin\left\{\frac{(a-b)^2\omega t}{a^2+b^2}\right\} + (a-b)\cos\left\{\frac{(a+b)^2\omega t}{a^2+b^2}\right\}\right].\ \textbf{Proved.}$$

Example 10: *If the liquid is contained between the elliptic cylinders* $\frac{x^2}{a^2}+\frac{y^2}{b^2} = 1$ *and* $\frac{x^2}{a^2}+\frac{y^2}{b^2} = K^2$ *and the whole rotates about OZ with angular velocity* ω, *prove that the velocity potential* ϕ *referred to the axes OX, OY is given by*

$$\phi = -\,\omega\,\frac{a^2-b^2}{a^2+b^2}xy,$$

and that the surfaces of equal pressure are hyperbolic cylinders

$$\frac{x^2}{3a^2+b^2} - \frac{y^2}{a^2+3b^2} = C.$$

Solution: For the first part see Example 47.

$$\text{Since } \phi = -\,\omega\,\frac{a^2-b^2}{a^2+b^2}xy. \qquad ...(1)$$

When the axes are rotating the pressure is given by

$$\left[\frac{p}{\rho}+\frac{1}{2}\left(\frac{\partial\phi}{\partial x}\right)^2+\left(\frac{\partial\phi}{\partial y}\right)^2\right]+\omega\left(x\frac{\partial\phi}{\partial y}-y\frac{\partial\phi}{\partial x}\right) = \text{const.} \qquad ...(2)$$

From (1) and (2), we have

$$\frac{p}{\rho}+\frac{1}{2}\omega^2\left(\frac{a^2-b^2}{a^2+b^2}\right)^2(x^2+y^2)+\omega^2\left(\frac{a^2-b^2}{a^2+b^2}\right)(-x^2+y^2) = \text{const.}$$

$$\text{or}\quad \frac{p}{\rho}+\frac{1}{2}\omega^2\left(\frac{a^2-b^2}{a^2+b^2}\right)\left\{\frac{a^2-b^2}{a^2+b^2}(x^2+y^2)+2(y^2-x^2)\right\}=\text{const.}$$

The surfaces of equal pressure are obtained by substituting p = const.,

$$\text{i.e.}\quad \frac{a^2-b^2}{a^2+b^2}(x^2+y^2)+2(y^2+x^2)=\text{const.}$$

$$\text{or}\quad x^2\,\frac{a^2+3b^2}{a^2+b^2}-y^2\,\frac{3a^2+b^2}{a^2+b^2}=\text{const.}$$

$$\text{or}\quad \frac{x^2}{3a^2+b^2}-\frac{y^2}{a^2+3b^2}=\text{const.},$$

which represent the hyperbolic cylinders. **Proved.**

Example 11: *Prove that if 2a. 2b are the axes of the cross-section of an elliptic cylinder placed across a stream in which the velocity an infinity is U parallel to the major axis of the cross-section, the velocity at a point (a cosη, b sin η) on the surface is*

$$U(a+b)\sin\eta\,(b^2\cos\eta+a^2\sin\eta)^{-1/2},$$

and that, in consequence of the motion of the liquid, the resultant thrust (per unit length) on that half cylinder on which the stream impinges is diminished by

$$\frac{2b^2\rho U^2}{a-b}\left\{1-\left(\frac{a+b}{a-b}\right)^{1/2}\tan^{-1}\left(\frac{a-b}{a+b}\right)^{1/2}\right\},$$

where ρ is the density of the liquid.

Solution: The velocity at any point is given by

$$q^2=U^2\frac{a+b\sinh^2(\xi-\alpha)+\sin^2\eta}{\sinh^2\xi+\sin^2\eta}.$$

The point (a cos η, b sin η) lies on the boundary of the elliptic cylinder at $\xi=\alpha$, it reduces to

$$q^2=U^2\,\frac{(a+b)^2}{c^2}\,\frac{\sin^2\eta}{\sinh^2\alpha+\sin^2\eta}=\frac{U^2(a+b)^2\sin^2\eta}{c^2\sinh^2\alpha+c^2\sin^2\eta}$$

$$q^2=\frac{U^2(a+b)^2\sin^2\eta}{b^2+(a^2-b^2)\sin^2\eta}$$

$$\text{or}\quad q^2=U^2\frac{(a+b)^2\sin^2\eta}{b^2\cos^2\eta+a^2\sin^2\eta},\quad c^2=a^2-b^2 \qquad ...(1)$$

Thus the velocity at a point (a cos η, b sin η) on the surface is

$q^2 = U^2 (a + b)^2 \sin^2 \eta (b^2 \cos^2\eta + a^2 \sin^2\eta)^{-1}$

or $q = U (a + b) \sin \eta (b^2 \cos^2\eta + a^2 \sin^2 \eta) - 1/2.$ **Proved.**

In a steady state, the pressure p is given by

$$\frac{p}{\rho} + \frac{1}{2}q^2 = \text{const.} \quad ...(2)$$

Since $P_\infty = \Pi$, $q = U$; const. $= \dfrac{\Pi}{\rho} + \dfrac{1}{2} U^2$,

$$\text{or } \Pi - p = \frac{1}{2}\rho (q^2 - U^2). \quad ...(3)$$

Thus diminution in the pressure at a point on the surface of the elliptic cylinder is

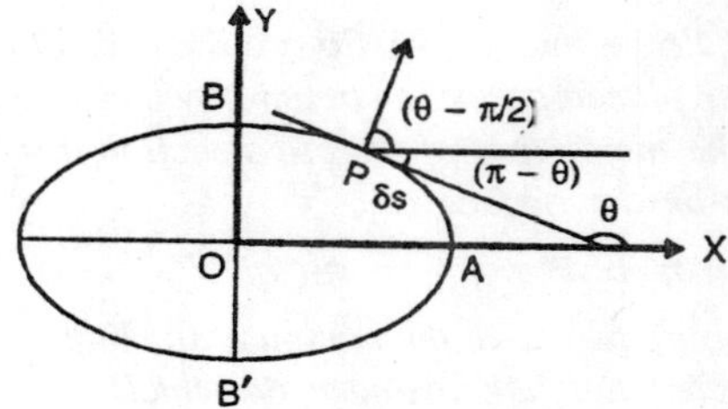

Fig. 2.17

$$= (\Pi - p) = \frac{1}{2}\rho (q^2 - U^2)$$

$$= \frac{1}{2}\rho\left\{\frac{U^2(a+b)^2 \sin^2\eta}{b^2\cos^2\eta + a^2\sin^2\eta} - U^2\right\}$$

$$= \frac{1}{2}\rho U^2\left\{\frac{(a+b)^2\sin^2\eta - (b^2\cos^2\eta + a^2\sin^2\eta)}{b^2\cos^2\eta + a^2\sin^2\eta}\right\}$$

Consider an element δs at any point P on the surface of the elliptic boundary. Let the tangent to the point P makes an angle θ with the major axis. Total diminution in the resultant thrust on the half cylinder BAB′ on which the stream impinges is

$$= \int(\Pi - p)\cos\left(\theta - \frac{\pi}{2}\right)ds = \int(\Pi - p)dy$$

$$= \frac{1}{2}\rho U^2 b\int_{-\pi/2}^{+\pi/2}\left\{\frac{(a+b)^2\sin^2\eta}{b^2(1-\sin^2\eta) + a\sin^2\eta} - 1\right\}\cos\eta \, d\eta$$

Let sin η = λ, cos η dη = dλ

$$= \frac{1}{2}\rho U^2 b \int_{-1}^{+1} \left\{ \frac{(a+b)^2 \lambda^2}{b^2 + (a^2 - b^2)\lambda^2} - 1 \right\} d\lambda$$

$$= \rho U2b \int_0^1 \left\{ \frac{2b}{a-b} - \frac{b^2(a+b)}{a-b} \frac{1}{b^2 + (a^2 - b^2)\lambda^2} \right\} d\lambda$$

$$= \frac{2\rho U^2 b^2}{a-b} \left\{ 1 - \frac{a+b}{2\sqrt{(a^2 - b^2)}} \tan^{-1}\left(\frac{\sqrt{(a^2 - b^2)}\lambda}{b} \right) \right\}_0^1$$

$$= \frac{2\rho U^2 b^2}{a-b} \left\{ 1 - \left(\frac{a+b}{a-b} \right)^{1/2} \tan^{-1}\left(\frac{a-b}{a+b} \right)^{1/2} \right\}.$$ Proved.

Example 12: *A circular cylinder is placed in a uniform stream, find the forces acting on the cylinder.*

Solution: The complex potential for undisturbed motion is given by

w = (u – iv) z

By circle's theorem, we have

$$w = (u - iv)\, z + (u + iv) \frac{a^2}{z}$$

or $$\frac{dw}{dz} = (u - iv) - (u + iv) \frac{a^2}{z^2}.$$

Let (X,Y) be the forces acting on the contour of a fixed circular cylinder, then by Blasius's theorem, we have

$$X - iY = \frac{1}{2} i\rho \int_c \left(\frac{dw}{dz} \right)^2 dz$$

or $$X - iY = \frac{1}{2} i\rho \int \left\{ (u - iv) - (u + iv) \frac{a^2}{z^2} \right\}^2 dz$$

or X – iY = 0 ⇒ X = 0 = Y

Now $$\frac{1}{2}\rho \int_c z \left(\frac{dw}{dz} \right)^2 dz$$

$$N = \text{Real part of} - \frac{1}{2}\rho \int_c \left\{ (u - iv)^2 + 2(u^2 + v^2) \frac{a^2}{z^2} + \ldots \right\} z\,dz$$

N = Real part of – 1/2 ρ {– 2 (u^2 + v^2) a^2} 2πi

N = Real part of $\{2\pi\rho\, a^2 i\, (u^2 + v^2)\}$ = zero

If follows that no force or couple acts on the cylinder. **Answer.**

Example 13: *The circle $(x + a)^2 + y^2 = a^2$ is placed in an oncoming wind of velocity U and there is a circulation $2\pi k$. Find the complex potential and show that moment about the origin is $2\pi k\rho a U$.*

Solution : The complex potential for the uniform stream is given by

$$w = Uz + \frac{Ua^2}{z-a} + ik\log(z-a) \quad \text{...(1)}$$

or $$\frac{dw}{dz} = U - \frac{Ua^2}{(z-a)^2} + \frac{ik}{(z-a)}$$

or $$\frac{dw}{dz} = U - \frac{Ua^2}{z^2}\left[1-\frac{a}{z}\right]^{-2} + \frac{ik}{z}\left[1-\frac{a}{z}\right]^{-1}$$

or $$\frac{dw}{dz} = U - \frac{Ua^2}{z^2}\left[1+\frac{2a}{z}\right]^{-2} + \frac{ik}{z}\left[1+\frac{a}{z}\right]. \quad \text{...(2)}$$

To determine the moment about the origin, equating the coeffcients of $1/z^2$ in (2), we have

$= - k^2 - 2U^2a^2 + 2Uika$

or $$\int z\left[\frac{dw}{dz}\right]^2 dz = (- k^2 - 2U^2a^2 + 2Uika)\, 2\pi i$$

Thus N = Real part of $-\dfrac{1}{2}\rho \int_c z\left[\dfrac{dw}{dz}\right]^2 dz$

or N = Real part of $-\dfrac{1}{2}$r [– k2 –2U2a2 + 2Ukai]2pi

or N = $\dfrac{1}{2}$r (2Uka) 2p = 2pkraU. **Proved.**

Example 14: *An elliptic cylinder, semi-axis a and b, is held with its length perpendicular to, and its major axis making an angle q with, the direction of a stream of velocity V. Prove that the magnitude of the couple per unit length of cylinder due to the fluid pressure is*

$\pi\, \rho\, (a^2 - b^2)\, V^2 \sin\theta \cos\theta,$

and determine its sense.

Solution: The complex potential is given by

$w = V\,(a + b)\cosh(\zeta - \zeta_0)$,

where $\zeta = \xi + i\eta$, $\zeta_0 = a + i\theta$ and $z = c\cosh\zeta$,

Now $\dfrac{dw}{dz} = \dfrac{dw}{d\zeta}\dfrac{d\zeta}{dz} = \dfrac{V(a+b)}{c}\dfrac{\sinh(\zeta-\zeta_0)}{\sinh\zeta}$

or $\dfrac{dw}{dz} = \dfrac{V(a+b)}{c}\dfrac{\sinh\zeta\cosh\zeta_0 - \cosh\zeta\sinh\zeta_0}{\sinh\zeta}$

or $\dfrac{dw}{dz} = \dfrac{V(a+b)}{c}\left\{\cosh\zeta_0 - \dfrac{z}{\sqrt{(z^2-c^2)}}\sinh\zeta_0\right\}$

Again $\dfrac{z}{\sqrt{(z^2-c^2)}} = \left(1-\dfrac{c^2}{z^2}\right)^{-1/2} = 1+\dfrac{1}{2}\dfrac{c^2}{z^2}+\ ...,$

when z is large.

$$\frac{dw}{dz} = \frac{V(a+b)}{c}\left\{\cosh\zeta_0 - \left(1+\frac{c^2}{2z^2}\right)\sinh\zeta_0\right\}$$

or $\left(\dfrac{dw}{dz}\right)^2 = \dfrac{V^2(a+b)^2}{c^2}\left\{e^{-\zeta_0}\ \dfrac{c^2}{2z^2}\sinh\zeta_0\right\}^2.$ ...(1)

If N is the couple about the origin then by Blasius's theorem, we have

$$N = \text{Real part of } -\frac{1}{2}\rho\int_c z\left(\frac{dw}{dz}\right)^2 dz$$

$N = \text{Real part of } -\dfrac{1}{2}r\left\{2\pi i \times \text{sum of the residues of } z\left(\dfrac{dw}{dz}\right)^2.\right.$

inside the contour} ...(2)

or $z\left(\dfrac{dw}{dz}\right)^2 = \dfrac{V^2(a+b)^2}{c^2} - z\left(e^{-\zeta_0} - \dfrac{c^2}{2z^2}\sinh\zeta_0\right)^2$ ...(3)

The only pole lies inside the contour is at origin, therefore equating the coefficients of (1/z) in (3), we have

$$= -\frac{V^2(a+b)^2}{c^2}c^2\sinh\zeta_0 e^{-\zeta_0}$$

From (2) and (3), we have

$N = \text{Real part of } \{\pi\rho i\ V^2\ (a+b)^2 \sinh \zeta_0 e - \zeta_0\}$

$= \text{Real part of } \{\pi\rho i\ V^2\ (a+b)^2 \sinh(\alpha + i\theta)\ e^{-(a+i\theta)}\}$

$= \text{Real part of } \{\pi\rho i\ V^2\ (a+b)^2 (\sinh\alpha\cos\theta$

$+ i \cosh \alpha \sin \theta) e^{-\alpha} (\cos \theta - i \sin \theta)\}$

$= - \pi\rho V^2 (a + b)^2 e^{-\alpha} (\cosh \alpha - \sinh \alpha) \sin \theta \cos \theta$

$= - \pi\rho V^2 (a^2 + b^2)^2 e^{-2\alpha} \sin \theta - \cos \theta$ [as ea = $\sqrt{\{(a+b)/(a-b)\}}$]

$= \pi\rho V^2 (a^2 + b^2)^2 \sin \theta \cos \theta,$ **Proved.**

negative sign shows that the couple tends to set the cylinder broadside to the stream.

Example 15: *Liquid of density ρ is circulating irrotationally between two confocal elliptic cylinders $\xi = \alpha$, $\xi = \beta$, $x + iy = c \cosh (\xi + i\eta)$.*

Prove that, if k is the circulation, the kinetic energy per unit length of cylinders is

$(1/4) \rho k^2 [(\beta - a)/\pi]$.

Solution: The complex potential for irrotational cyclic motion of circulation k round the elliptic cylinder

$$w = \frac{ik}{2\pi}(\xi + i\eta) \Rightarrow \phi + i\psi = \frac{ik}{2\pi}(\xi + i\eta).$$

Equating real real and imaginary parts, we have

$\phi = (k\eta/2\pi)$ and $\psi = (k\xi/2\pi)$. ...(1)

Let T be the kinetic energy per unit length of cylinder, which is given by

$$T = \frac{1}{2}\rho \int q^2 ds = \frac{1}{2}\rho \iint \left\{ \left(\frac{\partial \phi}{\partial \xi}\right)^2 + \left(\frac{\partial \psi}{\partial \eta}\right)^2 \right\} d\xi d\eta$$

As $ds = d\xi\, d\eta$ and $q^2 = \left(\frac{\partial \phi}{\partial \xi}\right)^2 + \left(\frac{\partial \psi}{\partial \eta}\right)^2$

$$\text{or } T = \frac{1}{2}\rho \frac{k^2}{4\pi^2} \int_{\xi=\alpha}^{\beta} \int_{\eta}^{2\pi} d\xi d\eta = \frac{1}{4}\rho k^2 \frac{(\beta - \alpha)}{\pi}.$$ **Proved.**

Example 16: *Show that the motion of a liquid streaming past the elliptic disc $x^2/a^2 + y^2/b^2 = 1$, the velocity at infinity being parallel to the axis of X and equal to U can be expressed by the relation*

$$\phi + i\psi = [U/(a - b)] \{az - b\sqrt{(z^2 - c^2)}\},$$

where $c^2 = a^2$ and $z = x + iy$.

Solution: The velocity potential f and the stream function ψ are given by

$$\phi = Ux + Ub \sqrt{(a+b)} / (a - b) e^{-\xi} \cos \eta,$$

and ψ = Uy + Ub $\sqrt{(a+b)}$ / (a – b) $e^{-\xi}$ sin η,

or $\phi + i\psi$ = Uz + Ub $\sqrt{(a+b)}$ / (a – b) $e^{-\zeta}$

$$\text{or } f + iy = U\sqrt{\left(\frac{a+b}{a-b}\right)}\left\{\frac{a-b}{\sqrt{(a^2-b^2)}}z + b\cosh\zeta\right.$$

$$\left. - b\sqrt{(\cosh^2\zeta - 1)}\right\}$$

$$\text{or } \phi + i\psi = U\sqrt{\left(\frac{a+b}{a-b}\right)}\left(\left(\frac{a-b}{c}\right)z + b\frac{z}{c} - b\sqrt{\left(\frac{z^2}{c^2} - 1\right)}\right),$$

$$\text{or } \phi + i\psi = U\sqrt{\left(\frac{a+b}{a-b}\right)}\left\{\left(\frac{a-b}{c}\right)z + b\frac{z}{c} - \frac{b}{c}\sqrt{(z^2-c^2)}\right\},$$

$$\text{or } \phi + i\psi = \frac{U}{c}\sqrt{\left(\frac{a+b}{a-b}\right)}\{az - b\sqrt{(z^2-c^2)}\},$$

$$\text{or } \phi + i\psi = \frac{U}{(a-b)}\{az - b\sqrt{(z^2-c^2)}\}.$$ **Proved.**

Example 17: *A circle | z | = a is transformed into a thin aerofoil by the transformation* $\zeta = z + (a^2/z)$. *Obtain the lift force and show the moment M about the centre is*

$M = 2\pi\rho\, b^2\, U^2 \sin 2\,(a + \mu)$

where a is the angle of attack and b, μ are constants of the transformation.

Solution : Let U be the velocity of the stream at infinity, whose direction makes an angle α with the negative direction of the X-axis. k is the circulation round the circle of radius b whose centre is at z_0 ($z_0 = ae^{i\beta}$). The complex potential becomes

$$w = Ue^{i\alpha}(z - z_0) + \frac{Ub^2e^{i\alpha}}{(z-z_0)} + \frac{ik}{2\pi}\log(z - z_0). \quad ...(1)$$

$$\frac{dw}{dz} = Ue^{i\alpha}\ \frac{Ub^2e^{-i\alpha}}{(z-z_0)^2} + \frac{ik}{2\pi(z-z_0)}. \quad ...(2)$$

Consider the stagnation point at $z - z_0 = be^{i}(\pi + \beta) = -be^{i\beta}$.

For stagnation point dw/ dz = 0, so

$$Ue^{i\alpha} - \frac{Ub^2e^{-i\alpha}}{(z-z_0)^2} + \frac{ik}{2\pi(z-z_0)} = 0.$$

or $2\pi bU\ [e^{i(\alpha+\beta)} - e^{-i(\alpha+\beta)}] = ik.$

or $4\pi bU \sin(\alpha+\beta) = k.$

Since the transformation

$$\zeta = z + \frac{a^2}{z}, \qquad \text{...(4)}$$

transforms a circle into a thin aerofoil. Then

$$\frac{dw}{d\zeta} = \frac{dw}{dz}.\frac{dz}{d\zeta} = \left(1-\frac{a^2}{z^2}\right)^{-1}\frac{dw}{dz},$$

or $$\left(\frac{dw}{d\zeta}\right)^2 d\zeta = \left(1-\frac{a^2}{z^2}\right)^{-1}\left(\frac{dw}{dz}\right)^2 dz$$

By Blasius's theorem, we have

$$X - iY = \frac{1}{2} i\rho \int_c \left(\frac{dw}{d\zeta}\right)^2 d\zeta. \qquad \text{...(5)}$$

From (2) and (5), we have

$$X - iY = \frac{1}{2} i\rho \int_c \left(1-\frac{a^2}{z^2}\right)^{-1}\left(Ue^{i\alpha} - \frac{Ub^2e^{-i\alpha}}{(z-z_0)^2} + \frac{ik}{2\pi(z-z_0)}\right)^2 dz.$$

$$\text{or } X - iY = \frac{1}{2} i\rho \int_c \left(1+\frac{a^2}{z^2}+...\right)\left[Ue^{i\alpha} - \frac{Ub^2e^{-i\alpha}}{z^2}\times\left(1+\frac{2z_0}{z}+...\right)\right.$$
$$\left. + \frac{ik}{2\pi z}\left(1+\frac{z_0}{z}+...\right)\right]^2 dz$$

The sum of the residues at the pole $z = 0$ is

$$X - iY = \frac{1}{2} i\rho\left(\frac{Ue^{i\alpha}ik}{\pi}\right)2\pi i = -\ i\rho k\ Ue^{i\alpha}$$

$$X - iY = -\ i\rho kU(\cos\alpha + i\sin\alpha)$$

$\Rightarrow X = \rho kU \sin\alpha,\ Y = \rho kU \cos\alpha.$

Thus the lift force L is given by

$$L = \sqrt{(X^2+Y^2)} = \rho kU = 4\pi b\ \rho U^2 \sin(\alpha+\beta), \qquad \text{...(6)}$$

which is perpendicular to the main stream. **Ans.**

Again, M is the moment about the centre z_0, then

$$M = N + xY - yX = N + a\cos\beta\, Y - a\sin\beta X, \quad ...(7)$$

where $N = \text{Real part of } -\frac{1}{2}\rho\int_c z\left(\frac{dw}{dz}\right)^2 dz$

$$N = 2\pi\rho U^2 [a^2 \sin 2(\alpha+\beta) + a_1 \sin 2\alpha]. \quad ...(8)$$

From (7) and (8), we have

$M = 2\pi\rho U^2 [a^2 \sin 2(\alpha+\beta) - a_1 \sin 2\alpha]$

$+ [a\cos\beta\, \rho k U \cos a - a\sin\beta\, \rho k U \sin\alpha]$

or $M = 2\pi\rho U^2 [a^2 \sin 2(\alpha+\beta) + a_1 \sin 2\alpha]$

$+ 2\pi\rho a^2 U^2 \sin 2(\alpha+\beta)$.

or $M = 2\pi\rho U^2 [2a^2 \sin 2(\alpha+\beta) + a_1 \sin 2\alpha]$

or $M = 2\pi\rho U^2 [(2a^2 \cos 2\beta + a_1)\sin 2\alpha + 2a^2 \sin 2\beta \cos 2\alpha]$

or $M = 2\pi\rho U^2 [b^2 \cos 2\mu \sin 2\alpha + b^2 \sin 2\mu \cos 2\alpha]$

or $M = 2\pi\rho U^2 b^2 \sin 2(\alpha+\mu)$. **Proved.**

Example 18: *A flat plate of infinite length and width l is placed in a current of incompressible fluid with its plane at an angle α to the undisturbed stream lines and its edge perpendicular to them. Determine the resulting flow on the circulation theory, assuming the velocity at the trailing edge is finite. By considering the pressures and velocities over a large cylinder whose axis is the median line of the plate, show that the forces on the plate are equivalent to a force $k\rho U^2 l \sin\alpha$ per unit length perpendicular to the current, acting at a distance l/4 from the leading edge.*

Solution: The circle of radius a transforms into the line of length a by the transformation

$$\zeta = z + \frac{a^2}{z}, \; l = 4a.$$

The complex potential in z- plane is given by

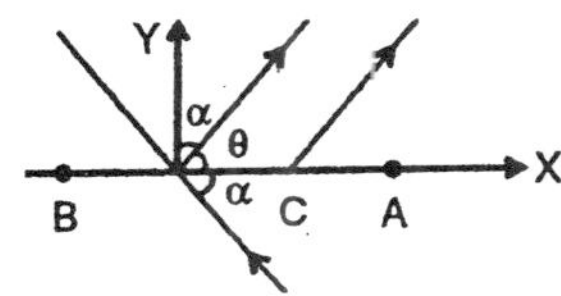

Fig. 2.18

$$w = Ueia\, z + \frac{Ua^2 e^{-i\alpha}}{z} + \frac{ik}{2\pi}\log z,$$

where centre of the circle is taken as origin.

$$\frac{dw}{dz} = Ue^{i\alpha} - \frac{Ua^2e^{i\alpha}}{z^2} + \frac{ik}{2\pi z}. \qquad ...(1)$$

the stagnation points corresponding to z = – a are given by dw/dz = *i.e.*,

$$Ue^{i\alpha} - \frac{Ua^2e^{-i\alpha}}{a^2} + \frac{ik}{2\pi a} = 0$$

or $iK = 2\pi a\ U\ (e^{i\alpha} - e^{-i\alpha}) = 4\ \pi a\ U\ i \sin\alpha$

$\Rightarrow K = 4\pi a\ U \sin\alpha = \pi l\ U \sin\alpha$. ...(2)

By Blasius's theorem, we have

$$X - iY = \frac{1}{2}i\rho\int_c\left(\frac{dw}{d\zeta}\right)^2 d\zeta$$

$$X - iY = \frac{1}{2}i\rho\int_c\left(1-\frac{a^2}{z^2}\right)^{-1}\left(\frac{dw}{dz}\right)^2 dz$$

$$X - iY = \frac{1}{2}i\rho\int_c\left(1-\frac{a^2}{z^2}\right)^{-1}\left(Ue^{\alpha i} - \frac{Ua^2e^{-i\alpha}}{z^2} + \frac{ik}{2\pi z}\right)^2 dz$$

The sum of the residues at the pole z = 0 is

$$X - iY = \frac{1}{2}i\rho\left(\frac{2iUke^{i\alpha}}{2\pi}\right)2\pi i$$

$X - iY = -\ i\rho\ Uk\ e^{i\alpha} = -\ i\rho\pi\ lU^2 \sin\alpha\ e^{i\alpha}$

$\Rightarrow X = \pi\rho\ lU^2 \sin^2\alpha$,

and $Y = \rho\pi\ lU^2 \sin\alpha \cos\alpha$.

The resulting force R on the circulation theory becomes

$$R = \pi\ \rho\ lU^2\ \sqrt{(\sin^4\alpha + \sin^2\alpha\cos^2\alpha)} = \pi\rho\ lU^2 \sin\alpha,$$

at an angle $\tan\theta = Y/X = \cot\alpha \Rightarrow \theta = \pi/2 - \alpha$.

It follows that the resulting flow is perpendicular to the uniform stream.

Ans.

Again N = Real part of $-\ \frac{1}{2}\rho\int_c\left(\frac{dw}{d\zeta}\right)^2 \zeta d\zeta$.

$$= \text{Real part of } -\ \frac{1}{2}\rho\int_c\left(1-\frac{a^2}{z^2}\right)^{-1}\left(z + \frac{a^2}{z}\right)\left(\frac{dw}{dz}\right)^2 dz$$

$$= \text{Real part of } -\frac{1}{2}\rho\int_c \left(z+\frac{2a^2}{z}\right) \times \left[Ue^{i\alpha}\frac{Ua^2e^{-i\alpha}}{r^2}+\frac{ik}{2\pi z}\right]^2 dz.$$

The sum of residues at the pole z = 0 (i.e., equating the coefficients of (1/z)) is given by

$$= \left[U^2e^{2i\alpha}.2a^2 - \frac{k^2}{4\pi^2} - 2U^2a^2\right] \times 2\pi i$$

Thus $N = 2\pi\ \rho\ U^2\ a^2 \sin 2\alpha = 1/8\ \pi\rho l^2\ U^2 \sin 2\alpha$.

Hence the system ruduces to a single resultant force R together with a couple N, then

$R\ x \cos\alpha = N$,

where x is the distance from origin to the point of application of R.

or $\pi\rho l U^2 \sin\alpha\ x \cos\alpha = \frac{1}{8}\ \pi\rho l^2\ U^2 \sin 2\alpha$.

or $\pi\rho l U^2 \sin 2\alpha \left(x - \frac{1}{4}l\right) = 0$. **Proved.**

Example 19: *A source is placed midway between two planes whose distance from one another is 2a. Find the equation for the streamlines when the motion is in two dimensions. Show that those particles which at an infinite distance are distance a/2 from one of the boundaries, issured from the source in a direction making an angle p/4 with it.*

Solution : The infinite strip $P_\infty\ Q_\infty R_\infty S_\infty$ in the Z-plane is transformed into the upper half of ζ-plane by the transformation

$\zeta = i \exp.(\pi z/2a)$

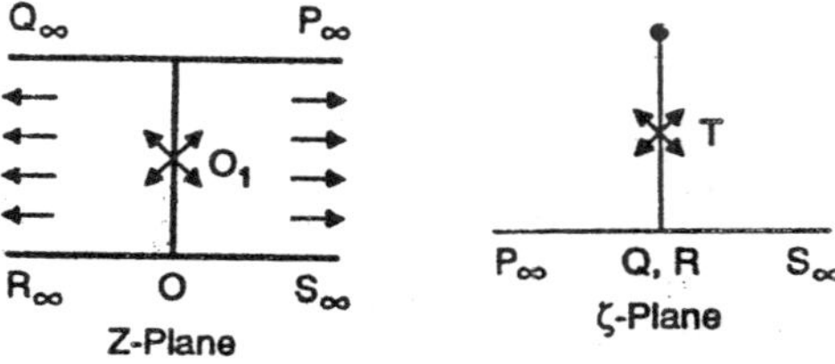

Fig. 2.19

Let O_1 be the origin in the Z-plane being midway between the two walls. The points $Q_\infty\ R_\infty$ coincide with Q, R, ζ = 0 in ζ-plane when z = 0, ζ = i. Thus a source of strength m at O_1 and an equal sink of same strength at infinite distance in Z-plane correspond to a source of strength m at T and sink (–m) at Q, R and hence an image source m at the point $\zeta = -i$. Thus the complex potential is given by

$$w = -m \log(\zeta - i) - m \log(\zeta + i) + m \log \zeta$$

$$\text{or } w = -m \log(\zeta^2 + 1)/\zeta\} = -m \log(\zeta + \zeta^{-1})$$

$$\text{or } w = -m \log i\,(e^{\pi z/2a} - e^{-\pi z/2a}) = -m \log \sinh(\pi z/2a)$$

$$\text{or } w = -m \log (e^{\pi z/2a} - e^{-\pi z/2a})$$

$$\text{or } w = -m \log [e^{\pi x/2a} \{ \cos(\pi y/2a) + i \sin(\pi y/2a)$$

$$- e^{-\pi x/2a} \{ \cos(\pi y/2a) + i \sin(\pi y/2a)\}]$$

$$\text{or } \psi = \tan^{-1}\left\{\frac{\cosh(\pi x/2a)\sin(\pi y/2a)}{\sinh(\pi x/2a)\cos(\pi y/2a)}\right\} - \tan^{-1}\left\{\frac{\tan(\pi y/2a)}{\tanh(\pi x/2a)}\right\}.$$

The stream lines are given by ψ = const.

$$\tan(\pi y/2a) = C \tanh(\pi x/2a).$$

Since $x = \infty$, $y = a/2$; $C = 1$.

Thus $\tan(\pi y/2a) = C \tanh(\pi x/2a)$.

Differentiating with regard to x, we have

$$\sec^2(\pi y/2a)\,\frac{dy}{dx} = \text{sech}^2(\pi x/2a).$$

The source is situated at the origin in Z-plane *i.e.*, $x = 0 = y$

$$\Rightarrow \frac{dy}{dx} = 1 \Rightarrow \tan\theta = \pi/4.$$ **Proved.**

Example 20: *The irrotational motion in two dimensions of a fluid bounded by the lines y = 0, y = a is due to a doublet of strength μ at the origin, the axis of the doublet being in the positive direction of the axis of X. Prove that the motion is given by w = (πμ/a) coth (πz/2a). Find the stream lines and show that those points where the fluid is moving parallel to the axis of Y lie on the curve*

$$\text{sech}(\pi x/a)\sec(\pi y/a) = 1.$$

Solution : The infinite strip in the Z-plane is transformed into the upper half of real axis in ζ-plane by the transformation

$$\zeta = \exp.(\pi z/a).$$

Let μ' be the strength of the doublet in ζ-plane then

$$\mu' = m\mu = \left|\frac{d\zeta}{dz}\right|\mu = \frac{\pi}{a}\mu \text{ along the real axis } z = 0.$$

The complex potential w is given by

$$w = \frac{\mu'}{\zeta - 1}.$$

(origin in Z-plane corresponds to $\zeta = 1$ in ζ-plane)

$$\text{or } w = \frac{\pi\mu}{a}\frac{1}{\zeta - 1} = \frac{\pi\mu}{a}\left[\frac{1}{e^{\pi z/a} - 1}\right]$$

$$\text{or } \frac{\pi\mu}{a}\left[\frac{1}{e^{\pi z/a} - 1} + \frac{1}{2} - \frac{1}{2}\right]$$

$$\text{or } w = \frac{\pi\mu}{2a}\left[\frac{e^{\pi z/a} + 1}{e^{\pi z/2a} - 1}\right] - \frac{\pi\mu}{2a}$$

$$\text{or } w = \frac{\pi\mu}{2a}\left[\frac{e^{\pi z/2a} + e^{-\pi z/2a}}{e^{\pi z/2a} - e^{-\pi z/2a}}\right] - \frac{\pi\mu}{2a} = \frac{\pi\mu}{2a}\coth\frac{\pi z}{2a} + \text{const. } \textbf{Proved.}$$

$$\text{or } w = \frac{\pi\mu}{2a}\left[\frac{\cosh\{\pi(x+iy)/2a\}\sinh\{\pi(x-iy)/2a\}}{\sinh\{\pi(x+iy)/2a\}\sinh\{\pi(x-iy)/2a\}}\right] + \text{const.}$$

$$\text{or } \phi + i\psi = \frac{\pi\mu}{2a}\left[\frac{\sinh(\pi x/a) - i\sin(\pi y/a)}{\cosh(\pi x/a) - \cos(\pi y/a)}\right], \text{ constant is neglected.}$$

$$\text{or } \psi = -\frac{\pi\mu}{2a}\frac{\sin(\pi y/a)}{\cosh(\pi x/a) - \cos(\pi y/a)}.$$

The streamlines are given by y = const. The points where the motion is parallel to Y-axis *i.e.*, $u = 0 \Rightarrow \partial\psi/\partial y = 0$.

or $\{\cosh(\pi x/a) - \cos(\pi y/a)\}\cos(\pi y/a) - \sin(\pi y/a)\sin(\pi y/a) = 0$

$\cosh(\pi x/a)\cos(\pi y/a) = \cos^2(\pi y/a) + \sin^2(\pi y/a) = 1$

or $\text{sech}(\pi x/a)\sec(\pi y/a) = 1$. **Proved.**

Example 21: *Fluid motion is taking place in the part of the plane bounded by the real axis and the lines x = ± a, which is due to a source at one corner and a sink at the other corner of the strip, each of strength m, show that the motion is given by*

$$\tanh(w/8m) = \tan(\pi z/4a).$$

and that the streamline which leaves the source at an angle $\pi/4$ to the side is

$$\cos(\pi x/2a) = \sinh(\pi y/2a).$$

Solution : Consider a semi-infinite strip $P_\infty QRS_\infty$ in Z-plane bounded by the lines $x = \pm a$, $y = 0$, with two vertices at infinity. By transforming $P_\infty Q$,

R on the points $\zeta = -\infty$, $\zeta = \pm 1$ of the real axis in the ζ-plane , then in the accordance with the Schwarz and Christoffel theorem, we have

$$\frac{dz}{d\zeta} = z\frac{K}{\sqrt{(\zeta^2 - 1)}} \Rightarrow z = K\cos^{-1}\zeta + A.$$

When $z = a$, $\zeta = 1 \Rightarrow a = K \cosh^{-1} 1 + A \Rightarrow A = a$.

When $z = -a$, $\zeta = -1$

$\Rightarrow -a = K \cosh^{-1} (-1) + a$

or $-2a = K \cosh^{-1} (-1)$

or $\cosh\left(-\frac{2a}{K}\right) = -1 \Rightarrow K = \frac{2ai}{\pi}$.

Therefore $z = \frac{2ai}{\pi} \cosh{-1}\ \zeta + a$.

or $\zeta = \cosh \frac{\pi}{2ai} (z - a)$

$$= \cosh\frac{\pi i}{2a}(a - z) = \sin\frac{\pi z}{2a}. \quad ...(1)$$

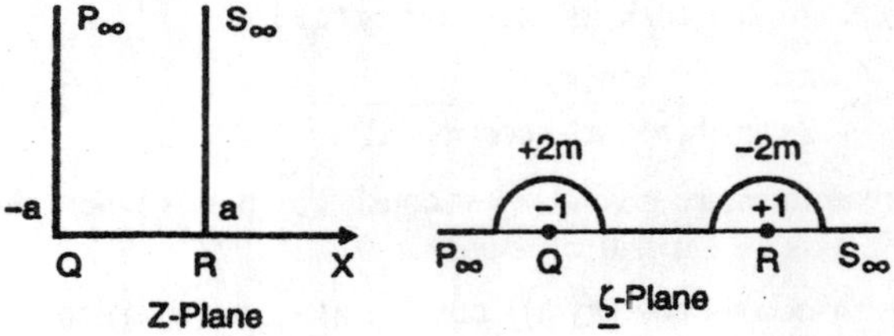

Fig. 2.20

Again, a source of strength m at x = a and an equal sink at x = – a in Z-plane correspond to a source of strength 2m at $\zeta = +1$ and an equal sink at $\zeta = -1$ in ζ -p plane. The complex potential is given by

$w = -2m (\zeta - 1) + 2m \log (\zeta + 1)$

or $w = -2m \log \frac{\sin(\pi z/2a) - 1}{\sin(\pi z/2a) + 1}$

w $w = -4m \log \frac{\cos(\pi z/4a) - \sin(\pi z/4a)}{\cos(\pi z/4a) + \sin(\pi z/4a)}$

or $w = -4m \log \frac{1 - \tan(\pi z/4a)}{1 + 4\tan(\pi z/4a)}$ ...(2)

$$e^{-w/4m} = \frac{1 - \tan(\pi z/4a)}{1 + 4\tan(\pi z/4a)},$$

or $\dfrac{1 - e^{-(w/4m)}}{1 + e^{-(w/4m)}} = \tan(\pi z/4a)$

or $\dfrac{e^{(w/8m)} - e^{-(w/8m)}}{e^{(w/8m)} + e^{-(w/8m)}} = \tan(\pi z/4a)$

or $\tanh\left(\dfrac{w}{8m}\right) = \tan\left(\dfrac{\pi z}{4a}\right)$. **Proved.**

Again from(2), we have

$$w = -4m \log \tan\left(\frac{\pi}{4} - \frac{\pi z}{a}\right)$$

or $w = -4m \log \tan \dfrac{\pi}{4a}(a - x - iy)$

or $w = -4m \log \dfrac{\sin\{(\pi/4a)(a - x - iy)\}\cos\{(\pi/4a)(a - x + iy)\}}{\cos\{(\pi/4a)(a - x - iy)\} - i\cosh\{(\pi/4a)(a - x + iy)\}}$

or $w = -4m \log \dfrac{\sin\{\pi(a - x)/2a\} - i\sinh(\pi y/2a)}{\cos\{\pi(a - x)/2a\} - i\cosh(\pi y/2a)}$

or $w = -4m \left[\log\left\{\sin\dfrac{\pi}{2a}(a - x) - i\sinh\dfrac{\pi y}{2a}\right\}\right]$

$$- \log\left\{\cos\frac{\pi}{2a}(a - x) - \cosh\frac{\pi y}{2a}\right\}$$

or $\psi = -4m \tan^{-1}\left\{-\dfrac{\sinh(\pi y/2a)}{\sin[\pi(a - x)/2a]}\right\} = 4m \tan^{-1}\dfrac{\sinh(\pi y/2a)}{\cos(\pi x/2a)}$

The streamlines are given by y = constant i.e.

$$\sinh\left(\frac{\pi y}{2a}\right) = B\cos\left(\frac{\pi x}{2a}\right) \Rightarrow \cosh\left(\frac{\pi y}{2a}\right).\frac{dy}{dx} = -B\sin\left(\frac{\pi x}{2a}\right). \quad ...(3)$$

For the stream leaving the source at $\pi/4$ $\left(\dfrac{dy}{dx}\right)_{\substack{x=0\\y=0}} = -1$

From (3), we have $\cosh\left(\dfrac{\pi}{2}\right).\dfrac{dy}{dx} = -B \Rightarrow B = 1.$

Therefore, the stream lines are given by

$\sinh(\pi y/2a) = \cos(\pi x/2a)$. **Proved.**

Example 22: *Prove that for the complex potential tan^{-1} z the stream lines and equi-potentials are circles. Find the velocity at any points and examine the singularities at z = ± i.*

Solution: The complex potential is given by

$w = \phi + i\psi = \tan^{-1}z,$ ...(1)

Also $\overline{w} = \phi + i\psi = \tan^{-1}\overline{z},$ ...(2)

By subtracting (1) and (2), we have

$$2i\psi = \tan^{-1}z - \tan^{-1}\overline{z} = \tan{-1}\frac{z-\overline{z}}{1+z\overline{z}}$$

$$\text{or } \tan 2i\,\psi = \frac{2iy}{1+x^2+y^2}$$

$\Rightarrow x^2 + y^2 + 1 = 2y \coth 2\psi.$

The stream lines ψ = constant represent the circles

$x^2 + y^2 + 1 = 2y \coth 2\psi.$

Similarly, by adding (1) and (2), we have

$$2\phi = \tan^{-1}z + \tan^{-1}\overline{z} = \tan^{-1}\frac{z-\overline{z}}{1+z\overline{z}}$$

or $1 - x^2 - y^2 = 2x \cot 2\phi.$

The equi-potentials φ = const. also represent circles which are orthogonal to the streamlines ψ = const. and form a co-axial system with limit points at z = ± i. The velocity component (u, v) are given by

$$\frac{dw}{dz} = -m + iv = \frac{1}{z^2+1},$$

The denominator vanishes at z = ± i, therefore, it represents the singularities at these points.

At z = + i, substitute $z = i + z_1$, where $|z_1|$ is very small

$$-u + iv = \frac{dw}{dz} = \frac{dw}{dz_1} = \frac{1}{1+(-1+2iz_1)} = \frac{1}{2iz_1}.$$

by integrating, we have $w = -\frac{1}{2}i \log z_1$

⇒ that the singularity at z = i is a vortex of strength k = – 1/2 with circulation – πk.

Similarly, the singularity at z = – i is a vortex of strength k = 1/2 with circulation πk.

$$= \frac{\mu e^{i\alpha}}{z-b} + \frac{\mu z e^{i(\pi-\alpha)}}{b\left(z - \frac{a^2}{b}\right)}$$

$$= \frac{\mu e^{i\alpha}}{z-b} + \frac{\mu e^{i(\pi-\alpha)}}{b} \frac{[z - a^2/b + a^2/b]}{z - a^2/b}$$

$$= \frac{\mu e^{i\alpha}}{z-b} + \frac{\mu a^2}{b^2} \frac{e^{i(\pi-\alpha)}}{(z-a^2/b)} + \frac{\mu}{b} e^{i(\pi-\alpha)}.$$

Ignoring the last term which is constant,

$$w(z) = \frac{\mu e^{i\alpha}}{z-a} + \frac{\mu a^2}{b^2} \frac{e^{i(\pi-\alpha)}}{z-a^2/b}.$$

This is the complex potential due to

(i) a doublet of strength m at z = b, inclined at an angle α to the x-axis.

(ii) a doublet of strength $\frac{\mu a^2}{b^2}$ at $z = \frac{a^2}{b}$ and inclined at an angle (π – α) to the x-axis.

The doublet (ii) is the image of doublet (i) in the circle $|z| = a$.

Example 23: *Find the lines of flow in two-dimensional fluid motion given by* $\phi + i\psi = -\frac{1}{2}\pi\ (x + iy)^2 e^2 int.$

Prove or verify that the paths of the particles of the fluid (in polar co-ordinates) may be obtained by eliminating t from the equations $r \cos(nt + \theta) - x_0 = r \sin(nt + \theta) - y_0 = nt(x_0 - y_0)$.

Solution: We have $\phi + i\psi = -\frac{1}{2} n (x + iy)^2 e^{2int}$.

Putting $x = r\cos\theta$, $y = r\sin\theta$, i.e., $x + iy = re^{i\theta}$,

$$\phi + i\psi = -\frac{1}{2} nr^2 e^{2i\theta} e^{2int} = -\frac{1}{2} nr^2 e^{2i(\theta + nt)}.$$

Equating real and imaginary parts,

$$\phi = \frac{1}{2} nr^2 \cos(2\theta + 2nt)$$

and $\psi = -\frac{1}{2} nr^2 \sin(2\theta + 2nt)$ ...(1)

Now the lines of flow are given by ψ = const.

i.e. $r^2 \sin(2\theta + 2nt)$ = const.

For the second par, we have

$$r = n \frac{dr}{dt} = -\frac{\partial \phi}{\partial r} = nr \cos(2\theta + 2nt)$$

$$\text{and } r\theta = r \frac{d\theta}{dt} = \frac{1}{r}\frac{\partial \phi}{\partial r} = -nr \sin(2\theta + 2nt) \qquad ...(2)$$

Now by actual differentiation,

$$\frac{d}{dt}\{r \cos(nt + \theta) = r \cos(nt + \theta) - r\theta \sin(nt + \theta) - rn \sin(nt + \theta)$$

$$= nr [\cos(nt + \theta) \cos(2nt + 2\theta) + \sin(nt + \theta) \sin(2nt + 2\theta) - \sin(nt+\theta)]$$

putting values of r and rq from (2)

$$= nr [\cos(nt + \theta) - \sin(nt + \theta)]. \qquad ...(3)$$

Similarly, $\frac{d}{dt}[r \sin(nt + \theta)]$

$$= nr [\cos(nt + q) - \sin(nt + q)]. \qquad ...(4)$$

Thus from (3) and (4),

$$\frac{d}{dt}[r \cos(nt + \theta) = \frac{d}{dt}[r \sin(nt + \theta)],$$

i.e. $r \cos(nt + \theta) = r \sin(nt + \theta) + A$,

where A is the constant of integration

i.e., $r [\cos(nt + \theta) - \sin(nt + \theta) = A$. ...(5)

To determine A, we have when $t = 0$,

$x = r \cos \theta = x_0$, $y = r \sin \theta = y_0$.

$\therefore [x_0 - y_0] = A$.

Thus (5) becomes

$r [\cos(nt + \theta) - \sin(nt + 0)] = x_0 - y_0$.

Putting this value on the right hand side of (3), it becomes

$$\frac{d}{dt}\{r \cos(nt + \theta)\} = n (x0 - y0).$$

Integrating, $r \cos(nt + q) + C1 = nt (x0 - y0)$.

Again when $t = 0$, $x = r \cos q = x0$.

$\therefore x_0 + C_1 = 0$, i.e., $C_1 = -x_0$.

Thus $r \cos(nt + \theta) - x_0 = nt (x_0 - y_0)$.

Similarly from (4), $r \sin(nt + \theta) - y_0 = nt (x_0 - y_0)$; so we have finally

$r \cos(nt + \theta) - x_0 = r \sin(nt + \theta) - y_0 = nt (x_0 - y_0)$.

Example 24: *In the case of the motion of liquid in a part of a plane bounded by a straight line due to a source in the plane prove that if mρ is the mass of fluid (of density ρ) generated at the source per unit of time the pressure on the length 2l of the boundary immediately opposite to the source is less than that on an equal length at a great distance by*

$$\frac{1}{2}\frac{m^2\rho}{\pi^2}\left\{\frac{1}{c}\tan^{-1}\frac{l}{c}-\frac{l}{l^2+c^2}\right\},$$

where c is the distance of the source from the boundary.

Solution: Let the bounding line oy be taken as y- axis and $\frac{m}{2\pi}$ be the source at A (c, 0).

The equivalent image system consistsof

(i) a source $\frac{m}{2\pi}$ at A (c, 0),

(ii) a source $\frac{m}{2\pi}$ at A′ (– c, 0).

Hence the complex potential is given by

$$w = -\frac{m}{2\pi}\log(z-c) - \frac{m}{2\pi}\log(z+c)$$

$$= -\frac{m}{2\pi}\log(z^2-c^2).$$

$$\text{Speed, } q = \left|\frac{dw}{dz}\right| = \frac{m}{2\pi}\left|\frac{2z}{z^2-c^2}\right|.$$

If P (z = iy) is a point on y-axis, then speed at this point is given by

$$q = \frac{m}{\pi}\frac{y}{y^2+c^2},$$

Also by Bernoulli's theorem,

$$\frac{p}{\rho} = \frac{p_0}{\rho}-\frac{1}{2}q^2,$$

where p_0 is the pressure on OY when y is the infinite, where velocity is zero.

$$\therefore \quad \frac{p_0-p}{\rho}\frac{1}{2}q^2 = \frac{1}{2}\frac{m^2}{\pi^2}\frac{y^2}{(y^2+c^2)^2}.$$

Therefore the required difference in pressure

$$= \int_{-l}^{l}(p_0-p)dy = \frac{1}{2}\frac{m^2\rho}{\pi^2}\int_{-l}^{l}\frac{y^2}{(y^2+c^2)}dy$$

$$\frac{m^2}{\pi^2}\rho\int_0^l \frac{y^2}{(y^2+c^2)^2}dy$$

$$\frac{m^2\rho}{\pi^2 c}\int_0^l \frac{c^2\tan^2\theta}{c^4\sec^4\theta.}c\sec^2\theta d\theta,\ \text{putting } y = c\tan\theta$$

$$= \frac{m^2\rho}{\pi^2 c}\int_0^l \sin^2\theta d\theta = \frac{m^2\rho}{2\pi^2 c}\int_0^l (1-\cos 2\theta)\,d\theta\quad \text{putting } y = c\tan\theta$$

$$= \frac{m^2\rho}{\pi^2 c}\left[\theta - \frac{\sin 2\theta}{2}\right] = \frac{1}{2}\frac{m^2\rho}{\pi^2}\left[\frac{1}{c}\theta - \frac{1}{c}\sin\theta\cos\theta\right]_0^l$$

$$= \frac{1}{2}\frac{m^2\rho}{\pi^2}\left[\frac{1}{c}\tan^{-1}\frac{y}{c} - \frac{y}{y^2+c^2}\right]_0^l \quad \text{as } y = c\tan\theta$$

$$= \frac{1}{2}\frac{m^2\rho}{\pi^2}\left[\frac{1}{c}\tan^{-1}\frac{l}{c} - \frac{l}{l^2-c^2}\right]_0^l.$$

This proves the result.

Example 25: An area A is bounded by that part of the x-axis for which $x > a$ and by that branch of $x^2 - y^2 = a^2$ which is in the positive quadrant. There is a two-dimensional unit source at (a, 0) which sends out liquid uniformly in all directions. Shew by means of the transformation $w = \log(z^2 - a^2)$ that in steady motion the stream lines of the liquid within the area A are portions of rectangular hyperbolas. Draw the stream lines corresponding.

Solution: Equating imaginary parts,

$$y = -\,m\tan^{-1}\frac{2xy}{x^2-y^2-a^2} + 2m\tan^{-1}\frac{y}{x}$$

$$= -\tan^{-1}\frac{2xy}{x^2-y^2-a^2} + m\tan^{-1}\frac{2xy}{x^2-y^2}$$

$$\text{as } 2\tan^{-1}A = \tan^{-1}\frac{2A}{1-A^2}$$

$$= -\,m\tan^{-1}\frac{2a^2xy}{(x^2+y^2)^2 - a^2(x^2-y^2)}.$$

The streamlines are given by y = const.

$$\text{i.e.,}\quad \frac{2a^2xy}{(x^2+y^2)^2 - a^2(x^2-y^2)} = \frac{2}{\lambda}\ \text{(say)}$$

or $(x^2 + y^2)^2 = a^2(x^2 - y^2 + \lambda xy)$;

by properly adjusting the constant, which is the parameter in this case and for different values of λ, we get different streamlines.

$$\text{speed} = \left|\frac{dw}{dz}\right| = \left|\frac{2m}{z^2 - a^2} - \frac{2m}{z}\right|$$

$$= \frac{2ma^2}{|z||z-a||z+a|} = \frac{2ma^2}{r_3 r_1 r_2}.$$

This proves the result.

Example 26: *If a homogeneous liquid is acted on by a repulsive force from the origin, them magnitude of which at distance r from the origin is μr per unit mass, shew that it is possible for the liquid to move steadily, without being constrained by any boundaries, in the space between one branch of the hyperbola $x^2 - y^2 = a^2$ and the asymptotes, and find the velocity potential.*

Solution: The liquid moves steadily between the space given by one branch of

$$x^2 - y^2 = a^2 \qquad \text{...(1)}$$

and its asymptotes given by $x^2 - y^2 = 0$. ...(2)

Thus $y = A(x^2 - y^2) = Ar^2 \cos^2 2\theta = Ar^2 \sin\left(\frac{\pi}{2} + 2\theta\right)$,

where A is some constant.

The harmonic conjugate of ψ which is given by φ is clearly

$$f = Ar^2 \cos\left(\frac{\pi}{2} + 2\theta\right),$$

$$\text{so that } w = \phi + i\psi = Ar^2 e^{i(\frac{1}{2}\pi + 2\theta)}$$

$$= Ae^{\frac{1}{2}i\pi r^2 e^2 i\theta} = Aiz^2 \text{ as } z = re^{i\theta}.$$

$$\therefore \quad \frac{dw}{dz} = Ai.\ 2z.$$

Since the only (repulsive) force is along the radius vector, the velocity is a function of r alone.

So if v is the velocity at a distance r, v = 2 Air.

The motion is steady ; hence the equation of motion becomes

$$\frac{p}{p} + \frac{1}{2}v^2 + V = \text{const.},$$

where as given $-\frac{\partial V}{\partial r} = \mu r$ or $V = -\frac{1}{2}\mu r^2$,

so that $\frac{p}{\rho} - 2A^2r^2 - \frac{1}{2}\mu r^2 = \text{const.}$

On the free surface p = const. $\therefore\ 2A^2 - \frac{1}{2}\mu = 0.$

Hence $v = 2Ai.\ r = -\sqrt{\mu r}$

and velocity potential,

$$\phi = Ar^2 2^{\frac{1}{2}i\pi} \cos 2\theta = Air^2 \cos 2\theta$$

$$= -\frac{1}{2}\sqrt{\mu . r^2} \cos 2\theta.$$

Example 27: *In a two-dimensional liquid motion ϕ and ψ are the velocity potential and current function; show that a second fluid motion exists in which ψ is the velocity potential and $-\phi$ the current function; and prove that if the first motion be due to sources and sinks, the second motion can be built up by replacing a source and an equal sink by a line of doublets uniformly distributed along any curve joining them.*

Solution: Since ϕ and ψ are the velocity and stream function for the two-dimensional motion, therefore we have the conditions

$$\frac{\partial\phi}{\partial x} = \frac{\partial\psi}{\partial y}, \frac{\partial\phi}{\partial y} = -\frac{\partial\psi}{\partial x}. \quad ...(1)$$

Now the fluid motion with ψ as velocity potential and $-\phi$ as current function will exist if conditions of the of (1) are satisfied when we put ψ for ϕ and $-\phi$ for ψ,

i.e., if $\frac{\partial\psi}{\partial x} = \frac{\partial(-\phi)}{\partial y}$ and $\frac{\partial\psi}{\partial y} = -\frac{\partial(-\phi)}{\partial x}$,

i.e., if $\frac{\partial\psi}{\partial x} = -\frac{\partial\phi}{\partial y}$ and $\frac{\partial\psi}{\partial y} = \frac{\partial\phi}{\partial x}$.

which hold because of (1). Hence such a motion exists.

Thus if $w = \phi + i\psi$ exists,

$w' = +\psi - i\phi = iw$ also exists.

Second part : Let there be a source m at B (a, 0) and a sink – m at (–a, 0). The complex potential for this is

$$w = -m \log (z - a) + m \log (z + a)$$

$$= m \log\left(\frac{z+a}{z-a}\right). \qquad ...(1)$$

Now let us join AB by any curve. The axis of the doublet on this curve is normal to AB.

If w_1 is the complex potential due to this line of doublets,

$$\text{then } w_1 = \int_A^B \frac{me^{i\pi/2}}{z-u}du = me^{i\pi/2}\log\frac{z-a}{z+a}$$

$$= mi \log\frac{z-a}{z+a} = -\, iw.$$

The result follows from the first part now.

Example 28: *Show that velocity potential*

$$\phi = \frac{1}{2}\log\frac{(x+a)^2+y^2}{(x-a)^2+y^2},$$

gives a possible motion. Determine the form of streamlines and the curves of equal speed.

Solution: We have

$$\phi = \frac{1}{2}\log\,[(x+a)^2+y^2] - \frac{1}{2}\log\,[(x-a)^2+y^2],$$

$$\text{so that } \frac{\partial\phi}{\partial x} = \frac{x+a}{(x+a)^2+y^2} - \frac{x-a}{(x-a)^2+y^2},$$

$$\text{and } \frac{\partial\phi}{\partial y} = \frac{y}{(x+a)^2+y^2} - \frac{y}{(x-a)^2+y^2}.$$

$$\text{Now } \frac{\partial u}{\partial x} = \frac{\partial}{\partial x}\left(-\frac{\partial\phi}{\partial x}\right) = -\frac{y^2-(x+a)^2}{[(x+a)^2+y^2]^2} + \frac{y^2-(x-a)^2}{[(x-a)^2+y^2]^2}$$

$$\text{and } \frac{\partial v}{\partial y} = \frac{\partial}{\partial y}\left(-\frac{\partial\phi}{\partial y}\right) = -\frac{(x+a)^2-y^2}{[(x+a)^2+y^2)]^2} + \frac{(x-a)^2-y^2}{[(x-a)^2+y^2]^2}.$$

Clearly the equation of continuity, *i.e.*,

$$\frac{\partial u}{\partial x} + \frac{\partial v}{\partial y} = 0 \text{ is satisfied.}$$

$$\text{Therefore } \phi = \frac{1}{2}\log\frac{(x+a)^2+y^2}{(x-a)^2+y^2} \text{ gives a possible motion.}$$

Now to find streamlines, we should determine the stream function ψ, which holds :

$$\frac{\partial\phi}{\partial x} = \frac{\partial\psi}{\partial y} \text{ and } \frac{\partial\phi}{\partial y} = -\frac{\partial\psi}{\partial x}. \qquad ...(3)$$

so $$\frac{\partial\psi}{\partial y} = \frac{(x+a)}{(x+a)^2+y^2} - \frac{x-a}{(x-a)^2+y^2}.$$

Integrating w.r.t. y, we get

$$\psi = \tan{-1}\frac{y}{x+a} - \tan^{-1}\frac{y}{x-a} + f(x)$$

to determine f (x), we have

$$\frac{\partial\psi}{\partial x} = \frac{y}{(x+a)^2+y^2} + \frac{y}{(x-a)^2+y^2} + f'(x)$$

$$= -\frac{\partial\phi}{\partial y} = \frac{y}{(x+a)^2+y^2} + \frac{y}{(x-a)^2+y^2} \text{ from (2).}$$

$\therefore$ f′ (x) = 0, i.e. f (x) = absolute constant and may be omitted.

$$\therefore \ \psi = \tan^{-1}\frac{y}{x+a} \tan^{-1}\frac{y}{x-0}$$

$$= \tan^{-1}\left[\frac{-2ay}{x^2+y^2-a^2}\right].$$

Thus streamlines are given by ψ = const.

when ψ = 0 and infinity, the streamlines are

$y = 0$ and $x^2 + y^2 = a^2$.

Now $w = \phi \ i\psi = \frac{1}{2}\log\{(x+a)^2 + y^2\} - \frac{1}{2}\log\{(x-a)^2 + y^2\}$

$$+ i\tan^{-1} x\frac{y}{x+a} - i\tan^{-1}\frac{y}{x-a}$$

$= \log\{(x + a) + iy\} - \log\{(x - a) + iy\}$

$= \log(z + a) - \log(z - a)$ where $z = x + iy$.

$$\text{Speed} = \left|\frac{dw}{dz}\right| = \left|\frac{1}{z+a} - \frac{1}{z-a}\right| = \frac{2a}{|z+a||z-a|}.$$

If $|z + a| = r'$ and $|z-a| = r$, then

$$\text{Speed} = \frac{2a}{rr'}$$

and putting speed equal to constant, the curves of equal speed are given by

$$\frac{2a}{rr'} = \text{constant, } i.e. \ rr' = \text{const.}$$

which are Cassini ovals.

Example 29: *In irrotational motion in two dimensions, prove that*

$$\left(\frac{\partial q}{\partial x}\right)^2 + \left(\frac{\partial q}{\partial y}\right)^2 = q\nabla^2 q. \qquad ...(1)$$

Solution: Since the motion is irrotational, the velocity potential ϕ exists and satisfies the equation

$$\frac{\partial^2\phi}{\partial x^2} + \frac{\partial^2\phi}{\partial y^2} = 0.$$

Also $q^2 = \left(\frac{\partial\phi}{\partial x}\right)^2 + \left(\frac{\partial\phi}{\partial y}\right)^2$

Differentiating it partially w.r.t. x and y respectively, we get

$$q\frac{\partial q}{\partial x} = \frac{\partial\phi}{\partial x}\frac{\partial^2\phi}{\partial x^2} + \frac{\partial\phi}{\partial y}\frac{\partial^2\phi}{\partial x\partial y} \qquad ...(2)$$

and $q\frac{\partial q}{\partial y} = \frac{\partial\phi}{\partial x}\frac{\partial^2\phi}{\partial x\partial y} + \frac{\partial\phi}{\partial y}\frac{\partial^2\phi}{\partial y^2}$. ...(3)

Differentiating these again partially w.r.t. x and y respectively, we get

$$q\frac{\partial^2 q}{\partial x^2} + \left(\frac{\partial q}{\partial x}\right)^2 = \left(\frac{\partial^2\phi}{\partial x^2}\right)^2 + \frac{\partial\phi}{\partial x}\frac{\partial^3\phi}{\partial x^3} + \left(\frac{\partial^2\phi}{\partial x\partial y}\right)^2 + \frac{\partial\phi}{\partial y}.\frac{\partial^3\phi}{\partial x^2\partial y} \qquad ...(4)$$

$$q\frac{\partial^2 q}{\partial y^2} + \left(\frac{\partial q}{\partial y}\right)^2 = \left(\frac{\partial^2\phi}{\partial x\partial y}\right)^2 + \frac{\partial\phi}{\partial x}\frac{\partial^3\phi}{\partial y^2\partial x} + \left(\frac{\partial^2\phi}{\partial y^2}\right)^2 + \frac{\partial\phi}{\partial y}.\frac{\partial^3\phi}{\partial y^3}. \qquad ...(5)$$

Adding (4) and (5), we get

$$q\nabla^2 q + \left(\frac{\partial q}{\partial x}\right)^2 + \left(\frac{\partial q}{\partial y}\right)^2 = \left(\frac{\partial^2\phi}{\partial x^2}\right)^2 + 2\left(\frac{\partial^2\phi}{\partial x\partial y}\right)^2 + \left(\frac{\partial^2\phi}{\partial y^2}\right)^2$$

$$+ \frac{\partial\phi}{\partial x}\frac{\partial}{\partial x}\left\{\frac{\partial^2\phi}{\partial x^2} + \frac{\partial^2\phi}{\partial y^2}\right\} + \frac{\partial\phi}{\partial y}\frac{\partial}{\partial y}\left\{\frac{\partial^2\phi}{\partial x^2} + \frac{\partial^2\phi}{\partial y^2}\right\}$$

$$= \left(\frac{\partial^2\phi}{\partial x^2}\right)^2 + \left(\frac{\partial^2\phi}{\partial x\partial y}\right)^2 + \left(\frac{\partial^2\phi}{\partial y^2}\right)^2 \text{ from (1)}$$

$$= 2\left(\frac{\partial^2\phi}{\partial x^2}\right)^2 + \left(\frac{\partial^2\phi}{\partial x\partial y}\right)^2 \text{ as } \frac{\partial^2\phi}{\partial x^2} = -\frac{\partial^2\phi}{\partial y^2}. \quad ...(6)$$

Also squaring and adding (2) and (3), we get

$$q^2\left[\left(\frac{\partial q}{\partial x}\right)^2 + \left(\frac{\partial q}{\partial y}\right)^2\right] = \left(\frac{\partial\phi}{\partial x}\right)^2\left[\left(\frac{\partial^2\phi}{\partial x^2}\right)^2 + \left(\frac{\partial^2\phi}{\partial x\partial y}\right)^2\right]$$

$$+\left(\frac{\partial\phi}{\partial y}\right)^2\left[\left(\frac{\partial^2\phi}{\partial x\partial y}\right)^2 + \left(\frac{\partial^2\phi}{\partial y^2}\right)^2\right] + 2\frac{\partial\phi}{\partial x}\frac{\partial\phi}{\partial y}\frac{\partial^2\phi}{\partial x\partial y}\left[\frac{\partial^2\phi}{\partial x^2} + \frac{\partial^2\phi}{\partial y^2}\right]$$

$$= \left[\left(\frac{\partial\phi}{\partial x}\right)^2 + \left(\frac{\partial\phi}{\partial y}\right)^2\right]\left[\left(\frac{\partial^2\phi}{\partial x^2}\right)^2 + \left(\frac{\partial^2\phi}{\partial x\partial y}\right)^2\right] \quad \text{using (1)}$$

$$= q^2\left[\left(\frac{\partial^2\phi}{\partial x^2}\right)^2 + \left(\frac{\partial^2\phi}{\partial x\partial y}\right)^2\right],$$

$$i.e. \left(\frac{\partial^2\phi}{\partial x^2}\right)^2 + \left(\frac{\partial^2\phi}{\partial x\partial y}\right)^2 = \left(\frac{\partial q}{\partial x}\right)^2 + \left(\frac{\partial q}{\partial y}\right)^2$$

Putting this in (6), we get

$$q\nabla^2 q + \left(\frac{\partial q}{\partial x}\right)^2 + \left(\frac{\partial q}{\partial y}\right)^2 = 2\left[\left(\frac{\partial q}{\partial x}\right)^2 + \left(\frac{\partial q}{\partial y}\right)^2\right]$$

$$\text{or } q\nabla^2 q = \left(\frac{\partial q}{\partial x}\right)^2 + \left(\frac{\partial q}{\partial y}\right)^2.$$

This proves the result.

Example 30: *A single source is placed in an infinite perfectly elastic fluid, which is also a perfect conductor of heat ; shew that if the motion be steady, the velocity V at a distance r from the source satisfies the equation*

$$\left(V - \frac{k}{V}\right)\frac{\partial V}{\partial r} = \frac{2k}{r};$$

and hence that $r = \frac{1}{\sqrt{V}}e^{V^2/4k}$ ...(1)

Solution: The fluid will obey Boyle's law as the liquid extends to infinity and is perfectly elastic, there being hardly any change in temperature.

$\therefore$ p = kρ.

Again since the motion is steady and is due to a single source, the flow is readial. The equation of continuity is

$$\frac{\partial}{\partial r}(\rho r^2 V) = 0.$$

$$\text{or } Vr^2\frac{\partial \rho}{\partial r} + \rho\left(r^2\frac{\partial V}{\partial r} + 2rV\right) = 0 \qquad ...(2)$$

and the equation of motion is

$$V\frac{\partial V}{\partial r} = -\frac{1}{\rho}\frac{\partial p}{\partial r},$$

$$\textit{i.e.,}\ V\frac{\partial V}{\partial r} + \frac{K}{\rho}\frac{\partial \rho}{\partial r} = 0 \text{ by (1).} \qquad ...(3)$$

Eliminating $\frac{\partial \rho}{\partial r}$ from (2) and (3), we get

$$Vr^2\left(-\frac{\rho V}{K}\frac{\partial V}{\partial r}\right) + \rho\left(r^2\frac{\partial V}{\partial r} + 2rV\right) = 0$$

$$\text{or } \frac{\partial V}{\partial r}.(k - V^2) + 2k\frac{V}{r} = 0$$

$$\text{or } \left(V - \frac{k}{V}\right)\frac{\partial V}{\partial r} = \frac{2k}{r}.$$

This proves the first result.

The above equation after separating the variables can be written as

$$V\ dV - \frac{k}{V}dV = \frac{2k}{r}dr.$$

Integrating $\frac{1}{2}V^2 - k\log V = 2k\log r + \log C'$

$$\text{or } \frac{V^2}{2k} = \log(Vr^2C)$$

or $Vr^2C = eV^2/2k$,

where $\log C' = k\log C$

$$\text{or } r = \frac{1}{\sqrt{(VC)}}eV^2/4k$$

Giving C the particular value 1, this becomes

$$r = \frac{1}{\sqrt{V}} eV^2 / 4k.$$

This proves the second result.

Example 31: *Parallel line -sources (perpendicular to the xy-plane) of equal strength m are place at the points z = nia, where n = ...–2, –1, 0, 1, 2, 3,...; prove that the complex potential is* $w = -m \log \sin h\frac{\pi z}{a}$.

Hence show that the complex potential for two-dimensional doublets (line doublets), with their axes parallel to the x-axis, of strength μ at the some points, is given by

$$w = \mu \coth\left(\frac{\pi z}{a}\right).$$

Solution: Hence we have source at points

$z = 0$; ai, –ai; 2ai, – 2ai; 3ai, – 3ai; ...etc.

Hence the complex potential of the system at any point z is given by

$$w = -m \log z - m \log (z - ai) - m \log (z + ai)$$

$$- m \log (z - 2ai) - m\log (z + 2ai)$$

$$- m \log (z - 2ai) - m\log (z + 3ai)$$

$$= - m \log z - m \log (z^2 + a^2) - m \log (z^2 + 2^2 a^2) + ...$$

$$= - m\log \left\{\frac{\pi z}{a}\left(1+\frac{z^2}{a^2}\right)\left(1+\frac{z^2}{2^2 a^2}\right)\right\}$$

$$- m \log \left\{\frac{a}{\pi} a^2 (2^2 a^2)(3^2 a^2)...\right\}$$

$$= - m \log \sin h \frac{\pi z}{a} + \text{const.}$$

$$= - m \log \sin h\frac{\pi z}{a}, \text{ leaving the constant} \qquad ...(1)$$

$$\text{Since } \sin h\, x = x\left(1+\frac{x^2}{\pi^2}\right)\left(1+\frac{x^2}{2^2\pi^2}\right)\left(1+\frac{x^2}{3^2\pi^2}\right)-...$$

This proves the first part of the result.

The complex potential due to doublets at these points can be obtained by differentiating (1) which is therfore given by

$$w = -\frac{m\pi}{a}\coth\left(\frac{\pi z}{a}\right)$$

$= \mu \coth\left(\frac{\pi z}{a}\right)$ numerically.

This proves the result.

Example 32: *λ denoting a variable parameter, and f a given function, find the condition that f (x, y, l) = 0 should be a possible system of stream lines for steady irrotational motion in two dimension.*

Solution : We know that stream lines are given by

$\psi = \text{const.}$ ...(1)

If $f(x, y, l) = 0$ ...(2)

is a system of stream lines for different values of λ, then for $\lambda = \lambda_1$, (2) must represent a stream line which must be given by (1), also for some value of the constant say c_1. Thus ψ is a function of λ alone (λ of course depending on x and y),

i.e. $y = y(\lambda)$ ...(3)

Now for irrotational steady motion, ψ satisfies the equation

$$\frac{\partial^2\psi}{\partial x^2} + \frac{\partial^2\psi}{\partial y^2} = 0. \quad ...(4)$$

The required condition will be obtained with the help of (3) and (4). We have

$$\frac{\partial\psi}{\partial x} = \frac{d\psi}{d\lambda}.\frac{\partial\lambda}{\partial y}.$$

Again $\frac{\partial^2\psi}{\partial x^2} = \frac{\partial}{\partial x}\left[\frac{\partial\psi}{\partial x}\right] = \frac{\partial}{\partial x}\left[\frac{d\psi}{d\lambda}.\frac{\partial\lambda}{\partial x}\right]$

$$= \frac{d^2\psi}{d\lambda^2}.\left(\frac{\partial\lambda}{\partial x}\right)^2 + \frac{d\psi}{d\lambda}.\frac{\partial^2\lambda}{\partial x^2}$$

and $\frac{\partial^2\psi}{\partial y^2} = \frac{\partial}{\partial y}\left[\frac{d\psi}{d\lambda}.\frac{\partial\lambda}{\partial y}\right]$

$$= \frac{d^2\psi}{d\lambda^2}.\left(\frac{\partial\lambda}{\partial y}\right)^2 + \frac{d\psi}{d\lambda}.\frac{\partial^2\lambda}{\partial y^2}.$$

Putting these in (4), we get

$$\frac{d^2\psi}{d\lambda^2}.\left(\frac{\partial\lambda}{\partial x}\right)^2 + \left(\frac{\partial\lambda}{\partial y}\right)^2 + \frac{d\psi}{d\lambda}.\left[\frac{\partial^2\lambda}{\partial x^2} + \frac{\partial^2\lambda}{\partial y^2}\right] = 0,$$

which is the required condition.

Example 33: *In two-dimensional irrotational fluid motion show that, if the stream lines are confocal ellipses,*

$$\frac{x^2}{a^2+\lambda}+\frac{y^2}{b^2+\lambda} = 1,$$

$$\psi = A \log [\sqrt{(a^2+\lambda)}+\sqrt{(b^2+\lambda)}] + B,$$

and the velocity at any point is inversely proportional to the square root of the rectangle under the focal radii of the point.

Solution: In example 13, we have shown that if f (x,y,λ) = 0 is a possible stream line, ψ is a function of λ alone and we have the condition

$$\frac{\left[\dfrac{\partial^2\lambda}{\partial x^2}+\dfrac{\partial^2\lambda}{\partial y^2}\right]}{\left(\dfrac{\partial\lambda}{\partial x}\right)^2+\left(\dfrac{\partial\lambda}{\partial y}\right)^2} = \frac{\dfrac{d^2\psi}{d\lambda^2}}{\dfrac{d\psi}{d\lambda}}. \qquad ...(1)$$

Here f (x, y, λ) = 0 is

$$\frac{x^2}{a^2+\lambda}+\frac{y^2}{b^2+\lambda}-1 = 0.$$

Differentiating it partially w.r.t. x, we get

$$\frac{2x}{a^2+\lambda}-\left[\frac{x^2}{(a^2+\lambda)^2}+\frac{y^2}{(b^2+\lambda)^2}\right]\frac{\partial\lambda}{\partial x}= 0, \qquad ...(2)$$

i.e. $2\xi - [\xi^2 + \eta^2]\dfrac{\partial\lambda}{\partial x}= 0,$

where $\xi = \dfrac{x}{a^2+\lambda}, \eta=\dfrac{y}{b^2+\lambda}$

or $$\frac{\partial\lambda}{\partial x} = \frac{2\xi}{\xi^2+\eta^2} \qquad ...(3)$$

Similarly, $\dfrac{\partial\lambda}{\partial y} = \dfrac{2\eta}{\xi^2+\eta^2}$.

Again differentiating (2), w.r.t.x, we get

$$\frac{2}{a^2+\lambda}-\frac{4x}{(a^2+\lambda)^2}\left(\frac{\partial\lambda}{\partial x}\right)+2\left[\frac{x^2}{(a^2+\lambda)^3}+\frac{y^2}{(b^2+\lambda)^3}\right]\left(\frac{\partial\lambda}{\partial x}\right)^2$$

$$= \frac{\partial^2\lambda}{\partial x^2}[\xi^2+\eta^2] \quad \text{i.e.,} \quad \frac{\partial^2\lambda}{\partial x^2}(\xi^2+\eta^2)$$

$$= \frac{2}{a^2+\lambda} - \frac{4x}{(a^2+\lambda)^2}\left(\frac{\partial\lambda}{\partial x}\right) + 2\left[\frac{\xi^2}{a^2+\lambda} + \frac{\eta^2}{b^2+\lambda}\right]\left(\frac{\partial\lambda}{\partial x}\right)^2.$$

Similarly differentiating twice w.r.t. y, we get

$$\frac{\partial^2\lambda}{\partial y^2}(\xi^2+\eta^2)$$

$$= \frac{2}{b^2+\lambda} - \frac{4y}{(b^2+\lambda)^2}\left(\frac{\partial\lambda}{\partial y}\right) + 2\left[\frac{\xi^2}{a^2+\lambda} + \frac{\eta^2}{b^2+\lambda}\right]\left(\frac{\partial\lambda}{\partial y}\right)^2.$$

Adding these, we get

$$\left(\frac{\partial^2\lambda}{\partial x^2} + \frac{\partial^2\lambda}{\partial y^2}\right)(\xi^2+\eta^2) = \frac{2}{a^2+\lambda} - \frac{2}{b^2+\lambda}$$

$$-4\left[\frac{x}{(a^2+\lambda)^2}\frac{\partial\lambda}{\partial x} + \frac{y}{(b^2+\lambda)^2}\frac{\partial\lambda}{\partial y}\right]$$

$$+2\left[\frac{\xi^2}{a^2+\lambda} + \frac{\eta^2}{b^2+\lambda}\right]\left[\left(\frac{\partial\lambda}{\partial x}\right)^2 + \left(\frac{\partial\lambda}{\partial y}\right)^2\right]$$

$$= \frac{2}{a^2+\lambda} + \frac{2}{b^2+\lambda};$$

other terms cancel when we put values of $\frac{\partial\lambda}{\partial x}\ \frac{\partial\lambda}{\partial y}$ and from (3).

Thus $\left(\frac{\partial^2\lambda}{\partial x^2} + \frac{\partial^2\lambda}{\partial y^2}\right) = \frac{2}{\xi^2+\eta^2}\left[\frac{1}{a^2+\lambda} + \frac{1}{b^2+\lambda}\right].$

Also from (3), $\left(\frac{\partial\lambda}{\partial x}\right)^2 + \left(\frac{\partial\lambda}{\partial y}\right)^2 = \frac{4}{\xi^2+\eta^2}.$

Thus, (1) gives $\frac{1}{2}\left[\frac{1}{a^2+\lambda} + \frac{1}{b^2+\lambda}\right] = -\frac{\psi''(\lambda)}{\psi'(\lambda)}.$

Integrating . $\frac{1}{2}[\log(a^2+\lambda) + \log(b^2+\lambda)] - \log A = -\log\psi'(\lambda)$

or $\psi'(\lambda) = \frac{A}{\sqrt{[(a^2+\lambda)(b^2+\lambda)]}}$ ...(4)

Integrating again,

$$\psi = A, \int \frac{1}{\sqrt{\{(a^2+\lambda)(b^2+\lambda)}} d\lambda + B$$

$$= A \log [\sqrt{(a^2+\lambda)} + \sqrt{(b^2+\lambda)}] + B .$$

This proves the result

Second part : If q is the velocity, then

$$q2 = \left(\frac{\partial\psi}{\partial x}\right)^2 + \left(\frac{\partial\psi}{\partial y}\right)^2$$

$$= \left(\frac{d\psi \partial\psi}{d\lambda \partial x}\right)^2 + \left(\frac{d\psi}{d\lambda}.\frac{\partial\psi}{\partial y}\right)^2$$

$$= [\psi'(\lambda)]^2 \left[\left(\frac{\partial\lambda}{\partial x}\right)^2 + \left(\frac{\partial\lambda}{\partial y}\right)^2\right]$$

$$= \frac{4A^2}{(a^2+\lambda)(b^2+\lambda)} \left[\frac{1}{\frac{x^2}{(a^2+\lambda)^2} + \frac{y^2}{(b^2+\lambda)^2}}\right]$$

$$= \frac{4A^2}{a^2+\lambda} \left[\frac{1}{\frac{(b^2+\lambda)}{(a^2+\lambda)^2}\left(1 - \frac{x^2}{a^2+\lambda}\right)}\right]$$

from the given equation

$$= \frac{4A^2}{(a^2+\lambda) - \frac{x^2(a^2-b^2)}{a^2+\lambda}}.$$

But from the property of the ellipse,

$a^2 - b^2 = (a^2 + \lambda) - (b^2 + \lambda) = (a^2 + \lambda)\, e^2.$

$$\therefore\ q^2 = \frac{4A^2}{(a^2+\lambda) - e^2x^2} = \frac{4A^2}{[\sqrt{(a^2+\lambda)} - ex][\sqrt{(a^2+\lambda)} + ex]}$$

$$= \frac{4A^2}{r_1 r_2},$$

where $r_1 = \sqrt{(a^2 + \lambda) - ex}$ and r2 = $\sqrt{(a^2 + \lambda) - ex}$ are the focal distances of the point.

Thus $q = \dfrac{2A}{\sqrt{(r_1)}\sqrt{(r_2)}}$

in inversely proportional to square - root of rectangle formed of focal radii to the point.

Example 34: *Find the lines of flow in the two dimensional fluid motion given by*

$\phi + i\psi = -(1/2)\, n\, (x + iy)^2 e^{2int}.$

Prove or verify that the paths of the particles of the fluid (in polar coordinates) may be obtained by eliminating t from the equations.

$r \cos (nt + \theta) - x_0 = r \sin (nt + \theta) - y_0 = nt\, (x_0 - y_0)$

Solution: Since $\phi + i\psi = -\dfrac{1}{2} n\, (x + iy)^2\, e^{2int}$ ...(1)

Substituting $x = r\cos\theta$ and $y = r \sin\theta$, we have

$$\phi + i\psi = -\frac{1}{2} n\, (r\cos\theta + ir\sin\theta)^2\, e^{2int}$$

$$\phi + i\psi = -\frac{1}{2} nr^2\, e^{2i\theta}\, e^{2int} = -\frac{1}{2} nr^2\, e^{2i}(\theta + nt)$$

$$\phi + i\psi = -\frac{1}{2} nr^2\, \{\cos 2\,(\theta + nt) + i \sin 2\,(\theta + nt)\}.$$

Separating into real and imaginary parts, we get

$$\phi = -\frac{1}{2} nr^2 \cos 2\,(\theta + nt),\ \psi = -\frac{1}{2} nr^2 \sin 2\,(\theta + nt).$$

The lines of flow are given by y = const. i.e.,

$r^2 \sin 2\,(\theta + nt) = \text{const.}$

Now, we shall determine the path of the particles of fluid

$$\frac{\partial r}{\partial t} = \frac{\partial \phi}{\partial r} = nr\cos(2\theta + 2nt) = nr\cos 2\lambda,\ \lambda = \theta + nt \qquad \text{...(2)}$$

$$\text{and } r\frac{\partial \theta}{\partial t} = -\frac{1}{r}\frac{\partial \phi}{\partial \theta} = -nr\sin(2\theta - 2nt) = -nr\sin 2\lambda \qquad \text{...(3)}$$

From (2), we have

$$nr\cos 2\lambda = \frac{dr}{dt} = \frac{dr}{d\lambda}\frac{d\lambda}{dt} = \frac{dr}{d\lambda}\left(\frac{d\theta}{dt} + n\right)$$

or $nr \cos 2\lambda = \frac{dr}{d\lambda}(n - n \sin 2\lambda)$

or $\frac{2dr}{r} = \frac{2\cos 2\lambda}{1-\sin 2\lambda} d\lambda$

By integrating, we have

$2 \log r + \log (1 - \sin 2\lambda, = \log C$

or $r^2 (1- \sin 2\lambda\ ) = C$

or $r (\cos \lambda - \sin \lambda) = D,$ ...(4)

where C and D are arbitrary constants.

At $t = 0; \lambda = \theta = \theta_1$ (initially), $r\ r_0$, so $D = x_0 - y_0$. ...(5)

Equation (4) reduces to

$r (\cos \lambda - \sin \lambda) = x_0 - y_0$

or $r \cos (\theta + nt) - x_0 = r \sin (\theta + nt) - y_0$ ...(6)

Again $\frac{d\lambda}{dt} = n - n \sin 2\lambda$

or $\int \frac{d\lambda}{1-\sin 2\lambda} = n \int dt$

or $\int \frac{d\lambda}{(\cos\lambda - \sin\lambda)^2} = n \int dt$

or $\int \frac{\sec^2 \lambda d\lambda}{(1-\tan\lambda)^2} = n \int dt$

or $\frac{1}{1-\tan\lambda} = nt + A.$ ...(7)

or $\frac{\cos\lambda}{\cos\lambda - \sin\lambda} = nt + A$

Using (5), the value of the constant becomes

$$A = \frac{\cos\theta_0}{\cos\theta_0 - \sin\theta_0} = \frac{x_0}{x_0 - y_0}$$

$$\Rightarrow \frac{r\cos\lambda}{r\cos\lambda - r\sin\lambda} = nt + \frac{x_0}{x_0 - y_0}$$

$\Rightarrow r \cos \lambda = nt (x_0 - y_0) + x_0$ . ...(8)

From (6) and (8), we have

$r \cos (\theta + nt) - x_0 = r \sin (\theta + nt) - y_0 = nt (x_0 - y_0)$. **Proved.**

Example 35: *In this case of the two dimensional fluid motion produced by a source of strength m placed at a point S outside a rigid circular disc of radius a whose centre is O, show that the velocity of slip of the fluid in contact with the disc is the greatest at the point where the line joining prove that its magnitude at these point is*

$$\frac{2m.OS}{(OS^2 - a^2)}.$$

Solution: Let S′ be an inverse point of

S with regard to the circular disc such that

OS OS′ = a^2

OS′ = a^2/c, OS = c.

The image system of a source of strength + m placed at a point S (c, 0) outside the circular disc consists of

(i) a source of strength + m at S (c, 0),

(ii) a source of strength + m at S′ (a^2/c, 0),

(iii) a sink of strength – m at O (0, 0).

The complex potential w for the motion of the fluid element at any point z is given by

$$w = -m \log (z - c) - m \log (z - (a^2/c)\} + m \log z.$$

Differentiating with regard to z, we have

$$\frac{dw}{dz} = -m\frac{1}{z-c} - m\frac{1}{z-(a^2/c)} + m\frac{1}{z}$$

Let q be the velocity given by

$$q = \left|\frac{dw}{dz}\right| = \left|-\frac{m}{z-c}\frac{m}{z-(a^2/c)} + \frac{m}{z}\right|$$

$$\text{or } q = m\left|\frac{z\{z-(a^2/c)\} + z(z-c) - (z-c)(z-(a^2/c)\}}{z(z-c)(z-a^2/c)\}}\right|$$

$$\text{or } q = m\left|\frac{(z-a)(z+a)}{z(z-c)\{z-(a^2/c)\}}\right|$$

The velocity at any point P, z = $ae^{\theta i}$, on the boundary of the circular disc reduces to

$$q = m\left|\frac{(ae^{\theta i} - a)(ae^{\theta i} + a)}{ae^{\theta i}(ae^{\theta i} - c)(ae^{\theta i} - (a^2/c)\}}\right|$$

$$\Rightarrow q = mc\left|\frac{(e^{\theta i}-1)(e^{\theta i}+1)}{e^{\theta i}(ae^{\theta i}-c)(ce^{\theta i}-a)}\right|$$

$$\Rightarrow q = mc\left|\frac{(e^{2\theta i}-1)}{e^{\theta i}(ae^{\theta i}-c)(ce^{\theta i}-a)}\right|$$

$$|(c\cos\theta - a) + ic\sin\theta|$$

$$\Rightarrow q = mc\,\frac{\sqrt{(\cos 2\theta - 1)^2 + \sin^2 2\theta}}{\sqrt{\cos^2\theta + \sin^2}\sqrt{(a\cos\theta - c)^2 + a^2\sin^2\theta}}$$

$$\sqrt{(c\cos\theta - a)^2 + c^2\sin^2\theta}$$

$$\text{or } q = \frac{2mc\sin\theta}{a^2 + c^2 - 2ac\cos\theta}$$

For q to be maximum of minimum $dq/d\theta = 0$, we have

$$\frac{dq}{d\theta} = 2mc\frac{(a^2+c^2)\cos\theta - 2ac}{(a^2+c^2) - 2ac\cos\theta)^2} = 0.$$

or $(a^2 + c^2)\cos\theta - 2ac = 0 \Rightarrow \cos\theta = 2ac/(a^2 + c^2)$.]

Now $\theta = 0$ gives minimum velocity as the expression vanishes at this point.

The maximum value of θ will be obtained to the corresponding value of $\cos\theta = 2ac/(a^2 + c^2)$. From (1), we have

$$q = 2mc\,\frac{\{(a^2-c^2)(a^2+c^2)\}}{(a^2+c^2) - \{4a^2c^2/(a^2+c^2)\}} = 2mc\,\frac{(a^2-c^2)}{(a^2+c^2) - 4a^2c^2}$$

$$\text{or } q = \frac{2mc}{c^2 - a^2} = \frac{2mOS}{OS^2 - a^2}.$$

The velocity will be along the direction of the tangent to the boundary and is equal to the velocity of slip as the boundary of the circular disc is a stream line.

Proved.

Example 36: *A source S and a sink T of equal strengths m are situated with in the space bounded by a circle whose centre is O. If S and T are at equal distance from O on opposite sides of it and on the same diameter AOB, shew that the velocity of the liquid at any point P is*

$$2m.\frac{OS^2 + OA^2}{OS}.\frac{PA.PB}{PSPS'.PTPT'},$$

where S′ and T′ are the inverse points of S and T with regards to the circle.

Solution: Consider OS = OT = f, let S′ and T′ be the inverse points of S and T regard to the circle that

OS. OS′ = OT. OT′ = a^2

OS′ = OT′ = a^2/f.

The image system of a source of a length + m and a sink of strength – m placed at the point S and T with in the circle consist of

(i) a source of strength + m at S (f, 0)

(ii) a source of strength + m at S'(a^2/f, 0)

(iii) a sink of strength – m at the origin O

(iv) a sink of strength – m at f (– f, 0)

(v) a sink of strength – m at T′ (– a^2/f, 0)

(vi) a source of strength + m at the origin O.

The source and sink of same strength cancels each other at the origin. Thus the complex potential w becomes

$$w = -m \log(z - f) - m \log\{z - (a^2/f)\} + m \log(z + f) + m \log\{z + (a^2/f)\}$$

The velocity q at any point P is given by

$$q = \left|\frac{dw}{dz}\right| = m\left|\frac{1}{z-f} - \frac{1}{z-(a^2/f)} + \frac{1}{z+f} + \frac{1}{z+(a^2/f)}\right|$$

$$\text{or } q = m\left|\frac{2f}{z^2-f^2} - \frac{2a^2/f}{z^2-(a^4/f^2)}\right|$$

$$\text{or } q = 2m.\frac{f^2+a^2}{f}\left|\frac{z^2-a^2}{(z^2-f^2)\{z^2-(a^4/f^2)\}}\right|$$

$$\text{or } q = 2m.\frac{f^2+a^2}{f}\left|\frac{(z-a)(z+a)}{(z-f)(z+f)\{z-(a^2/f)\}\{z+(a^2/f)\}}\right|$$

$$\text{or } q = 2m.\ \frac{f^2+a^2}{f}\left|\frac{|z-a|\ |z+a|}{|z-f||z+f||z-(a^2/f)||z+(a^2/f)|}\right|$$

$$\text{or } q = 2m.\ \frac{OS^2+OA^2}{OS}.\frac{PA.PB}{PSPS'.PTPT'}.$$ **Proved.**

Example 37(a): *In the parts of an infinite plane bounded by a circular quadrant AB and the productions of the radii OA and OB, there is a two-dimensional motion due to the production of liquid at A, and its absorption at B, at the uniform rate m. Find the velocity potential of the motion ; and shew that the fluid which issues from A in the direction making an angle m with OA follows the path whose polar equation is*

$$r = a\ sin^{1/2}\ 2q\ [cot\ \mu + \sqrt{\{\cot^2 \mu + \operatorname{cosec}^2 2\theta\}}]^{1/2},$$

the positive sign being taken for all square-roots.

Solution: Since there is a production of liquid at the point A of the quadrant, the image system of the source at A with regard to the circular boundary consists of:

(i) a source of strength + (m/2π) at A,

(ii) a source of strength + (m/2π) at A (since the point A is an inverse point of itself),

(iii) a sink of strength – (m/2π) at the origin O.

The image system of the source at A with regard to the circular boundary and the radii OA and OB will give rise to :

(i) two source of equal strength + (m/2π) at A *i.e.*, a source of strength + (m/π) at A,

(ii) two sources of equal strength + (m/2π) at A′ *i.e.*, a source of strength + (m/π) at A′. (This is the image of the source at A).

(iii) a sink of strength – (m/π) at the origin O.

Similarly there is an absorption of liquid at the point B of the quadrant, the image system of the sink at B with regard of the circular boundary consists of:

(i) a sink of strength – (m/2π) at B,

(ii) a sink of strength – (m/2π)at B, (since the point B is an inverse point of itself),

(iii) A source of strength + (m/2π) at the origin O.

The image system of the sink at B with regard to the circular boundary and the radii OA and OB will give rise to

(i) two sinks of equal strength – (m/2π) at B i.e., a sink of strength – (m/2π) at B

(ii) two sinks of equal strength – (m/2π) at B′ i.e., a sink of strength – (m/π) at B′. (This is the image of the sink at B)

(iii) a source of strength + (m/2π) at the origin O.

A source of strength + (m/2π) and a sink of strength – (m/2π) at the origin O neutralize each other. Hence the image system, finally, consists of

I. a source of strength + (m/π) at A,

II. a source of strength + (m/π) at A′,

III. a sink of strength – (m/π) at B,

IV. a sink of strength – (m/π) at B′.

The complex potential w at P due to the given system becomes

$$w = -\frac{m}{\pi}\log\ (z - a) - \frac{m}{\pi}\log\ (z + a) + \frac{m}{\pi}\log (z - ai) + \frac{m}{\pi}\log (z + ai)$$

$$\text{or } w = -\frac{m}{\pi}\log (z^2 - a^2) + \frac{m}{\pi}\log (z^2 + a^2) \qquad ...(1)$$

To determine the velocity potential of the motion, equating the real part from both the sides, we have

$$\phi = -\frac{m}{\pi}\log|z - a| - \frac{m}{\pi}\log|z + a| + \frac{m}{\pi}\log|z - ai| + \frac{m}{\pi}\log\ |z + ai|$$

$$\text{or } \phi = \frac{m}{\pi}\log AP - \frac{m}{\pi}\log A'P + \frac{m}{\pi}\log BP + \frac{m}{\pi}\log B'\ P.$$

$$\text{or } \phi = \frac{m}{\pi}\ [\log BP\ B'\ P - \log APA'P] = \frac{m}{\pi}\log\left(\frac{BPB'P}{APA'P}\right).\ \textbf{Ans.}$$

Since P (z = x + iy) be any point in the fluid, substituting $z = re^{\theta i}$ in (1), we have

$$\phi + i\psi = -\frac{m}{\pi}\log\ (r^2e^{2\theta i} - a^2) + \frac{m}{\pi}\ \log\ (r^2e^{2\theta i} + a^2)$$

$$\text{or } \phi + i\psi = -\frac{m}{\pi}\ \log\ (r^2\cos 2\theta - a^2) + ir^2 \sin 2\theta) + \frac{m}{\pi}\log\ \{(r^2 \cos 2\theta + at^2) + ir^2 \sin 2\theta\}$$

$$\text{or } \psi = \frac{m}{\pi}\tan^{-1}\left(\frac{r^2 \sin 2\theta}{r^2 \cos 2\theta - a^2}\right) + \frac{m}{\pi}\tan^{-1}\left(\frac{r^2 \sin 2\theta}{r^2 \cos 2\theta + a^2}\right)$$

$$\text{or } \psi = \frac{m}{\pi}\tan^{-1}\left[\frac{\dfrac{r^2\sin 2\theta}{r^2\cos 2\theta + a^2} - \dfrac{r^2\sin 2\theta}{r^2\cos 2\theta - a^2}}{1 + \dfrac{r^2\sin 2\theta}{r^2\cos 2\theta + a^2}\cdot\dfrac{r^2\sin 2\theta}{r^2\cos 2\theta - a^2}}\right]$$

$$\text{or } \psi = -\frac{m}{\pi}\tan^{-1}\left(\frac{2a^2r^2\sin 2\theta}{r^4 - a^4}\right) \qquad ...(2)$$

The streamline that leaves A at an inclination m with OA, is

$\psi = -(m/\pi)\,\mu$.

From (2), we have

$$-\frac{m}{\pi}\tan^{-1}\frac{2a^2r^2\sin 2\theta}{r^4 - a^4} = \frac{m}{\pi}\mu$$

or $r^4 - 2a^2r^2 \sin 2\theta \cot^2 \mu - a^4 = 0$.

or $r^2 = a^2 \sin 2\theta \cot \mu \pm \sqrt{(a^4 \sin^2 2\theta \cot^2 \mu + a^4)}$.

Neglecting the negative sign before the radical because it provides a relation $r^2 < 0$, so we have

$r = a\,(\sin 2\theta)^{1/2}\,[\cot \mu + \sqrt{\{(\cot^2 \mu + \operatorname{cosec}^2 2\theta)\}}]^{1/2}$ **Proved.**

Example 37(b): *Show that the velocity potential*

$$\phi = \frac{1}{2}\log\frac{(x+a)^2 + y^2}{(x-a)^2 + y^2},$$

gives a possible motion. Determine the form of streamlines and curves of equal speed.

Solution: $\phi = \dfrac{1}{2}\log\dfrac{(x+a)^2 + y^2}{(x-a)^2 + y^2}$ ...(1)

$$\frac{\partial \phi}{\partial x} = \frac{x+a}{(x+a)^2 + y^2} - \frac{x-a}{(x-a)^2 + y^2} \qquad ...(2)$$

$$\frac{\partial \phi}{\partial y} = \frac{y}{(x+a)^2 + y^2} - \frac{y}{(x-a)^2 + y^2}. \qquad ...(3)$$

$$\text{Also } \frac{\partial u}{\partial x} = \frac{\partial}{\partial x}\left\{\frac{\partial \phi}{\partial x}\right\} = \frac{\partial}{\partial x}\left\{\frac{x-a}{(x-a)^2 + y^2} - \frac{x+a}{(x+a)^2 + y^2}\right\}$$

$$\text{or } \frac{\partial u}{\partial x} = \frac{y^2-(x-a)^2}{[(x-a)^2+y^2]^2} - \frac{y^2-(x-a)^2}{[(x-a)^2+y^2]^2}$$

$$\text{and } \frac{\partial v}{\partial y} = \frac{\partial}{\partial y}\left(\frac{\partial \phi}{\partial y}\right) = \frac{\partial}{\partial x}\left\{\frac{y}{(x-a)^2+y^2} - \frac{y}{(x+a)^2+y^2}\right\}$$

$$\text{or } \frac{\partial v}{\partial y} = \frac{(x-a)^2-y^2}{[(x-a)^2+y^2]^2} - \frac{(x+a)^2-y^2}{[(x+a)^2+y^2]^2}$$

$\Rightarrow \dfrac{\partial u}{\partial x} + \dfrac{\partial v}{\partial y} = 0$ which satisfies the equation of continuity.

Hence the relation (1) gives a possible liquid motion.

Since the velocity potential f and the stream function y satisfies the property of conjugate functions, so we have

$$\frac{\partial \phi}{\partial x} = \frac{\partial \psi}{\partial y} \text{ and } \frac{\partial \phi}{\partial y} = -\frac{\partial \psi}{\partial x} \qquad ...(4)$$

$$\text{or } \frac{\partial \psi}{\partial y} = \frac{x+a}{(x+a)^2+y^2} - \frac{x-a}{(x-a)^2+y^2}$$

$$\text{or } y = \tan^{-1}\frac{y}{x+a} - \tan^{-1}\frac{y}{x-a} + f(x). \qquad ...(5)$$

To determine f (x), we shall differentiate (5) with regard to x,

$$\frac{\partial \psi}{\partial x} = -\frac{y}{(x+a)^2+y^2} + \frac{y}{(x-a)^2+y^2} + f'(x)$$

$$\therefore \frac{\partial \psi}{\partial x} = -\frac{\partial \phi}{\partial y} \Rightarrow f'(x) = 0 \text{ Þ } f(x) = \text{const.}$$

$$\text{or } \psi = \tan^{-1}\frac{y}{x+a} - \tan^{-1}\frac{y}{x-a} = \tan{-1}\left(\frac{2ay}{x^2+y^2-a^2}\right).$$

The lines of flow can be obtained by ψ = const.

$$\text{i.e. } \tan^{-1}\left\{\frac{2ay}{x^2+y^2-a^2}\right\} = \text{const.}$$

$$\text{or } \frac{2ay}{a^2-x^2-y^2} = \tan\mu = A = \text{const.}$$

The constant A = 0 gives the stream line y = 0 i.e., a real axis and A =∞ gives the stream line $x^2 + y^2 = a^2$, i.e., a circle. **Ans.**

Again $w = \phi + i\,\psi = \frac{1}{2}\log\{(x+a)^2+y^2\} - \frac{1}{2}\log\{(x-a)^2+y^2\}$

$$+\, i\tan^{-1}\frac{y}{x+a} - i\tan^{-1}\frac{y}{x-a}.$$

or $w = \left[\frac{1}{2}\log\{(x+a)^2+y^2\} + i\tan^{-1}\frac{y}{x+a}\right]$

$$-\left[\frac{1}{2}\log\{(x-a)^2+y^2\} + \tan^{-1}\frac{y}{x-a}\right]$$

or $w = \log\{(x + a) + iy\} - \log\{(x - a) + iy\}$

or $w = \log\{(z + a) - \log(z - a)\}$; $z = x + iy$

So $q = \left|\frac{dw}{dz}\right| = \left|\frac{1}{z+a} - \frac{1}{z-a}\right| = \frac{2a}{|z+a|\,|z-a|} = \frac{2a}{rr'}$,

where $|z + a| = r'$, $|z - a| = r$, r' and r are the distances from $(-a, 0)$ and $(a, 0)$.

Hence the curves of equal speeds are given by.

$\frac{2a}{rr'}$ = const., rr' = const., which are known as Cassini Ovals.

Example 37(c): *Between the fixed boundaries $\theta = \pi/6$ and $\theta = -\pi/6$, there is a two-dimensional liquid motion due to a source at the point ($r = c$, $\theta = \alpha$) and a sink at the origin, absorbing, water at the same rate as the source produces it. Find the stream function, and shew that one of the stream lines is a part of the curve*

$$r^3 \sin 3\alpha = c^3 \sin 3\theta.$$

Solution: Let the transformation be $t = z^3$,

where $t\ Re^{\phi i}$, $z = re^{\theta i}$.

or $Re^{\phi i} = r^3 e^{3\theta i}$,

$\therefore$ $R = r^3$ and $\phi = 3\theta$.

Now the boundaries $\theta = \pi/6$ and $\theta = -\pi/6$ in Z-plane transforms into $\phi = \pi/2$ and $\phi = -\pi/2$ (i.e., an imaginary axis) in t-plane.

A source of strength + m at the point ($r = c$, $\theta = \alpha$) in Z-plane transforms to a source of same strength at the point ($R = c^3$, $f = 3\alpha$) in t-plane. Also a sink of same strength – m at the origin in Z-plane corresponds to a sink of same strength at origin in t-plane. The image system consists of

(i) a source of strength + m at $\{c^3, 3\alpha\}$

(ii) a source of strength + m at $\{c^3, (\pi - 3\alpha)\}$

(iii) a source of strength – m at an origin.

Thus the complex potential becomes

$w = 2m \log t - m \log (t - c^3e^{3\alpha i}) - m \log (t - c^3e(\pi - 3\alpha) i)$

$w = 2m \log t - m \log (t - c^3e^{3\alpha i}) - m \log (t + c^3e^{-3\alpha i})$

$w = 2m \log z^3 - m \log (z^3 - c^3e^{3\alpha i}) - m \log (z^3 + c^3e - 3^{\alpha i}),$

Let P $(z = re^{\theta i})$ be a point in Z-plane, then

$w = 6m \log (re^{\theta i}) - m \log (r^3e^{3\theta i} - c^3e^{\alpha i}) - m \log (r^3e^{3\theta i} + c^3e - 3\alpha i)$

or $\phi + i\psi = 6m (\log r + \theta i) - m \log [(r^3e^{3\theta i} - c^3e^3\alpha i)(r^3e^{3\theta i} + c^3e - 3\alpha i)]$

or $\phi + i\psi = 6m (\log r + \theta i)$

$- m \log [(c^6 \cos 6\theta + 2c^3r^3 \sin 3\theta \sin 3\alpha - c^6)$

$+ i (r^6 \sin \theta - 2c^3r^3 \sin 3\alpha \cos 3\theta)]$

Equating the imaginary parts, we have

$$\psi = 6m\theta - m \tan^{-1}\left\{\frac{r^6 \sin 6\theta - 2c^3r^3 \sin 3\alpha \cos 3\theta}{r^6 \cos 6\theta + 2c^3r^3 \sin 3\theta \sin 3\alpha - c^6}\right\}$$

Now one of the stream lines is y = 0, which gives

$$6m\theta - m \tan^{-1}\left\{\frac{r^6 \sin 6\theta - 2c^3r^3 \sin 3\alpha \cos 3\theta}{r^6 \cos 6\theta + 2c^3r^3 \sin 3\theta \sin 3\alpha - c^6}\right\} = 0$$

$$\text{or } \frac{\sin 6\theta}{\cos 6\theta} = \frac{r^6 \sin 6\theta - 2c^3r^3 \sin 3\alpha \cos 3\theta}{r^6 \cos 6\theta + 2c^3r^3 \sin 3\theta \sin 3\alpha - c^6}$$

or $r^6 \sin 6\theta \cos 6\theta + 2c^3r^3 \sin 3\theta \sin 3\alpha \sin 6\theta - c^6 \sin 6\theta$

$= r^6 \sin 6\theta \cos 6\theta - 2c^3r^3 \sin 3\alpha \cos 3\theta \cos 6\theta$

or $c^6 \sin 6\theta = 2r^3c^3 \sin 3\alpha [\cos 6\theta \cos 3\theta + \sin 6\theta \sin 3\theta]$

or $c^6 \sin 6\theta = 2r^3c^3 \sin 3\alpha \cos (6\theta - 3\theta)$

or $2c^6 \sin 3\theta \cos 3\theta = 2r^3c^3 \sin 3a \cos 3\theta$

or $2c^3 \cos 3\theta [r^3 \sin 3\alpha - c^3 \sin 3\theta] = 0.$

Either $\cos 3\theta = 0$, or $r^3 \sin 3\alpha - c^3 \sin 3\theta = 0$

If $\cos 3\theta = 0$, then $3\theta = \pm \pi/2 \Rightarrow \theta = \pm \pi/6$, which are the given boundaries.

and $r^3 \sin 3\alpha - c^3 \sin 3\theta = 0$

or $r^3 \sin 3\alpha = c^3 \sin 3\theta,$

which is the required equation of curve.

Example 38: *Prove that the equation of motion is satisfied for an inviscid, incompressible, steady flow with negligible body force whose velocity components are given by*

$$q_r = U\left(1-\frac{A^3}{r^3}\right)\cos\theta,\ q_\theta = -U\left(1+\frac{A^3}{2r^3}\right)\sin\theta,\ q_\phi = 0,$$

where A is constant. Find the resultant velocity when $r \to \infty$.

Solution: The equations of motion for an inviscid, incompressible and steady flow with negligible external force, in spherical polar coordinates, are given as

$$q_r\frac{\partial p_r}{\partial r} + \frac{q\theta}{r}\frac{\partial q_r}{\partial \theta} + \frac{q\phi}{r\sin\theta} - \frac{\partial q_r}{\partial \phi} - \frac{q_\theta^2 + q_\phi^2}{r} = -\frac{1}{\rho}\frac{\partial p}{\partial r},$$

$$q_r\frac{\partial q_\theta}{\partial r} + \frac{q_\theta}{r}\frac{\partial q_\theta}{\partial \theta} + \frac{q\phi}{r\sin\theta}\frac{\partial q\theta}{\partial \phi} + \frac{q_r q_\theta}{r} - \frac{q_\phi^2\cot\theta}{r} = -\frac{1}{\rho}\frac{\partial p}{\partial \theta},$$

$$q_r\frac{\partial q_\phi}{\partial r} + \frac{q_\theta}{r}\frac{\partial q\phi}{\partial \theta} + \frac{q\phi}{r\sin\theta}\frac{\partial q\phi}{\partial \phi} + \frac{q_\phi q_r}{r} - \frac{q_\theta q_\phi\cot\theta}{r}$$

$$= -\frac{1}{\rho}\frac{1}{r\sin\theta}\frac{\partial p}{\partial \phi} \qquad \text{...(1, 2, 3)}$$

Here $q_r = U\left(1-\frac{A^3}{r^3}\right)\cos\theta,\ q_\theta = -U\left(1+\frac{A^3}{2r^3}\right)$

$\sin\theta,\ q_\phi = 0,$ ...(4)

From the relation (4), equations (1,2,3) reduce to

$$U\left(1-\frac{A^3}{r^3}\right)\cos\theta\left(\frac{3UA^3}{r^4}\right)\cos\theta + \frac{U}{r}\left(1+\frac{A^3}{2r^3}\right)\sin\theta$$

$$\times U\left(1-\frac{A^3}{r^3}\right)\sin\theta - \frac{U^2}{r}\left(1+\frac{A^3}{2r^3}\right)\sin^2\theta = -\frac{1}{\rho}\frac{\partial p}{\partial r},$$

$$U\left(1-\frac{A^3}{r^3}\right)\cos\theta\left(\frac{3UA^3}{r^4}\right)\sin\theta + \frac{U}{r}\left(1+\frac{A^3}{2r^3}\right)\sin\theta$$

$$\times U\left(1+\frac{A^3}{2r^3}\right)\cos\theta - \frac{U^2}{r}\left(1-\frac{A^3}{r^3}\right)\left(1+\frac{A^3}{2r^3}\right)\sin\theta\cos\theta$$

$$= -\frac{1}{\rho}\frac{\partial p}{r\partial\theta},$$

$$0 = \frac{1}{\rho}\frac{1}{r\sin\theta}\frac{\partial p}{r\phi}. \qquad ...(5, 6, 7)$$

Equation (7) shows that the pressure p is independent of ϕ, therefore p = p (r, θ). On simplifying the equation (5) and (6), we have

$$\frac{3U^2A^3}{r^4}\left(1-\frac{A^3}{r^3}\right)\cos^2\theta + \frac{U^2}{r}\left(1-\frac{A^3}{2r^3}-\frac{A^6}{2r^6}\right)\sin^2\theta$$

$$-\frac{U^2}{r}\left(1+\frac{A^3}{r^3}+\frac{A^6}{4r^6}\right)\sin^2\theta = -\frac{1}{\rho}\frac{\partial p}{\partial r},$$

or $$\frac{3U^2A^3}{r^4}\left(1-\frac{A^3}{r^3}\right)\cos^2\theta + \frac{U^2}{r}\left(-\frac{3A^3}{2r^3}-\frac{3A^6}{4r^6}\right)\sin^2\theta = \frac{1}{\rho}\frac{\partial p}{\partial r},$$

or $$\frac{3U^2A^3}{r^4}\left(1-\frac{A^3}{r^3}\right)\cos^2\theta - \frac{3U^2A^3}{2r^4}\left(1+\frac{A^3}{2r^3}\right)\sin^2\theta$$

$$= -\frac{1}{\rho}\frac{\partial p}{\partial r}. \qquad ...(8)$$

and $$\frac{3U^2A^3}{2r^4}\left(1-\frac{A^3}{r^3}\right)\sin\theta\cos\theta + \frac{U^2}{r}\left(1+\frac{A^3}{r^3}+\frac{A^6}{4r^6}\right)\sin\theta\cos\theta$$

$$-\frac{U^2}{r}\left(1-\frac{A^3}{2r^3}-\frac{A^6}{2r^6}\right)\sin\theta\cos\theta \ -\frac{1}{\rho}\frac{\partial p}{\partial\theta}.$$

or $$\frac{3U^2A^3}{2r4}\left(1-\frac{A^3}{r^3}\right)\sin\theta\cos\theta + \frac{3U^2A^3}{2r^4}\left(1+\frac{A^3}{2r^3}\right)\sin\theta\cos\theta =$$

$$-\frac{1}{\rho}\frac{\partial p}{r\partial\theta}. \qquad ...(9)$$

Differentiating equation (8) with regard to θ, we have

$$-\frac{6U^2A^3}{r^4}\left(1-\frac{A^3}{r^3}\right)\cos\theta\sin\theta - \frac{3U^2A^3}{r^4}\left(1+\frac{A^3}{2r^3}\right)\sin\theta\cos\theta$$

$$= -\frac{1}{\rho}\frac{\partial^2 p}{\partial r\partial\theta}$$

or $$\left(-\frac{9U^2A^3}{r^4}+\frac{9U^2A^6}{2r^7}\right)\sin\theta\cos\theta = -\frac{1}{\rho}\frac{\partial^2 p}{\partial r\partial\theta} \ - \qquad ...(10)$$

Differentiating equation (9) with regard to r, we have

$$\frac{3U^2A^3}{2}\left(-\frac{3}{r^4}+\frac{6A^3}{r^7}\right)\cos\theta\sin\theta$$

$$+\frac{3U^2A^3}{2}\left(-\frac{3}{r^4}+\frac{3A^3}{r^7}\right)\sin\theta\cos\theta = -\frac{1}{\rho}\frac{\partial^2 p}{\partial r\partial\theta}$$

or $$\frac{9}{2}\frac{U^2A^3}{2}\left(1-\frac{2A^3}{r^3}\right)\cos\theta\sin\theta$$

$$-\frac{9}{2}\frac{U^2A^3}{2}\left(1+\frac{A^3}{r^3}\right)\sin\theta\cos\theta = -\frac{1}{\rho}\frac{\partial^2 p}{\partial r\partial\theta}$$

or $$\left(-\frac{9U^2A^3}{r^4}+\frac{9U^2A^6}{2r^7}\right)\sin\theta\cos\theta = -\frac{1}{\rho}\frac{\partial^2 p}{\partial r\partial\theta} \quad ...(11)$$

Equation (10) and (11) are identical. Hence, the equation of motion is satisfied.

When r → ∞, the resultant velocity is equal to U. **Proved.**

Example 39(a): *Prove that the velocity components*

$$q_r\ (r,\theta) = -U\left(1-\frac{a^2}{r^2}\right)\cos\theta,\ q_\theta(r,\theta) = U\left(1+\frac{a^2}{r^2}\right)\sin\theta,$$

satisfy the equation of motion for a two-dimensional inviscid incompressible flow. Find the pressure associated with this velocity field. U and a are constants.

Solution: The equations of motion for a two-dimensional steady, inviscid, incompressible flow in the absence of external force, in spherical polar coordinates, are given by

$$q_r\frac{\partial q_r}{\partial r}+\frac{q_\theta}{r}\frac{\partial q_r}{\partial\theta}-\frac{q_\theta^2}{r} = -\frac{1}{\rho}\frac{\partial p}{\partial r},$$

$$q_r\frac{\partial q_\theta}{\partial r}+\frac{q_\theta}{r}\frac{\partial q_\theta}{\partial\theta}+\frac{q_r q_\theta}{r} = -\frac{1}{\rho}\frac{\partial p}{r\partial\theta},$$

$$0 = -\frac{1}{\rho}\frac{1}{r\sin\theta}\frac{\partial p}{\partial\phi}. \quad ...(1, 2, 3)$$

Here $q_r\ (r,\theta) = -U\left(1-\frac{a^2}{r^2}\right)\cos\theta,$

$$q_\theta\ (r,\theta) = U\left(1+\frac{a^2}{r^2}\right)\sin\theta. \quad ...(4)$$

From (3), it follows that the pressure p is independent of ϕ *i.e.*, p = p (r, θ).

Using the relation (4) into (1) and (2), we have

$$U\left(1-\frac{a^2}{r^2}\right)\cos\theta+\left(\frac{2Ua^2}{r^3}\right)\cos\theta+\frac{U}{r}\left(1+\frac{a^2}{r^2}\right)\sin\theta$$

$$\times U\left(1-\frac{a^2}{r^2}\right)\sin\theta-\frac{U^2}{r}\left(1+\frac{a^4}{r^2}\right)^2\sin^2\theta=-\frac{1}{\rho}\frac{\partial p}{\partial r},$$

or $$\frac{2U^2a^2}{r^3}\left(1-\frac{a^2}{r^2}\right)\cos^2\theta-\frac{2U^2a^2}{r^3}\left(1+\frac{a^2}{r^2}\right)\sin^2\theta$$

$$=-\frac{1}{\rho}\frac{\partial p}{\partial r}. \quad ...(5)$$

and $$\frac{2U^2a^2}{r^3}\left(1-\frac{a^2}{r^2}\right)\sin\theta\cos\theta+\frac{U^2}{r}\left(1+\frac{a^2}{r^2}\right)\sin\theta\cos\theta$$

$$-\frac{U^2}{r}\left(1-\frac{a^2}{r^4}\right)\sin\theta\cos\theta=-\frac{1}{\rho}\frac{\partial p}{r\partial\theta}$$

or $$\frac{2U^2a^2}{r^3}\left(1-\frac{a^2}{r^2}\right)\sin\theta\cos\theta+\frac{2U^2a^2}{r}\left(1+\frac{a^2}{r^2}\right)\sin\theta\cos\theta$$

$$=-\frac{1}{\rho}\frac{\partial p}{r\partial\theta}$$

$$\frac{4U^2a^2}{r^3}\sin\theta\cos\theta=-\frac{1}{\rho}\frac{\partial p}{r\partial\theta}. \quad ...(6)$$

Differentiating (5) with regard to θ, we have

$$-\frac{4U^2a^2}{r^3}\left(1-\frac{a^2}{r^2}\right)\cos\theta\sin\theta\frac{4U^2a^2}{r^2}\left(1-\frac{a^2}{r^2}\right)\sin\theta\cos\theta$$

$$=-\frac{1}{\rho}\frac{\partial^2 p}{\partial r\partial\theta},\ \frac{8U^2a^2}{r^3}\cos\theta\sin\theta=\frac{1}{\rho}\frac{\partial^2 p}{\partial r\partial\theta}. \quad ...(7)$$

Differentiating (6) with regard to r, we have

$$\frac{8U^2a^2}{r^3}\cos\theta\sin\theta=\frac{1}{\rho}\frac{\partial^2 p}{\partial r\partial\theta}. \quad ...(8)$$

Equation (7) and (8) are identical. Hence, the equation of motion are satisfied.

Again, p is a function or r and θ, we have

$$dp = \left(\frac{\partial p}{\partial r}\right) dr + \left(\frac{\partial p}{\partial \theta}\right) d\theta$$

$$\text{or } dp = -\ 2r\ U2a2 \left[\left(\frac{1}{r^3} - \frac{a^2}{r^5}\right)\cos^2\theta - \left(\frac{1}{r^3} + \frac{a^2}{r^5}\right)\sin^2\theta\right] dr$$

$$- \frac{4\rho U^2 a^2}{r^2}(\sin\theta\cos\theta)\ d\theta$$

By integrating, we have

$$p = -2\rho U^2 a^2 \left[\left(-\frac{1}{2r^2} + \frac{a^2}{4r^4}\right)\cos^2\theta - \left(-\frac{1}{2r^2} - \frac{a^2}{4r^4}\right)\sin^2\theta\right] + C$$

$$\text{or } p = \frac{4\rho U^2 a^2}{r}(\cos^2\theta - \sin^2\theta) + C,$$

where C is an integration constant.

or p = $2\rho U^2 a^2$

$$\left[\left(\frac{1}{2r^r} - \frac{a^2}{4r^4} - \frac{2}{r^2}\right)\cos^2\theta - \left(\frac{1}{2r^2} - \frac{a^2}{4r^4} - \frac{2}{r^2}\right)\sin^2\theta\right] + C$$

$$\text{or } p = 2\rho U^2 a^2 \left[-\frac{3}{2r^2}\cos 2\theta - \frac{a^2}{4r^4}\right] + C$$

$$\text{or } p = -\ \rho U^2 a^2 \left[-\frac{3}{r^2}\cos 2\theta + \frac{a^2}{2r^4}\right] + C,$$

Which gives the required pressure distribution.

Example 39(b): *An elastic fluid, the weight of which is neglected, obeying Boyle's law is in motion in a uniform straight tube; show that on the hypothesis of parallel sections the velocity at any time t at a distance r from a fixed point in the tube is defined by the equation*

$$\frac{\partial^2 v}{\partial t^2} + \frac{\partial}{\partial r}\left(2v\frac{\partial v}{\partial t} + v^2\frac{\partial v}{\partial r}\right) = k\ \frac{\partial^2 v}{\partial r^2}.$$

Solution: Since the fluid obeys Boyle's law then

$$p = k\rho. \qquad \text{...(1)}$$

The equation of continuity and the equation of motion is given by

$$\frac{\partial \rho}{\partial t}+\frac{\partial}{\partial r}(\rho v)=0, \qquad ...(2)$$

and
$$\frac{\partial v}{\partial t}+v\frac{\partial v}{\partial r}=-\frac{1}{\rho}\frac{\partial p}{\partial r} \qquad ...(3)$$

Using the relation (1), we have

$$\frac{\partial v}{\partial t}+v\frac{\partial v}{\partial r}=-\frac{k}{\rho}\frac{\partial \rho}{\partial r}. \qquad ...(4)$$

Differentiating (4) partially with regard to t, we have

$$\frac{\partial^2 v}{\partial t^2}+\frac{\partial}{\partial t}\left(v\frac{\partial v}{\partial r}+\frac{k}{\rho}\frac{\partial \rho}{\partial r}\right)=0,$$

$$\Rightarrow \quad \frac{\partial^2 v}{\partial t^2}+\frac{\partial}{\partial r}\left(v\frac{\partial v}{\partial t}+\frac{k}{\rho}\frac{\partial \rho}{\partial t}\right)=0$$

$$\Rightarrow \quad \frac{\partial^2 v}{\partial t^2}+\frac{\partial}{\partial r}\left\{v\frac{\partial v^*}{\partial t}+\frac{k}{\rho}\left(-\frac{\partial}{\partial r}(\rho v)\right)^t\right\}=0$$

$$\Rightarrow \quad \frac{\partial^2 v}{\partial t^2}+\frac{\partial}{\partial r}\left\{v\frac{\partial v}{\partial t}+\frac{k}{\rho}\left(\rho\frac{\partial}{\partial r}+v\frac{\partial \rho}{\partial r}\right)^t\right\}=0$$

$$\Rightarrow \quad \frac{\partial^2 v}{\partial t^2}+\frac{\partial}{\partial r}\left\{v\frac{\partial v}{\partial t}-k\frac{\partial v}{\partial r}-\frac{k}{\rho}\frac{\partial \rho}{\partial r}v\right\}=0$$

$$\Rightarrow \frac{\partial^2 v}{\partial t^2}+\frac{\partial}{\partial r}\left\{v\frac{\partial v}{\partial t}-k\frac{\partial v}{\partial r}+\left(\frac{\partial v}{\partial t}+v\frac{\partial v}{\partial r}\right)v\right\}=0$$

$$\Rightarrow \quad \frac{\partial^2 v}{\partial t^2}+\frac{\partial}{\partial r}\left(2v\frac{\partial v}{\partial t}+v^2\frac{\partial v}{\partial r}-k\frac{\partial v}{\partial r}\right)=0$$

$$\Rightarrow \quad \frac{\partial^2 v}{\partial t^2}+\frac{\partial}{\partial r}\left(2v\frac{\partial v}{\partial t}+v^2\frac{\partial v}{\partial r}\right)=k\frac{\partial^2 v}{\partial r^2}.$$ **Proved.**

Example 40: *An area A is bounded by that part of the X-axis for which $x > a$ and by that branch of $x^2 - y^2 = a^2$ which is in the positive quadrant. There is a two dimensional unit source at (a, 0) which sends out liquid uniformly in all direction. Show by means of the transformation $w = \log(z^2 - a^2)$ that in steady motion the stream lines of the liquid within the area A are portions of rectangular hyperbolas. Determine the stream lines* *corresponding to* $\psi = 0$, $\pi/4$ *and* $\pi/2$. *If* ρ_1 *and* ρ_2 *are the distance of a point P within the fluid from the points* ($\pm$ *a,* 0). *Shew that the velocity of fluid at P is measured by* $\frac{2OP}{\rho_1\rho_2}$, *O being the origin.* ...(1)

Solution: The complex potential is given by

$$w = \log(z^2 - a^2) \qquad ...(2)$$

or $\phi + i\psi = \log\{(x + iy)^2 - a^2\} = \log\{(x^2 - y^2 - a^2) + 2ixy\}$

$$\text{or } \psi = \tan^{-1}\frac{2xy}{x^2 - y^2 - a^2} \Rightarrow \tan\psi = \frac{2xy}{x^2 - y^2 - a^2} \qquad ...(3)$$

The stream lines are given by y = const.

$$\text{i.e., } \frac{2xy}{x^2 - y^2 - a^2} = \text{const.} = k \text{ (let)}$$

If k is zero, the stream lines will be x = 0 and y = 0.

If k is infinite, the stream lines will be $x^2 - y^2 = a^2$.

Thus the liquid flows in the area A bound by x = 0, y = 0, and $x^2 - y^2 = a^2$ *i.e.*, the portion of a rectangular hyperbola in the positive quadrant.

Again, the complex potential (1) can be written as

$$w = \log(z - a) + \log(z + a),$$

which shows that the image of a unit source at the point (a, 0) consists a unit source at the point (– a, 0) with regard to Y-axis. **Proved.**

The velocity of the fluid at any point is

$$q = \left|\frac{dw}{dz}\right| = \left|\frac{1}{z-a} + \frac{1}{z+a}\right| = \frac{|2z|}{|z-a|\,|z+a|} = \frac{2OP}{\rho_1\rho_2}. \qquad \textbf{Proved.}$$

The stream lines corresponding to $\psi = 0$ and $\psi = \pi/2$ are x = 0, y = 0 and $x^2 - y^2 = a^2$.

Also, the stream lines corresponding to $\psi = \pi/4$ are

$$\frac{2xy}{x^2 - y - a^2} = \tan\pi/4 = 1$$

or $x^2 - y^2 - a^2 = 2xy$. **Ans.**

Example 41: *The space on one side of an infinite plane wall y = 0, is filled with inviscid, incompressible fluid, moving at infinity with velocity U in the direction of the axis of X. The motion of the fluid is wholly two dimensional, in the (x, y) plane. A doublet of strength m is at a distance a from the wall and points in the negative direction of the axis of X. Shew that* *if μ is less than $4a^2U$, the pressure of the fluid on the wall is a maximum at points distance $\sqrt{3}a$ from O, the foot of the perpendicular from the doublet on the wall, and is minimum at O.*

If μ is equal to $4a^2U$, find the points where the velocity of the fluid is zero, and shew that the stream lines include the circle

$x^2 + (y-a)^2 = 4a^2$,

where the origin is taken at O.

Solution: We know that the complex potential for a doublet of strength μ at $z = z_0$ inclined at an angle α to the real axis is given by

$w = e^{i\alpha}./(z - z_0)$

The doublet points in the negative direction of the X-axis will make an angle π with it. Since the doublet of strength m is placed at a distance a from the real axis then the image system consists of an equal doublet on the other side of that axis. The complex potential for the system is

$$w = \frac{\mu e^{i\pi}}{z-ia} + \frac{\mu e^{i\pi}}{z+ia} - Uz = -\frac{2\mu z}{z^2+a^2} - Uz \quad ...(1)$$

$$\text{or} \quad \frac{dw}{dz} = -\frac{2\mu(a^2-z^2)}{(a^2+z^2)^2} - U \quad ...(2)$$

Let q be the velocity on the wall then

$$q = \left|\frac{dw}{dz}\right| = \left|\frac{2\mu(a^2-z^2)}{(a^2+z^2)} + U\right|$$

Consider p be any point on the wall. Substituting z = x, we have

$$q = \left[2\mu\frac{a^2-x^2}{(a^2+x^2)^2} + U\right]$$

By Bernoulli's theorem, we get $\frac{p}{\rho} + \frac{1}{2}q^2 = \text{const.}$

As $p = \Pi$, $q = U$; const. $= \frac{\Pi}{\rho} + \frac{1}{2}U^2$

$$\Rightarrow \frac{p}{\rho} + \frac{1}{2}q^2 = \frac{\Pi}{\rho} + \frac{1}{2}U^2 \Rightarrow \quad \Pi - p = \frac{1}{2}\rho(q^2 - U^2)$$

$$\Rightarrow \Pi - p = \frac{1}{2}\rho\left\{\left(2\mu\frac{(a^2-x^2)}{(a^2+x^2)^2} + U\right)^2 - U^2\right\}$$

$$\Rightarrow \Pi - p = 2\mu\rho\left[\mu\frac{(a^2 - x^2)}{(a^2 + x^2)^4} + U\frac{a^2 - x^2}{(a^2 + x^2)^2}\right]$$

The pressure will be maximum or minimum if $\frac{dp}{dx} = 0$

i.e. $2\mu^2\left\{\frac{4x(a^2 - x^2)}{(a^2 + x^2)^4} + \frac{8x(a^2 - x^2)^2}{(a^2 + x^2)^5}\right\}$

$$+ 2\mu U\left\{\frac{2x}{(a^2 + x^2)^2} + \frac{4x(a^2 - x^2)^2}{(a^2 + x^2)^3}\right\} = 0$$

or $4\mu x\ [2\mu\ (a^2 - x^2) + U\ (a^2 + x^2)^2]\frac{3a^2 - x^2}{(x^2 + a^2)^5} = 0$

Either $x\ (3a^2 - x^2) = 0$ or $[2\mu\ (a^2 - x^2) + U\ (a^2 + x^2)] = 0$.

If x (3a2 – x2) = 0 then $x = 0, \sqrt{3a}, -\sqrt{3a}$.

Now on the wall ($y = 0$) at $x = +\sqrt{3a}$

$$\frac{dw}{dx} = -U - \frac{2\mu(a^2 - 3a^2)}{(a^2 + 3a^2)^2} = \frac{\mu}{4a^2} - U.$$

If $\mu < 4a^2U$, the value of $(d^2p/dx^2)x = a\sqrt{3}$ is negative, then the pressure of the fluid at the wall is maximum. Also if $(d^2p/dx^2)_x = 0$ is positive then the pressure at the wall is minimum, and if $m = 4a^2U$, then $(dw/dz) = 0$.

From (2), we have

$$8a^2U\frac{a^2 - z^2}{(a^2 + z^2)^2} + U = 0$$

or $z^4 - 6a^2z^2 + 9a^4 = 0 \Rightarrow z = \pm\ a\sqrt{3}$.

The stagnation points are $(a\sqrt{3}, 0)$ $(-\ a\sqrt{3}, 0)$.

Again, to determine the stream function, equating the imaginary parts from (1) and substituting $\mu = 4a^2U$, we have

$$\psi = \frac{2\mu(x^2 + y^2 - a^2)}{(x^2 + y^2)^2 + 2a^2(x^2 - y^2) + a^4} - Uy$$

$$\text{or}\quad \psi = \frac{8a^2Uy(x^2 + y^2 - a^2)}{(x^2 + y^2)^2 + 2a^2(x^2 - y^2) + a^4} - Uy$$

The stream lines are given by $\psi = 0$

i.e., $-\ y\ \{(x^2 + y^2)^2 + 2a^2\ (x^2 - y^2) + a^4\} + 8a^2y\ (x^2 + y^2 - a^2) = 0$

or $(x^2 + y^2)^2 - 6a^2x^2 - 10\, a^2y^2 + 9a^4 = 0$.

or $(x^2 + y^2 - 2ay - 3a^2)\,(x^2 + y^2 + 2ay - 3a^2) = 0$.

Obviously which includes the circle

$x^2 + y^2 - 2ay - 3a^2 = 0$,

or $x^2 + (y - a)^2 = 4a^2$. **Proved.**

Example 42: *A rectangle open at infinity in the X-direction has solid boundaries along $x = 0$, $y = 0$ and $y = a$. Fluid of amount $2\pi m$ flows into and out of the rectangle at the corners $x = 0$, $y = 0$ and $x = 0$, $y = a$ respectively. Prove that the motion of the fluid is given by*

$w = 4m \log \tanh (\pi y/2a)$

Also show that half the streamlines lies between $x = 0$ and the streamline which cuts $y = a/2$ at the point $x = (a/\pi) \log (1 + \sqrt{2})$.

Solution: The transformation becomes

$z = k \cosh^{-1} \zeta + A$.

$z = ai,\ \zeta = -1 \Rightarrow ai = k \cosh^{-1} (-1) + A$.

$z = 0,\ \zeta = +1 \Rightarrow 0 = k \cosh^{-1} (1) + A$, ...(1)

which $= 0$ and $k = a/\pi$.

$z = (a/\pi) \cosh{-1}\zeta \Rightarrow \zeta = \cosh (\pi z/a)$. ...(2)

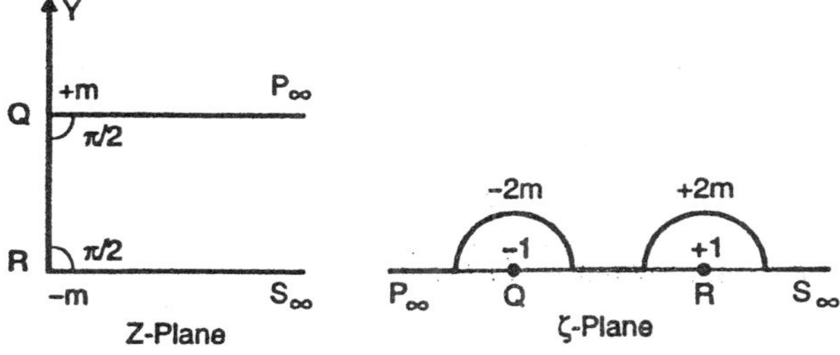

Fig. 2.21

Again, the complex potential is given by

$w = -2m \log (\zeta + 1) + 2m \log (\zeta + 1) = 2m \log [(\zeta - 1)/(\zeta + 1)]$

or $w = 2m \log \left[\dfrac{\cosh(\pi z/a) - 1}{\cosh(\pi z/a) + 1}\right] = 4m \log \tanh (\pi z/2a)$ **Proved.**

or $w = 4m \log \left[\dfrac{\sin\{(\pi/2a)(x + iy)\}\cos\{(\pi/2a)(x - iy)\}}{\cos\{(\pi/2a)(x + iy)\}\cosh\{(\pi/2a)(x - iy)\}}\right]$

or $\phi + i\psi = 4m \log \dfrac{\sin(\pi x/a) - i\sin(\pi y/a)}{\cosh(\pi x/a) - \cos(\pi y/a)}$

or $\phi + i\psi = 4m \log [\log \{\sinh (\pi x/a) - i \sin(\pi/a)\}$

$+ \log \{\cosh (\pi x/a) - \cos(\pi y/a)\}]$

or $\psi = 4m \tan^{-1}\left[\dfrac{\sin(\pi y/a)}{\sinh(\pi x/a)}\right]$.

At $x = 0$, $\psi_0 = 4m \tan^{-1} \infty = 2m\pi$,

and $x = \infty$, $\psi_\infty = 4m \tan^{-1} 0 = 0$.

Thus, total flow $\psi_0 - \psi_\infty = 2m\pi$.

The corresponding streamline due to half of the flow is

$m\pi = 4m \tan^{-1}\left[\dfrac{\sin(\pi y/a)}{\sinh(\pi x/a)}\right]$.

or $\sin (\pi y/a) = \sinh (\pi x/a)$.

The condition when the streamline cuts at $y = a/2$ is

$\dfrac{\pi x}{a} = \sinh^{-1} 1 = \log (1 + \sqrt{2})$

or $x = (a/\pi) \log (1 + \sqrt{2})$. **Proved.**

EXERCISES

1. An elliptic cylinder is placed in a steady stream which at infinity makes an angle a with the major axis of the cylinder. Show that on the ellipse, pressure is the greatest at the points where the stream divides, and least at the points where the fluid is moving parallel to the stream as it meets the ellipse.
2. The space between two confocal co-axial elliptic cylinders is filled with liquid which is at rest. Prove that if the outer cylinder be moved with a velocity U parallel to the major axis, the inner will begin to move in the same direction with a velocity

   $$\frac{\rho U \sinh\beta \operatorname{cosech}(\beta-\alpha)}{\rho \sinh\alpha \coth(\beta-\alpha) + \sigma\cosh\alpha},$$

   where c cosh α, c sinh α are semi-axes of the inner cylinder c cosh β, c sinh β those of the outer and σ is the density of the inner cylinder.
3. An elliptic cylinder whose semi-axes are c cosh α, c sinh α is divided in two by a plane through the axis of the cylinder and the major axis

of its cross-section. An infinite liquid of density r streams past to the major axis of the cross-section of the cylinder. Shew that in consequence of the motion of the liquid the pressure between the two portions of the cylinder is diminished by

$$\rho c\ U^2\ e^{\alpha} \sinh \alpha \left\{2\cosh\alpha + e^{\alpha}\sinh\alpha \log\tanh\frac{1}{2}\alpha\right\},$$

per unit length of the cylinder.

4. An elliptic cylinder, semi-axes a and b is held with its length perpendicular to, and its major axis making an angle ψ with, the direction of a stream of velocity U. Prove that the magnitude of the couple per unit length on the cylinder due to the fluid pressure is $\pi\rho\ c^2\ U^2 \sin\theta \cos\theta$.

5. Verify the stream function for uniform streaming parallel to the axis past a solid, bounded by those parts of the circles

   $(x + 1)^2 + y^2 = 2;\ (x - 1)^2 + y^2 = 2$

which are external to each other, are

$$\phi = y\left\{1 + \frac{1}{x^2+y^2} - \frac{2}{(x+1)^2} - \frac{2}{(x-1)^2+y^2}\right\},$$

$$\text{and } \psi = -x + \frac{x}{x^2+y^2} + \frac{2(x+1)}{(x+1)^2+y^2} + \frac{2(x-1)}{(x-1)^2+y^2}$$

Hint : We know that

$|z - 1|^2 = (x - 1)^2 + y^2$, and $|z + 1|^2 = (x + 1)^2 + y^2$.

Let the transformation be $t = \{2z/(z^2 - 1)\}$, so that the circle $|t| = 1$ in t-plane is same as $|2z/(z^2 - 1)| = 1$ in Z-plane.

$$\text{or } \frac{4|z|^2}{|z-1|^2|z+1|^2} = \text{in Z - plane.}$$

$$\text{or } \frac{4(x^2+y^2)}{\{(x-1)^2+y^2\}\{(x+1)^2+y^2\}} = 1.$$

or $\{(x - 1)^2 + y^2 - 2\}\ (x + 1)^2 + y^2 - 2\} = 0$.

Thus the part of plane outside the given circles in Z-plane transforms into part outside the circle $|t| = 1$ in t-plane.

$$\text{Let } \phi + i\psi = i\left\{-(x+iy) + \frac{x-iy}{x^2-y^2} + 2\frac{(x+1)-iy}{(x+1)^2+y^2}\right.$$

$$2\left.\frac{(x-1)-iy}{(x-1)^2+y^2}\right\}\text{(Given)}$$

$$\Rightarrow w = i\left\{-z+\frac{1}{z}+\frac{2}{z+1}+\frac{2}{z-1}\right\}$$

$$\Rightarrow w = 2i\left\{\frac{2z}{z^2-1}+\frac{z^2-1}{2z}\right\} = 2i\left\{t-\frac{1}{t}\right\},$$

which is a complex potential at a point t when the liquid is extending upto infinity.

6. In the two-dimensional irrotational motion of a liquid streaming past a fixed elliptic disc $(x^2/a^2) + (y^2/b^2) = 1$, the velocity at infinity being parallel to the major axis and equal to V, prove that if

   $x + iy = c \cosh (x + ih)$, $a^2 - b^2 = c^2$ and

   $a = c \cosh a$, $b = c \sinh a$,

   the velocity at any point is given by

   $$q^2 = V^2 \frac{a + b\sinh^2(\xi-\alpha) + \sin^2\eta}{\sinh^2\xi + \sin^2\eta}$$

   and that it has its maximum value $[V(a+b)]/a$ at the end of the minor axis.

7. An elliptic cylinder, semi axis 2a and 2b, is held with its length perpendicular to, and its major axis making an angle f with the direction of a stream of velocity U. Find the magnitude of the couple per unit length on the cylinder due to the fluid pressure.

3

Irrotational Motion in Three Dimension

(Motion of a Sphere)

Here we discuss the theory of irrotational motion in three dimensions with a particular emphasis to the motion of the sphere. The treatment is based more or less to the motion of a circular cylinder.

BUTLER'S SPHERE THEOREM

Let a rigid sphere $r = a$ be introduced into a flow field on an axis-symmetric irrotational flow in an incompressible inviscid fluid with no rigid boundaries, characterised by the stream function ψ_0 (r, θ) all of whose singularities are at a distance greater than a from the origin, where $\psi_0 =$ (r^2) at the origin, then the stream function become.

$$\psi = \psi_0 - \psi_2(r,\theta) - \left(\frac{r}{a}\right)\psi_0\left(\frac{a^2}{r},\theta\right).$$

The following conditions for the flow field are to be satisfied :

(i) The flow given by the stream function y *must be irrotation.*

(ii) The steam function is constant on the boundary of the sphere.

(iii) ψ_1 has no singularities outside the sphere r = *a*.

(iv) The velocity due to ψ_1 tend to zero as $r \to \infty$ and ψ_1 must introduce no net flux over the sphere at infinity.

Since the motion is irrotational, the velocity potential ϕ exists at any point in the fluid such that the velocity components in terms of ϕ and ψ are connected by

$$q_r = -\frac{\partial\phi}{\partial r} = -\frac{1}{r^2 \sin}\frac{\partial\psi}{\partial\theta} \text{ and } q_\theta = -\frac{1}{r}\frac{\partial\phi}{\partial\theta} = \frac{1}{r^2\sin\theta}\frac{\partial\psi}{\partial r},$$

$$\text{or } \frac{\partial}{\partial\theta}\left(\frac{\partial\phi}{\partial r}\right) = \frac{\partial}{\partial\theta}\left(\frac{1}{r^2\sin}\frac{\partial\psi}{\partial\theta}\right);\ \frac{\partial}{\partial r}\left(\frac{\partial\phi}{\partial\theta}\right) = -\frac{\partial}{\partial r}\left(\frac{1}{\sin\theta}\frac{\partial\psi}{\partial r}\right)$$

$$\text{or } \frac{\partial}{\partial r}\left(\frac{1}{\sin\theta}\frac{\partial\psi}{\partial r}\right)+\frac{\partial}{\partial\theta}\left(\frac{1}{r^2\sin\theta}\frac{\partial\psi}{\partial\theta}\right)=0$$

$$\text{or } r^2\frac{\partial^2\psi}{\partial r^2}+\sin\theta\frac{\partial}{\partial\theta}\left(\frac{1}{\sin\theta}\frac{\partial\psi}{\partial\theta}\right)=0 \qquad \text{...(1)}$$

The equation (1) is satisfied by direct differentiating the function ψ_0 and ψ_1 (at origin). Therefore the flow given by the stream function ψ is irrotational and hence satisfies the condition (i). It is that the steam function vanishes at the surface of the sphere. This satisfies the condition (ii).

Since r and (a^2/r) are the inverse points with regard to the sphere r = *a*, if one point is inside the sphere then the other point must lies outside the sphere. Thus all the singularities of ψ_0 lie outside the sphere and consequently all the singularities of ψ_1 (at origin) will be inside. Therefore the condition (iii) is satisfied.

As ψ_0 is analytic inside the sphere r = *a* and near the origin $\psi_0 = O(r^2)$, therefore at infinity ψ_1 (at origin) = O (1 / r).

Since $q_r = -\dfrac{1}{r^2\sin\theta}\dfrac{\partial\psi}{\partial\theta}$, it follows that the velocity at infinity due to ψ_1 (at origin) is of $O(1/r^3)$ which tens to zero as *r* tends to infinity.

Again the flux $\int q_r\, ds = O(1/r)$ which also tends to zero as r tend to infinity. Hence the condition (iv) is also satisfied.

SOLUTION OF LAPLACE EQUATION

The most important method for obtaining the solution of the Laplace equation $\nabla^2\phi = 0$ in three dimensions is that of spherical harmonics when the suitable boundary conditions are prescribed to spherical surfaces. Since the Laplacian operator ∇^2 is homogeneous with respect to (x, y, z) it follows that the function ϕ must satisfy the Laplace equation separately. Thus any homogeneous solution is called a spherical solid harmonic. If ϕ_n be a spherical solid harmonic of degree n, then assuming $\phi_n = r^n S_n$, where S_n is a function of the direction only in which the point (x, y, z) lies with regard to the origin. This is known a spherical surface harmonic of order n.

Again to any solid harmonic ϕ_n of degrade n there corresponds another harmonic of degree (– n – 1). Therefore the two fundamental solutions of Laplace equation in a three dimensional field are $r^n S_n$ and $r^{-n-1} S_n$, where r = | x | and S_n is the spherical surface harmonics of order n (a positives integer).

The Laplace equation may be expressed in spherical polar co-ordinates (r, θ, ω) as

$$\sin\theta\frac{\partial}{\partial r}\left(r^2\frac{\partial\phi}{\partial r}\right)+\frac{\partial}{\partial\theta}\left(\sin\theta\frac{\partial\phi}{\partial\theta}\right)+\frac{1}{\sin\theta}\frac{\partial^2\phi}{\partial\omega^2}=0$$

Consider that ϕ is a homogeneous function of degree n; substituting $\phi = r^n S_n$ and $\phi = r^{-n-1} S_n$ in (1), we get

$$\frac{1}{\sin\theta}\frac{\partial}{\partial\theta}\left(\sin\theta\frac{\partial S_n}{\partial\theta}\right)+\frac{1}{\sin^2\theta}\frac{\partial^2 S_n}{\partial\omega^2}+n(n+1)\,S_n=0, \quad ...(2)$$

which is the general differential equation of spherical surface harmonic.

The term $\partial^2 S_n/\partial\omega^2$ will be neglected in the axis-symmetric flow. Let $\pi = \cos\theta$, the equation (2) becomes

$$\frac{d}{d\mu}\left[(1-\mu^2)\frac{dS_n}{d\mu}\right]+n(n+1)S_n=0,.$$

Whose solution is given by

$$S_n = A\left[1-\frac{n(n+1)}{1.2}\mu^2+\frac{n(n+1)(n+2)(n+3)}{1.2.3.4}\mu^4-...\right]$$

$$+B\left[\mu-\frac{(n-1)(n+2)}{1.2.3}\mu^3+\frac{(n-3)(n-1)(n+2)(n+4)}{1.2.3.4.5}\mu^5-...\right] \quad ...(3)$$

The first and the second series terminates when n is an even or odd integer respectively. Both the series are convergent for values of π between $\pi = \pm 1$; but if $\gamma - \alpha - \beta = 0$, it becomes infinite so the series diverges between the values of $\pi = \pm 1$. It follows that the terminating series corresponding to the integral values of n are the only zonal surface harmonics which are finite over the unit sphere. Thus both these cases are included in the expansion.

$$P_n(\mu)=\frac{1}{2^n n!}\frac{d^n}{d\mu^n}(\mu^2-1)^n,$$

where $\quad P_0(\mu)=1, P_1(\mu)=\mu, P_2(\mu)=\frac{1}{2}(3\mu^2-1)$ etc.

MOTION OF A SPHERE IN AN INFINITE MASS OF LIQUID AT REST AT INFINITY

Consider the origin *O* at the centre of the sphere, which is moving along the X-axis in the direction of motion with velocity *V*. The motion of the liquid is symmetric about this line. In the absence of the viscous terms in the equation of fluid motion, a velocity potential ϕ is defined such that

$$q = -\nabla\phi \Rightarrow \nabla.\, q = -\nabla.\,(\nabla\phi)$$

$$\Rightarrow \nabla^2\phi = 0$$

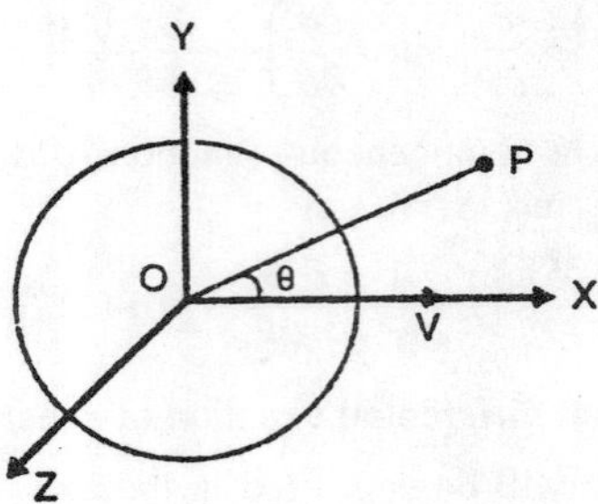

Fig. 3.1

$$\Rightarrow \frac{\partial}{dr}\left(r^2 \frac{d\phi}{dr}\right)+\frac{1}{\sin\theta}\frac{\partial}{d\theta}\left(\sin\theta\frac{\partial\phi}{d\theta}\right)=0 \qquad ...(1)$$

(1) $\left(-\frac{\partial\phi}{dr}\right)_{r\to\infty}=0$ ...(2)

(ii) Normal velocity $\left(-\frac{\partial\phi}{dr}\right)_{r=a}=V\cos\theta$ ...(3)

The surface condition can be satisfied for all θ if ϕ is proportional to the axis symmetric surface harmonics or Legendre polynomial of order one. The surface condition (3) suggests that the solution of (1) is of the form

$$\phi = f(r)\cos\theta$$

i.e., $\frac{\partial}{dr}\left[r^2 f'(r)\cos\theta\right]+\frac{1}{\sin\theta}\frac{\partial}{d\theta}\left[-\sin\theta f(r)\sin\theta\right]=0$

$$\Rightarrow \quad \cos\theta\left[\frac{\partial}{dr}\{r^2 f'(r)\}-2f\right]=0$$

$$\Rightarrow \quad r^2\frac{\partial^2 f}{dr^2}+2r\frac{\partial f}{dr}-2f=0,\ \cos\theta\neq 0$$

$$\Rightarrow \quad f(r)-Ar+\frac{B}{r^2}$$

Thus f assuming the form of the function ϕ as

$$\phi=\left(Ar+\frac{B}{r^2}\right)\cos\theta$$

$$\frac{\partial\phi}{\partial r}=\left(A-\frac{2B}{r^3}\right)\cos\theta \qquad ...(4)$$

Since the space derivative of ϕ must vanish at infinity, so

$A\cos\theta = 0 \Rightarrow A = 0$

From the relation (3) and (4), we have

$$\frac{2B}{a^3}\cos\theta = V\cos\theta \ \forall\ \theta$$

$$\Rightarrow B = \frac{1}{2}a^3V.$$

Thus the required solution becomes

$$\phi = \frac{1}{2}\frac{Va^3}{r^2}\cos\theta \text{ and } \psi = -\frac{1}{2}\frac{Va^3}{r}\sin\theta. \qquad ...(5)$$

Lines of flow : Referring to the centre of sphere as origin, the differential equation of the lines of flow are given by

$$\frac{dr}{\partial\phi/\partial r}\ \frac{rd\theta}{\partial\phi/r\partial\theta}$$

$$\Rightarrow \frac{dr}{-(Va^3/r^3)\cos\theta} = \frac{rd\theta}{-(Va^3/r^3)\sin\theta}$$

$$\Rightarrow \frac{dr}{\cos\theta} = \frac{2rd\theta}{\sin\theta} \Rightarrow \frac{dr}{r} = 2\cot\theta\, d\theta$$

By integrating, we have

log r = log c + log $\sin^2\theta$, *where C* is an arbitrary constant.

$\Rightarrow$r = *C* $\sin^2\theta$,

represents the equation of the lines of flow.

IDEAL FLUID FLOW AROUND A SPHERE

When the viscous terms are neglected in the equation of fluid motion, a velocity potential ϕ is defined such that

$$q = -\nabla\phi \Rightarrow \nabla^2\phi = 0.$$

The Laplace equation may be written in spherical polar coordinates as

$$\frac{\partial}{\partial r}\left(r^2\frac{\partial\phi}{dr}\right) + \frac{1}{r^2\sin\theta}\frac{\partial}{\partial\theta}\left(\sin\theta\frac{\partial\phi}{\partial\theta}\right) = 0. \qquad ...(1)$$

The fluid flows with velocity U_0 along the axis of symmetry $\theta = 0$ towards the origin at the centre of the sphere. The boundary conditions are

(i) $(\partial\phi/\partial r)_{r=a} = 0$

(ii) $(\partial\phi/\partial r)_{r\to\infty} = U_0\cos\theta.$

$(\partial\phi/r\partial\theta)_{r\to\infty} = -U_0\sin\theta.$

The condition (i) prohibits flow through the surface of the sphere, and condition (ii) states that the flow at a large distance from the obstacle, in any direction, is an undisturbed parallel flow. In view of the boundary condition (ii), assuming a particular solution of the form

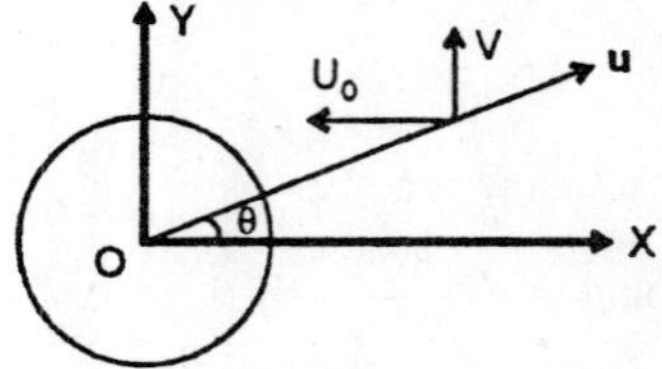

Fig. 3.2

$$\phi = f(r) \cos \theta \qquad ...(2)$$

From (1) and (2), we have

$$r^2 \frac{d^2 f}{dr^2} + 2r\frac{df}{dr} - 2f = 0 \Rightarrow \phi = \left(Ar + \frac{B}{r^2} \right) \cos\theta.$$

$$\frac{d\phi}{dr} = \left(A - \frac{2B}{r^3} \right) \cos\theta, \frac{d\phi}{d\theta} = -\left(Ar + \frac{B}{r^2} \right) \sin\theta. \qquad ...(3)$$

From the boundary conditions (i) 3.1, (ii) and the equation (3), we have

$A = U_0$ and $B = (1/2)\, U_0 a^3$.

Thus the solution becomes

$$\phi = U_0 \left(r + \frac{a^3}{2r^2} \right) \cos\theta, \; \psi = \frac{1}{2} U_0 r^2 \left(1 - \frac{a^3}{r^3} \right) \sin^2\theta. \qquad ...(4, 5)$$

We notice that the θ component of velocity does not vanish at the surface of the sphere. From (3), we have

$$-\frac{1}{r}\frac{\partial\phi}{\partial\theta} = \frac{2}{3} U_0 \sin\theta \qquad ...(6)$$

which shows that there is slipping at the surface of the sphere, physically, this is not reasonable and is caused by neglecting the viscous terms in the equation of motion. The difficulty can be over-come by assuming that there is a thin layer of fluid near any boundary in which viscous terms are important, which is the basis of *Prandtl's boundary layer theory.*

LIQUID STREAMING PAST A FIXED SPHERE

Consider the sphere to be fiexed and the liquid streaming past is with velocity V. By superposing a velocity $-V$ on the sphere and the liquid, the velocity potential becomes

$$\phi = V\left(r + \frac{1}{2}\frac{a^3}{r^2}\right)\cos\theta. \qquad ...(1)$$

Equation to the lines of flow are given by

$$\frac{dr}{\partial\phi / \partial r} = \frac{rd\theta}{\partial\phi / r\partial\theta}$$

or $$\frac{dr}{V\{1-(a^3/r^3)\}\cos\theta} = -\frac{rd\theta}{V\left\{1+\frac{1}{2}(a^3/r^3)\right\}\sin\theta}$$

or $$\frac{2r^3 + a^3}{r(r^3 - a^3)}dr = -\cot\theta\, d\theta$$

By integrating, we have

$$\log(r^3 - a^3) - \log r = -2\log\sin\theta + \log A,$$

where A is an integration constant,

or $$\frac{r^3 - a^3}{r} = \frac{A}{\sin^2\theta},$$

or $(r^3 - a^3)\sin^2\theta = Ar,$ ...(2)

which represent the lines of flow relative to the sphere.

CONCENTRIC SPHERES (INITIAL MOTION)

Determine the velocity potential of the initial motion when an impulsive force is so applied that the inner sphere starts moving with velocity U and the outer sphere with velocity V in the same direction.

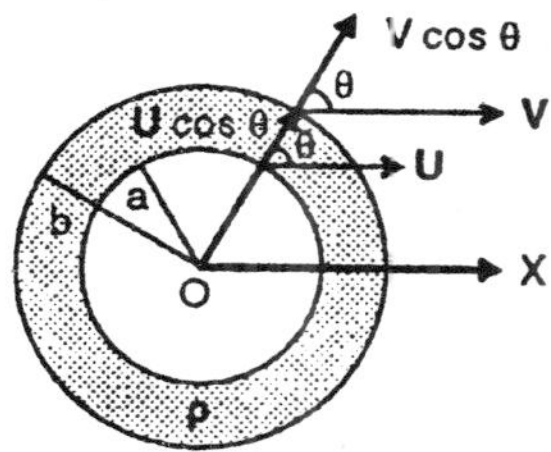

Fig. 3.3

Consider two concentric spheres of radii a and b (a < b), the intervening space being filled with liquid of density p. The motion is started by impulse and is consequently irrotational. Thus the velocity potential ϕ satisfies the Laplace equation $\nabla^2\phi = 0$. The boundary conditions become

(i) $(-\partial\phi/\partial r)_{r=a} = U\cos\theta,$

(ii) $(-\partial\phi/\partial r)_{r=b} = V\cos\theta$,

Assuming $\phi = \left(Ar + \dfrac{B}{r^3}\right)\cos\theta$,

or $\dfrac{\partial\phi}{\partial r} = \left(A - \dfrac{2B}{r^3}\right)\cos\theta$.

Using the condition (i) and (ii), we have

$$\left(A - \frac{2B}{a^3}\right)\cos\theta = -U\cos\theta. \left(A - \frac{2B}{b^3}\right)\cos\theta = -V\cos\theta. \qquad ...(2, 3)$$

These being true for all values of θ, so

$$A - \frac{2B}{a^3} = -U;\ A - \frac{2B}{b^3} = -V.$$

$$\Rightarrow A - \frac{Ua^3 - Vb^3}{b^3 - a^3};\ B = \frac{(U-V)a^3b^3}{2(b^3 - a^3)} = -V.$$

Substituting the value of the constants A and B, we get

$$\phi = \left\{\frac{Ua^3 - Vb^3}{b^3 - a^3}\right\} r\cos\theta + \left\{\frac{(U-V)a^3b^3}{2(b^3 - a^3)}\right\}\frac{\cos\theta}{r^2}. \qquad ...(4)$$

which determines the velocity potential at an instant of start.

Particular case : Let a given velocity is imparted to either of the spheres. Suppose the outer sphere be at rest (V = 0) then the velocity potential reduces.

$$\phi = \frac{Ua^3}{b^3 - a^3} r\cos\theta + \frac{Ua^3b^3}{2(b^3 - a^3)}\frac{\cos\theta}{r^2} = \frac{Ua^3}{b^3 - a^3}\left\{r + \frac{b^3}{2r^2}\right\}\cos\theta. \qquad ...(5)$$

Let M be the mass of the inner sphere and I be the impulse necessary to produce the velocity U, then by the principle of momentum, we have

$$I - MU = \int\int \varpi \cos\theta \, dS.$$

where ($\varpi = -\rho\phi$) denotes the impulsive pressure of the liquid at r = a.

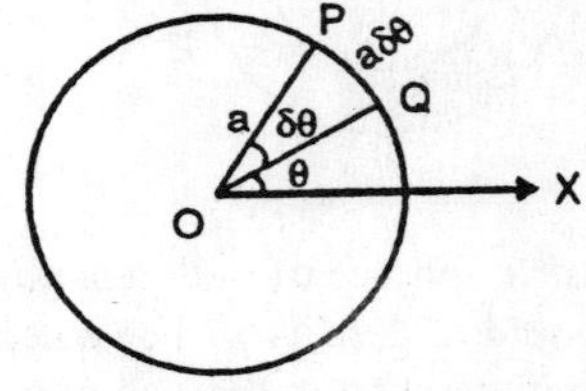

Fig. 3.4

or $I - MU = \dfrac{1}{2}\rho\dfrac{Ua(2a^3 + b^3)}{(b^3 - a^3)}$

$$\int_0^\pi \cos^2\theta(2\pi a\sin\theta)a\,d\theta$$

or $$I - MU = \frac{2}{3}\frac{\pi\rho U(2a^3+b^3)a^3}{(b^3-a^3)}.$$

Let M' be the mass of the liquid displaced then M' = $\frac{4}{3}\pi a^3\rho$.

or $$I = MU + \frac{1}{2}\frac{M'U(2a^3+b^3)}{(b^3-a^3)},$$

or $$I = MU + \frac{1}{2}M'U\left\{\frac{2(a^3/b^3)+1}{1-(a^3/b^3)}\right\}.$$

If the radius of the outer sphere (r = b) is increased indefinitely (*i.e.*, $b \to \infty$) then the impulse necessary to impart a velocity U to the inner sphere is given by

$$I = MU + \frac{1}{2}M'U = \left(M + \frac{1}{2}M'\right)U.$$ **Proved.**

EQUATION OF MOTION OF A SPHERE

Let a sphere of radius a moves with velocity v in a liquid at rest at infinity. Consider (x_0, y_0, z_0) be the coordinates of the centre of the moving sphere at any instant t referred to fixed axes and (U, V, W) are the components of the velocity of the centre along the coordinate axes respectively.

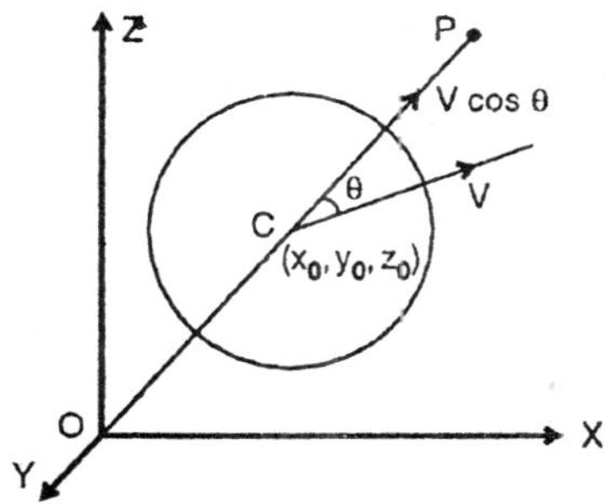

Fig. 3.5

$U = x,\ V = y,\ W = z.$

Assuming the velocity potential be of the form

$$\phi = \frac{1}{2}\frac{Va^3}{r^2}\cos\theta \qquad ...(1)$$

where v cosθ is the resolved part of the velocity v along CP.

$$= \left(U \frac{x - x_0}{r} + V \frac{y - y_0}{r} + W \frac{z - z_0}{r} \right),$$

where $r^2 = (x - x_0)^2 + (y - y_0)^2 + (z - z_0)^2$.

$$\Rightarrow rr = -(x - x_0)\ U - (y - y_0)\ V - (z - z_0)\ W. \quad \text{...(2)}$$

Thus the velocity potential at a fixed point of space (x, y, z) is

$$\phi = \frac{1}{2} \frac{a^3}{r^3} \left\{ U \frac{x - x_0}{r} + V \frac{y - y_0}{r} + W \frac{z - z_0}{r} \right\} \quad \text{...(3)}$$

Neglecting the extraneous forces, the pressure ρ is given by Bernoulli's equation as

$$\frac{p}{\rho} = F(t) + \frac{\partial \phi}{\partial t} - \frac{1}{2} q^2, \quad \text{...(4)}$$

where $q^2 = (\partial \phi / \partial x)^2 + (\partial \phi / \partial y)^2 + (\partial \phi / \partial z)^2$

$$\frac{\partial \phi}{\partial x} = \frac{a^3 U}{2r^3} - \frac{3a^3}{2r^5} (x - x_0)\ \{ U(x - x_0) + U(y - y_0) + W(z - z_0) \} + \dots$$

and $q^2 = \left(\frac{a^6}{4r^6} \right) (U^2 + V^2 + W^2) + (3a^6 / 4r^6)\ \{ U(x - x_0) + \dots + \dots \}^2$ ...(5)

Also $\frac{\partial \phi}{\partial t} = \left(\frac{a^3}{2r^3} \right) \{ U(x - x_0) + V(y - y_0) \dots + \}$

$$- \left(\frac{a^3}{2r^3} \right) \{ U^2 + V^2 + W^2 \} + \left(\frac{3a^3}{2r^5} \right) \{ U(x - x_0) + \dots + \dots \}^2$$

Substituting the value of $\frac{\partial \phi}{\partial t}$ and q^2 in (4), we have

$$\frac{p}{\rho} = F(t) + \left(\frac{a^3}{2r^3} \right) \{ U(x - x_0) + \dots + \dots \} - \left(\frac{a^3}{2r^3} \right) \{ U^2 + V^2 + W^2 \}$$

$$+ \left(\frac{3a^3}{2r^5} \right) \{ U(x - x_0) + \dots + \dots \}^2 - \left(\frac{a^6}{8r^6} \right) \{ U^2 + V^2 + W^2 \}$$

$$- \left(\frac{3a^6}{8r^6} \right) \{ U(x - x_0) + \dots + \dots \}^2 .$$

Thus the pressure on the sphere r = a is given by

$$\frac{p}{\rho} F(t) + (1/2) [U(x - x_0) + \dots + \dots] - (5/8) [U^2 + V^2 + W^2]$$

$$+ (9/8a^3) [U(x - x_0) + \dots + \dots]^2 \quad \text{...(6)}$$

Let f be the acceleration of the sphere and θ and θ_1 be the angles that CP makes with the direction of the velocity v and acceleration f then

$$v = \cos\theta = U\,(x - x_0)/\,a + V\,(y - y_0)/\,a + W\,(z - z_0)/\,a,$$

$$f\cos\theta_1 = U\,(x - x_0)/\,a + V\,(y - y_0)/\,a + W\,(z - z_0)/\,a. \qquad ...(7)$$

From (6) and (7), we have

$$\frac{p}{\rho} = F(t) + (1/2)af\cos\theta_1 - (5/8)v^2 + (9/8)v\cos^2\theta$$

Since p= ∏ at r = ∞; $F(t)$ = ∏/ρ.

or $$\frac{p}{\rho} = (\Pi/\rho) + (1/2)af\cos\theta_1 + (1/8)v^2 + (9\cos^2\theta - 5). \qquad ...(8)$$

Resultant thrust on the sphere due to the motion is

$$= -\iint p\cos\theta\, dS$$

$$= -\int_0^x \left[\Pi + (1/8)\rho v^2(9\cos^2\theta - 5)\right]\cos\theta\; 2\pi a\;\sin\theta\; ad\theta$$

$$-\int_0^x \frac{1}{2}af\rho\cos\theta_1\cos\theta_1.2\pi a\;\sin\theta_1.a\,a\theta_1$$

$$= 0 - \pi a^3\rho f\int_0^x \cos\theta_1\cos\theta_1\sin\theta_1\; d\vartheta_1 = -\frac{1}{2}M'f,$$

where $M' = \frac{4}{3}\pi a^3\rho$ is the mass of the liquid displaced.

Let X, Y, Z be the components of resultant thrust then

$$X = -\frac{1}{2}M'U, Y = -\frac{1}{2}M'V, \text{ and } Z = -\frac{1}{2}M'W.$$

Again if (X', Y', Z') are the components of the extraneous force on the sphere when no liquid is present and M represents the mass of the sphere of density σ then the equation of motion parallel to X-axis is given by

$$MU = -\frac{1}{2}M'U + \frac{\sigma - \rho}{\sigma}X'$$

or $$MU = \frac{M}{M + \frac{1}{2}M'}\frac{\sigma - \rho}{\sigma}X'$$

or $$MU = \frac{\frac{3}{4}\pi a^3\sigma}{\frac{1}{3}\pi a^3\sigma + \frac{1}{2}.\frac{3}{4}\pi a^3\sigma}\frac{\sigma - \rho}{\sigma}X'; M = \frac{3}{4}\pi a^3\sigma$$

$$MU = \frac{\sigma - \rho}{\sigma + \frac{1}{2}\rho} X'.$$

or ...(9)

Therefore the whole effect of the presence of the liquid is to reduce the extraneous forces in the ratio $\sigma - \rho : \sigma + \frac{1}{2}\rho.$

If S = σ / ρ, the specific gravity of the sphere compared with the liquid then the above ratio is expressed as $S - 1 : S + \frac{1}{2}.$

Let T be the kinetic energy of the liquid, then

$$T = -\frac{1}{2}\rho \int\int q^2 d\tau$$

where $$q^2 = \left(\frac{a^6}{4r^6}\right)v^2 + \left(\frac{3a^6}{4r^6}\right)v^2 \cos^2\theta.$$

or $$T = \frac{1}{8}\rho a^6 v^2 \int_{\theta=0}^{x} \int_{r=a}^{\infty} \left(\frac{1}{r^6} + \frac{3}{r^6}\cos^2\theta\right) 2\pi r \sin\theta \, r d\theta \, dr$$

or $$T = \frac{1}{4}\pi a^6 \rho v^2 . \frac{1}{3a^3} \int_0^{\pi} (1 + 3\cos^2\theta) \sin\theta \, d\theta$$

or $$T = \frac{1}{3}\pi \rho a^3 v^2 \frac{1}{4}.\left(\frac{4}{3}\pi a^3 \rho\right) v^2 = \frac{1}{4} M' v^2.$$

Hence the effect of the liquid is to increase the inertia of the sphere by half the mass of liquid displaced.

THREE DIMENSIONAL SOURCE AND SINK

If the motion of a liquid consists of symmetrical outward radial flow in all directions proceeding from a point then the point is called a *three dimensional source.* Let the source emits the volume 4π m per unit time then m is called the strength of the source.

Let there is a source of strength + m at the origin, the outward flux across a sphere of radius r, whose centre is at the source, is related to the radial velocity q_r by

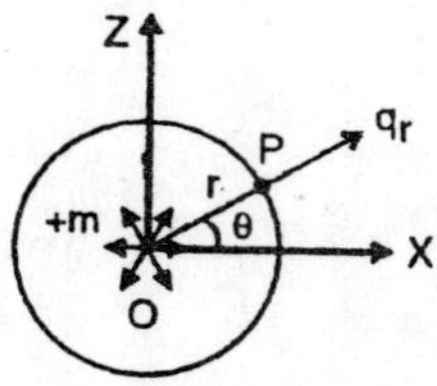

Fig. 3.6

$$4\pi r^2 q_r = 4\pi m \Rightarrow q_r = \frac{m}{r^2}$$

$$\Rightarrow q = \frac{m}{r^2}\hat{r}$$

Let ϕ be the velocity potential at P then

$$\nabla\phi = \phi'(r)\hat{r}$$

$$\Rightarrow \phi'(r) = -\frac{m}{r^2} \Rightarrow \phi(r) = \frac{m}{r}$$

and $$\psi = m\cos\theta = \frac{mx}{r}$$

Let the source is at the point A of the axis instead of at the origin, then

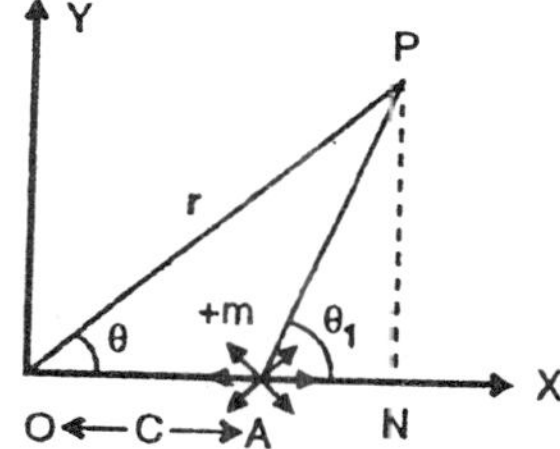

Fig. 3.7

$$\phi = \frac{m}{AP}\left(\text{since } \cos\theta = \frac{c^2 + r^2 AP^2}{2cr}\right)$$

$$\Rightarrow \phi = -\frac{m}{\sqrt{r^2 + c^2 - 2cr\cos\theta}}$$

and $$\psi = m\cos\theta_1 = \frac{m(r\cos\theta - c)}{\sqrt{r^2 + c^2 - 2cr\cos\theta}}$$

$(AN = ON - OA = r\cos\theta - c)$

If the fluid is directed radially symmetrically inwards to O from all Directions, then the fluid leaves the system at O which is called a sink.

THREE DIMENSIONAL DOUBLET

A three dimensional doublet is a combination consists of a three dimensional sink of strength - m and a source of strength + m at a distance δs apart. Assuming $m \to \infty$, $\delta s \to 0$ in such a way that the product $m\delta s$ ($=\mu$) remains finite and constant. The quantity μ is termed the strength of the doublet. The direction of the doublet is taken from sink to source.

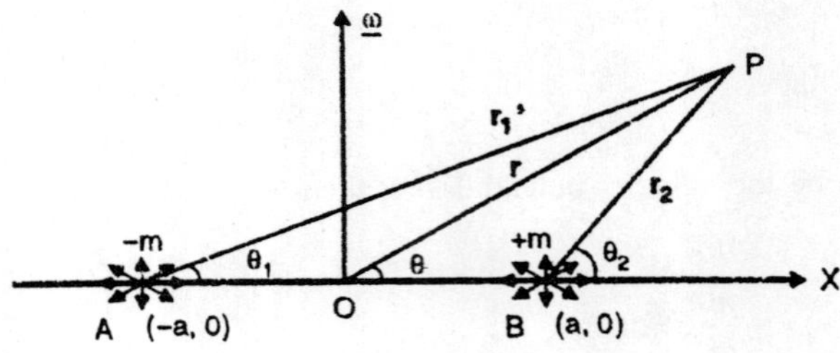

Fig. 3.8

Consider AB be the doublet with the sink of strength - m at A (- a, 0) and a source of strength + m at B (a, 0). Let P be any point in the fluid such that AP = r_1 and BP = r_2. The velocity potential ϕ due to the doublet at P is given by

$$\phi = -\frac{m}{r_1} + = \frac{m}{r_2}$$

Let the product 2ma = μ remains constant when m $\to \infty$ and 2a $\to$ 0, the combination becomes a double source or doublet.

From the sine rule in the ΔAPB, we have

$$\frac{r^2}{\sin\theta_2} = \frac{r^2}{\sin\theta_1} = \frac{2a}{\sin(\theta_2 - \theta_1)}$$

$$\Rightarrow r_1 - r_2 = \frac{2a(\sin\theta_2 - \sin\theta_1)}{\sin(\theta_2 - \theta_1)}$$

or $$r_1 - r_2 = \frac{2a.2\cos\frac{1}{2}(\theta_1 + \theta_2)\sin\frac{1}{2}(\theta_2 - \theta_1)}{2\sin\frac{1}{2}(\theta_2 - \theta_1)\cos\frac{1}{2}(\theta_2 + \theta_1)}$$

$$\Rightarrow r_1 - r_2 = \frac{2a\cos\frac{1}{2}(\theta_1 + \theta_2)}{\cos\frac{1}{2}(\theta_2 + \theta_1)}$$

Then $$\phi = m\frac{r_1 - r_2}{r_1 r_2} = \frac{\mu\cos\frac{1}{2}(\theta_1 + \theta_2)}{r_1 r_2 \cos\frac{1}{2}(\theta_2 + \theta_1)} \qquad ...(1)$$

Again $\psi = m(\cos\theta_2 - \theta_1)$

$$\Rightarrow \psi = \frac{m(x-a)}{r_2} - \frac{m(x+a)}{r_1}$$

$$\Rightarrow \psi = \frac{\mu x \cos\frac{1}{2}(\theta_2+\theta_1)}{r_1 r_2 \cos\frac{1}{2}(\theta_2-\theta_1)} - \frac{1}{2}\mu\left(\frac{1}{r_2}+\frac{1}{r_1}\right)$$

when a $\to$ 0, $\theta_2 \to \theta$, $r_2 \to r_1 \to r$, then, we have

$$\phi = \frac{\mu\cos\theta}{r_2}$$

and $$\psi = \frac{\mu x\cos\theta}{r_2} - \frac{\mu}{r} = \frac{\mu x^2}{r^3} - \frac{\mu}{r} = \frac{\mu(x^2-r^2)}{r^2} = \frac{\mu\omega^2}{r^3}$$

The stream lines for a doublet are shown as

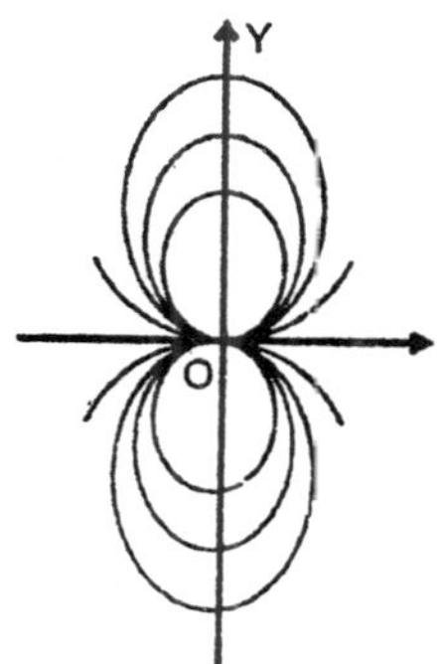

Fig. 3.9

IMAGE OF A SOURCE WITH REGARD TO A SPHERE

Let a be the radius of the sphere. Consider a source of strength + m at a point A such that OA=f, outside the sphere. Let ϕ_1 be the velocity potential due to the source alone when the sphere is not present and ϕ_1 due to the presence of the sphere. The velocity potential ϕ_1 at any point (r, o) is

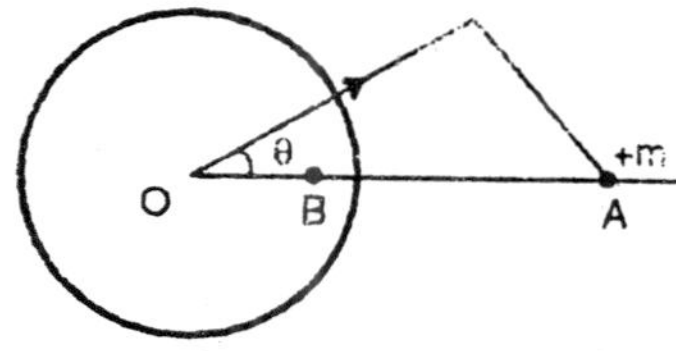

Fig. 3.10

$$\phi_1 = \frac{m}{AP}$$

$$\phi_1 = m(r^2 + f^2 - 2fr\cos\theta)^{-1/2}$$

$$\phi_1 = m/f[1-(2r/f)\cos\theta + (r^2/f^2)^{-1/2}$$

$$\phi_1 = \frac{m}{f}\left\{1+\sum_{n=1}^{\infty}\left(\frac{r}{f}\right)^n P_n(\mu)\right\}, r < f \quad ...(1)$$

where μ = cos θ, and P_n is the Legendre's coefficient of order n.

The motion is symmetrical about OA and the velocity potential satisfies Laplace equation. Let the velocity potential ϕ_2 be of the form

$$\phi_2 = \Sigma A_n \frac{a^n}{a^n+1} P_n(\mu) \quad ...(2)$$

Again, the velocity normal to the sphere is zero

i.e., $$\frac{\partial}{\partial r}(\phi_1 + \phi_2) = 0, at\ r = a.$$

Therefore $$\frac{m}{f}\sum_1^{\infty}\frac{na^{n-1}}{f^n}P_n(\mu) - \sum_0^{\infty}(n+1)\frac{A_n}{a^2}P_n(\mu) = 0$$

for all values of θ, so that

$$A_0 = 0, A_n = \Sigma\frac{nma^{n+1}}{(n+1)f^{n+1}}. \quad ...(3)$$

From (2) and (3), we have

$$\phi_2 = m\sum_1^{\infty}\frac{n}{n+1}\cdot\frac{a^{2n+1}}{r^{n+1}f^{n+1}}P_n(\mu)$$

or $$\phi_2 = m\sum_1^{\infty}\left(1-\frac{1}{n+1}\right)\frac{a^{2n+1}}{r^{n+1}f^{n+1}}P_n(\mu)$$

or $$\phi_2 = m\sum_1^{\infty}\frac{a^{2n+1}}{r^{n+1}f^{n+1}}P_n(\mu) - m\sum_1^{\infty}\frac{a^{2n+1}}{r^{n+1}f^{n+1}}\frac{P_n(\mu)}{n+1}. \quad ...(4)$$

Since B is an inverse point of A with regard to the sphere, so that

$$OA.OB = a^2 \Rightarrow ob = \frac{a^2}{f} = f'.$$

By adding and subtracting a term $\left(\frac{ma}{f}\frac{P_0}{r}\right)$ in (4), we have

$$\phi_2 = \frac{ma}{f}\sum_1^{\infty}\frac{f'^n}{r^{n+1}}P_n(\mu) - \frac{ma}{f}\sum_1^{\infty}\frac{f'^n}{r^{n+1}}\frac{P_n(\mu)}{n+1}.$$

or $$\phi_2 = \frac{ma}{f(r^2+f'^2-2rf'\cos\theta)^{1/2}} - \frac{ma}{f}\sum_0^{\infty}\frac{f'^n}{r^{n+1}}\frac{P_n(\mu)}{n+1}$$

$$\text{or } \phi_2 = \frac{(ma / f)}{BP} - \frac{ma}{f}\sum_0^{\infty}\frac{f'^n}{r^{n+1}}\frac{P_n(\mu)}{n+1}. \qquad ...(5)$$

The first term in (5) is the velocity potential due to a source of strength (+ ma/f) at B.

Again the velocity potential due to the source (+ ma / f) at any point on OB at assistance λ from the centre is given by

$$\phi = \frac{ma}{f}(r^2 + \lambda^2 - 2r\lambda\cos\theta)^{-1/2}\frac{ma}{f}\sum\frac{\lambda_2}{r^{n+1}}P_n(\mu),$$

$$\Rightarrow \phi = \frac{ma}{ff'}\int_0^{f'}\left(\sum\frac{\lambda_n}{r^{n+1}}P_n\right)d\lambda$$

$$\phi = \frac{ma}{ff'}\sum_0^{\infty} - \frac{f'^{n+1}}{(n+1)r^{n+1}}P_n.$$

Thus the second term in (5) is the velocity potential due to the continuous line distribution of sinks of strength (ma / ff') or (- m /a) per unit length extending from O to B.

Hence the image system of a source placed outside with regard to a sphere consists of a source of strength + ma / f at an inverse point and a line link of strength - m /a per unit length extending from the centre to the inverse point.

MOTION OF LIQUID INSIDE A ROTATING ELLIPSOIDAL SHELL

The equation to the surface is

$$\frac{x^2}{a^2} + \frac{y^2}{b^2} + \frac{z^2}{c^2} = 1. \qquad ...(1)$$

Let ω_x, ω_y, ω_z, are the components of the angular velocity referred to the coordinate axes fixed in space and coincident with the axes of the ellipsoid at any instant t. Then the component of linear velocities of a a point P (x y x) of the ellipsoidal shell are given by

$$z\omega_y - y\omega_z,\ x\omega_z - z\omega_z,\ y\omega_x - x\omega_y.$$

The direction cosines of the normal are proportional to x / a^2, y / b^2, and z / c^2. Let ϕ be the velocity potential of a liquid motion then the boundary condition is given by

$$-\frac{x^2}{a^2}\frac{\partial\phi}{\partial x} - \frac{y}{b^2}\frac{\partial\phi}{\partial y} - \frac{z}{c^2}\frac{\partial\phi}{\partial z} = \frac{x}{a^2}(z\omega_y - y\omega_z)$$

$$+\frac{y}{b^2}(x\omega_z - z\omega_x) + \frac{z}{c^2}(y\omega_x - x\omega_y). \qquad ...(2)$$

Let ϕ = Ayz + Bzx + Cxy. ...(3)

From (2) and (3), we have

$$Ayz\left(\frac{1}{b^2}+\frac{1}{c^2}\right)+Bzx\left(\frac{1}{c^2}+\frac{1}{a^2}\right)+Cry\left(\frac{1}{a^2}+\frac{1}{b^2}\right)$$

$$= yz\omega_x\left(\frac{1}{b^2}-\frac{1}{c^2}\right)+zx\omega_y\left(\frac{1}{c^2}-\frac{1}{a^2}\right)+zy\omega_z\left(\frac{1}{a^2}-\frac{1}{b^2}\right).$$

$$\Rightarrow A = -\frac{b^2-c^2}{b^2+c^2}\omega_x,\ B = -\frac{c^2-a^2}{c^2+a^2}\omega_y,\ \text{and } C = -\frac{a^2-b^2}{a^2+b^2}\omega_z.$$

Substituting the values of A, B and C into (3), we have

$$\phi = -\frac{b^2-c^2}{b^2+c^2}\omega_x yz - \frac{c^2-a^2}{c^2+a^2}\omega_y zx - \frac{a^2-b^2}{a^2+b^2}\omega_z xy. \qquad ...(4)$$

The velocity potential ϕ depends only on the mutual ratios of a, b, c and not on their absolute magnitudes, it follows that the motion is the same in all ellipsoids of the same shape rotating with the same angular velocity.

PATH OF THE PARTICLES RELATIVE TO THE ELLIPSOIDS

Let (λ, μ, v) are the coordinates of a particle P referred to the axes of the ellipsoid. Then the velocity components at P referred to the axes fixed in space are

$$\lambda - \mu\omega_z + v\omega_y,\ \mu - v\omega_x + \lambda\omega_y + \mu\omega_x,$$

$$\text{or } \lambda - \mu\omega_z + v\omega_y = -\frac{\partial\phi}{\partial x} = \frac{c^2-a^2}{c^2+a^2}v\omega_y + \frac{a^2-b^2}{a^2+b^2}\mu\omega_z$$

$$\text{or } \lambda = \left(\frac{(c^2-a^2)}{c^2+a^2}-1\right)v\omega_y + \left(\frac{a^2-b^2}{a^2+b^2}\right)\mu\omega_z$$

$$\text{or } \lambda = \frac{2a^2}{c^2+a^2}\omega_z\mu - \frac{2a^2}{c^2+a^2}\omega_y v = a^2(\gamma\mu - \beta v).$$

Similarly $\mu = b^2(\alpha v - \gamma\lambda)$ and $v = c^2(\beta\lambda - \alpha\mu)$, ...(5)

$$\text{where} \quad \alpha = \frac{2\omega_x}{b^2+c^2},\ \beta = \frac{2\omega_y}{c^2+a^2} \text{ and } \gamma = \frac{2\omega_z}{a^2+b^2}.$$

Multiplying equations (5) by α/a^2, μ/b^2, γ/c^2 and adding, we have

$$\lambda\frac{\alpha}{a^2} + \mu\frac{\beta}{b^2} + v\frac{\gamma}{c^2} = \alpha(\gamma\mu - \beta v) + \beta(\alpha v - \gamma\lambda) + \gamma(\beta\lambda - \alpha\mu)$$

or $\lambda\dfrac{\alpha}{a^2}+\mu\dfrac{\beta}{b^2}+v\dfrac{\gamma}{c^2}=0 \Rightarrow \dfrac{\lambda x}{a^2}+\dfrac{\mu\beta}{b^2}-v\dfrac{v\gamma}{c^2}=\text{const.}$ ...(6)

Again multiplying (5) by λ/a^2, μ/b^2 and v/c^2 and adding, we have

$$\frac{\lambda\lambda}{a^2}+\frac{\mu\mu}{b^2}+\frac{vv}{c^2}=\lambda(\gamma\mu-\beta v)+\mu(\alpha v-\gamma\lambda)+v(\beta\lambda-\alpha\mu)=0.$$

By integrating, we have

$$\frac{\lambda^2}{a^2}+\frac{\mu^2}{b^2}+\frac{v^2}{c^2}=\text{const.} \qquad ...(7)$$

Thus the path of the particle lies on the planes (6) and the ellipsoid (7), so that it is an ellipse.

Again assuming that the relation (5) have the solutions of the form $\lambda = Pe^{ipt}$, $\mu = Qe^{ipt}$ and $v = Re^{ipt}$. Substituting the value of λ, μ, *v in (5)* and eliminating P, Q, R, we have

$$\begin{vmatrix} ip/a^2 & -\gamma & \beta \\ \gamma & ip/b^2 & -\alpha \\ -\beta & \alpha & ip/c^2 \end{vmatrix}=0,$$

where $p = abc\left(\dfrac{\alpha^2}{a^2}+\dfrac{\beta^2}{b^2}+\dfrac{\gamma^2}{c^2}\right)^{1/2}$

Thus every particle of the liquid describes an ellipse relative to the ellipsoid as a particle moving under a law of force varying as the distance from a fixed point.

Periodic time for each particle is (2x/p), where

$$p=2abc\left\{\left(\frac{\omega_x/a}{b^2+c^2}\right)^2+\left(\frac{\omega_y/b}{c^2+a^2}\right)^2+\left(\frac{\omega_z/c}{a^2+b^2}\right)^2\right\}^{1/2}$$

For a sphere, a = b = c, we have

$$p=(\omega_x^2+\omega_y^2+\omega_z^2)^{1/2}.$$

It follows that the period of revolution of the liquid relative to the spherical shell is the same as the period of revolution of the shell which implies that the shell is revolving alone and the liquid is left at rest in space.

MOTION OF AN ELLIPSOID IN AN INFINITE MASS OF LIQUID

Equation to an ellipsoid is

$$\frac{x^2}{a^2}+\frac{y^2}{b^2}+\frac{z^2}{c^2}=1. \qquad ...(1)$$

We know that the potential of a solid homogeneous ellipsoid of unit density at an external point (x, y, z) is

$$V = \pi abc \int_{\lambda}^{\infty} \left(1 - \frac{x^2}{a^2+u} - \frac{y^2}{b^2+u} - \frac{z^2}{c^2+u}\right)$$

$$\times \frac{du}{(a^2+u)^{1/2}(b^2+u)^{1/2}(c^2+u)^{1/2}}, \qquad ...(2)$$

where λ is the positive root of the equation

$$\frac{x^2}{a^2+\lambda} + \frac{y^2}{b^2+\lambda} + \frac{z^2}{c^2+\lambda} = 1. \qquad ...(3)$$

Assuming $V = \pi(\delta - \alpha x^2 - \beta y^2 - \gamma z^2)$, ...(4)

where $\delta = abc \int_{\lambda}^{\infty} \frac{du}{\Delta}, \alpha = abc \int_{\lambda}^{\infty} \frac{du}{(a^2+u)\Delta},$

$$\beta = abc \int_{\lambda}^{\infty} \frac{du}{(b^2+u)\Delta} \text{ and } \gamma = abc \int_{\lambda}^{\infty} \frac{du}{(c^2+u)\Delta},$$

and $\Delta = (a^2+u)^{1/2}(b^2+u)^{1/2}(c^2+u)^{1/2}$. ...(5)

Let X, Y, Z are the components of attraction at an external point, then

$$X = \frac{\partial V}{\partial x} = -2\pi\alpha x,\ Y = 2\pi\beta y \text{ and } Z = -2\pi\gamma z. \qquad ...(6)$$

Here α, β, γ are the functions of λ or x, y, z. Since V is a solution of Laplace's equation and, therefore, so are X, Y, Z. Consider an ellipsoid moving with velocity U in the direction of X-axis. The boundary condition is

$$-\frac{x}{a^2}\frac{\partial \phi}{\partial x} - \frac{y}{b^2}\frac{\partial \phi}{\partial y} - \frac{z}{c^2}\frac{\partial \phi}{\partial z} = U\frac{x}{a^2} \qquad ...(7)$$

over the ellipsoid $\frac{x}{a^2} + \frac{y}{b^2} + \frac{z}{c^2} = 1$ *i.e.*, $\lambda = 0$.

In order to satisfy the above relation assuming a solution of the form $\phi = AX = A(-2\pi\alpha x)$

or $\frac{\partial \phi}{\partial x} = -2\pi A\left(\frac{\partial \alpha}{\partial \lambda}\frac{\partial \lambda}{\partial x}\right);\ \lambda = 0,\ \frac{\partial \alpha}{\partial \lambda} = -\frac{1}{a^2}.$ ...(8)

Differentiating (3) with regard to x, we have

$$\frac{2x}{a^2+\lambda} - \frac{\partial \lambda}{\partial x}\left\{\frac{x^2}{(a^2+\lambda)^2} + \frac{y^2}{(b^2+\lambda)^2} + \frac{z^2}{(c^2+\lambda)^2}\right\} = 0$$

or $\frac{\partial \lambda}{\partial x}=\frac{2p^2 x}{a^2+\lambda}; \frac{1}{p^2}=\frac{x^2}{(a^2+\lambda)^2}+\frac{y^2}{(b^2+\lambda)^2}+\frac{z^2}{(c^2+\lambda)^2}$.

Similarly $\frac{\partial \lambda}{\partial y}=\frac{2p^2 x}{b^2+\lambda}$ and $\frac{\partial \lambda}{\partial z}=\frac{2p^2 z}{c^2+\lambda}$.

Substituting the value of $\partial\lambda/\partial x$ and $\partial\alpha/\partial\lambda$ in (8), we have

$$\frac{\partial \phi}{\partial x}=-2\pi A\left\{\alpha-\frac{2p^2x^2}{a^2(a^2+\lambda)}\right\}=-2\pi A\left\{\alpha_0-\frac{2p^2x^2}{a^4}\right\}.$$

Similarly $\frac{\partial \phi}{\partial y}=-2\pi A\left(-\frac{2p^2xy}{a^2b^2}\right), \frac{\partial \phi}{\partial z}=-2\pi A\left(-\frac{2p^2xz}{a^2c^2}\right)$.

Substituting the values of $\partial\phi / \partial x$, $\partial\phi / \partial y$ and $\partial\phi / \partial z$ in (7), we have

$$-\frac{x}{a^2}\left\{-2\pi A\left(\alpha_0-\frac{2\pi^2x^2}{a^4}\right)\right\}-\frac{y}{b^2}\left\{-2\pi A\left(-\frac{2p^2xy}{a^2b^2}\right)\right\}$$

$$-\frac{z}{c^2}\left\{-2\pi A\left(-\frac{2p^2xy}{a^2c^2}\right)\right\}=U\frac{x}{a^2}$$

or $2\pi A\left\{\frac{\alpha_0 x}{a^2}+\frac{2p^2x}{a^2}\left(\frac{x^2}{a^4}+\frac{y^2}{b^4}+\frac{z^2}{c^4}\right)\right\}=\frac{Ux}{a^2}$

or $2\pi A\left\{\frac{\alpha_0 x}{a^2}+\frac{2p^2x}{a^2}\frac{1}{p^2}\right\}=\frac{Ux}{a^2}$

or $2\pi A(\alpha_0-2)(x/a^2)=(Ux/a^2)\Rightarrow A=U/\{2\pi(\alpha_0-2)\}$.

Thus $\phi=-2\pi\alpha.\frac{U}{2\pi(\alpha_0-2)}x=\frac{U\alpha x}{2-\alpha_0}$,

which gives the velocity potential of the liquid motion.

Hence if the ellipsoid has a velocity of which U, V, W are the components parallel to the axes; then the velocity potential ϕ becomes

$$\phi=\frac{U\alpha x}{2-\alpha_0}+\frac{V\beta y}{2-\beta_0}+\frac{W\gamma z}{2-\gamma_0}. \qquad \textbf{Ans.}$$

STOKES' STREAM FUNCTION

Let P be an arbitrary point and A be a fixed point on the axis of symmetry. Join the arbitrary point P (x, ω) to the fixed point A by curves AT_1P, AT_2P both lying in the meridian plane. Let the meridian curves AT_1P, AT_2P are rotated about the axis of symmetry then a closed surface will be formed into which as much liquid flows from right to left across the surface generated by AT_2P as flows out in the same time across the surface generated by AT_1P. The surface is free form the basic singularities i.e., sources or sinks.

Let $2\pi\psi$ be the flux across either of these surfaces then the function ψ is called the *Stoke's steam function. The* value of the steam function ψ remains unaltered if AT_1P is taken fixed and replacing AT_2P by any other meridian curve. The stream function ψ depends on the position of P, and perhaps on the fixed point A. Consider another fixed point B on the axis of symmetry and join the points by the meridian curves BT_3P and BT_4P. The flux across the surface generated by BT_3P will be the same as that across AT_1P. There is no flow across AB. It follows that the value of the stream function ψ does not depend on the particular fixed point, provided that this lies on the axis.

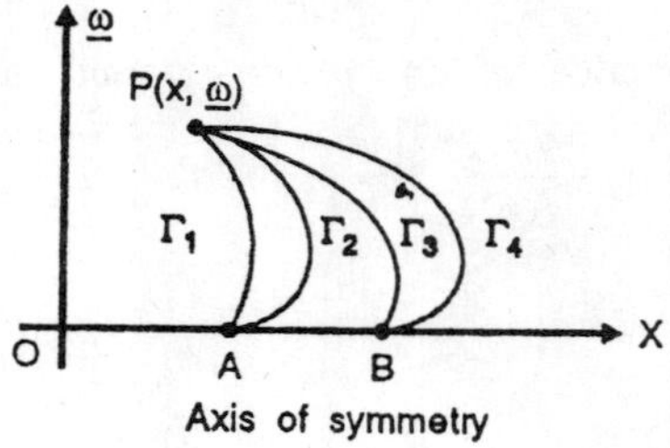

Fig. 3.11

Hence the value of the stream function at P depends solely on the position of P. When the point P is on the axis then $\psi = 0$.

In the case of axi-symmetric flows, the velocity components with regard to cylindrical coordinates are all independent of the azimuthal angle ϕ. Such a motion occurs, for example, in uniform flow past a stationary sphere, sphere moving with uniform velocity in a fluid at rest. This type of motion present some analogies with the two-dimensional case, in particular, a stream function can be defined. It represents the flux across any surface of revolution about the axis of symmetry. Consider a surface of revolution generated by rotating any curve joining a point P (r, θ) to any point A on the axis of symmetry. The flux of fluid from left to right across this surface will independent of the position of A.

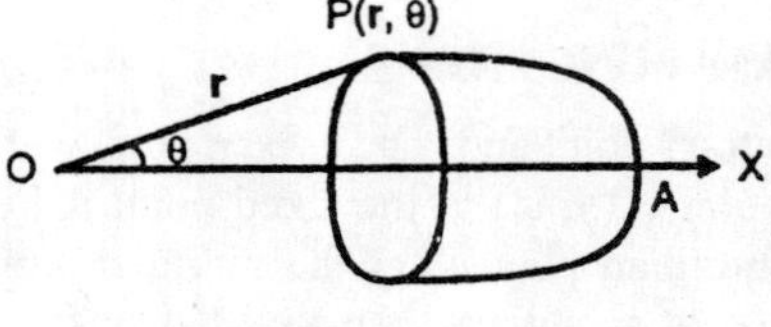

Fig. 3.12

Let a stream function ψ be defined such that if the position vector OP is rotated around the axis of symmetry-X-axis *i.e.,* ω is varied through 2π while r and θ are held fixed, the quantity of fluid which crosses the surface of revolution formed by the vector OP will be 2πψ. Let ψ + δψ be the stream function at a neighbouring point P'. If the line element PP' is rotated about the axis of symmetry the resulting surface will huge a quantity of fluid [2π (ψ + δψ) - 2πψ] (= 2πδψ) crossing it per unit time. Also, a quantity of fluid (q_θ dr - q_r rdθ) crosses a unit area of this surface so that

$$2\pi d\psi = 2\pi r\sin\theta(q_\theta dr - q_r r d\theta) \qquad ...(1)$$

$$d\psi = q_\theta r\sin\theta \; dr - q_r r^2 \sin\theta \; d\theta$$

Since ψ is a function of both r and θ, so

$$d\psi = (\partial\psi/\partial r)\; dr + (\partial\psi/\partial\theta)d\theta \qquad ...(2)$$

Comparing (1) and (2) for dψ, we have

$$q_r = -\frac{1}{r^2\sin\theta}\frac{\partial\psi}{\partial\theta} \text{ and } q_\theta = \frac{1}{r\sin\theta}\frac{\partial\psi}{\partial r}. \qquad ...(3)$$

The steam function ω defined, in this manner, is known as the *Stokes' stream function* in spherical polar coordinates.

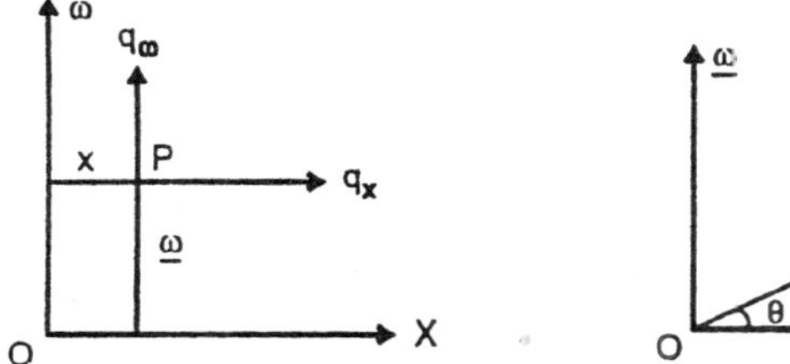

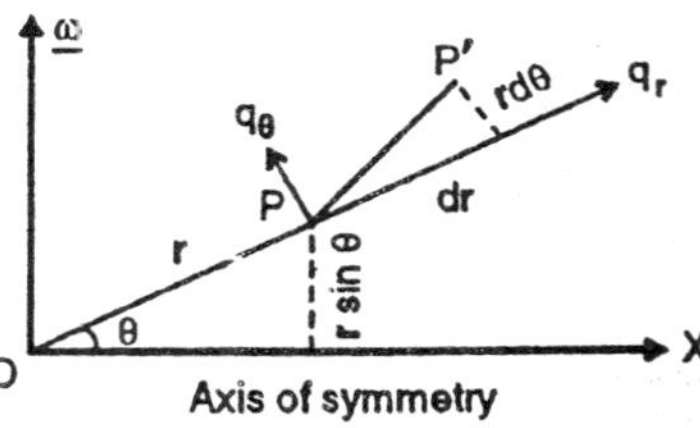

Fig. 3.13

Similarly, the velocity components in cylindrical polar coordinate are

$$q_x = -\frac{1}{\underline{\omega}}\frac{\partial\psi}{\partial\underline{\omega}},\; q_{\underline{\omega}} = \frac{1}{\underline{\omega}}\frac{\partial\psi}{\partial x}$$

The streamlines are given by ψ = const. for across such a line there is no flow. The dimensions of ψ are L^3T^{-1} and the dimensions of the velocity potential ϕ are L^2T^{-1}.

When the motion is irrotational a velocity potential ϕ of course always exist *i.e.,*

$$q_r = -(\partial\phi/\partial r),\; q_\theta = -(\partial\phi/r\,\partial\theta) \qquad ...(4)$$

From (3) and (4), we have

$$\frac{1}{r^2\sin\theta}\frac{\partial\psi}{\partial\theta} = \frac{\partial\phi}{\partial r} \text{ and } \frac{1}{r\sin\theta}\frac{\partial\psi}{\partial r} = -\frac{1}{r}\frac{\partial\phi}{\partial\theta}$$

Hence $\dfrac{\partial}{\partial\theta}\left(\dfrac{1}{r^2\sin\theta}\dfrac{\partial\psi}{\partial r}\right)=-\dfrac{\partial}{\partial r}\left(\dfrac{1}{\sin\theta}\dfrac{\partial\psi}{\partial r}\right)$

or $r^2\dfrac{\partial^2\psi}{\partial r^2}+\sin\theta\dfrac{\partial}{\partial\theta}\left(\dfrac{1}{\sin\theta}\dfrac{\partial\psi}{\partial\theta}\right)=0$

or $r^2\dfrac{\partial^2\psi}{\partial r^2}+(1-\mu^2)\dfrac{\partial^2\psi}{\partial\mu^2}=0,\ \mu=\cos\theta$ ...(5)

From the continuity equation, in spherical polar coordinates, we have

$$\frac{\partial}{\partial r}\left(r^2\frac{\partial\phi}{\partial r}\right)+\frac{1}{\sin\theta}\frac{\partial}{\partial\theta}\left(\sin\theta\frac{\partial\phi}{\partial\theta}\right)=0 \qquad ...(6)$$

or $\dfrac{\partial}{\partial r}\left(r^2\dfrac{\partial\phi}{\partial r}\right)+\dfrac{\partial}{\partial\mu}\left\{(1-\mu^2)\dfrac{\partial\phi}{\partial\mu}\right\}=0$ ...(7)

which is the Laplace's equation and has solution of the form

$$r^nP_n(\mu)=\phi_1 \text{ and } r^{-n-1}P_n(\mu)=\phi_2$$

Let ψ_1 and ψ_2 are the solution of equation (6) then from (5), we have

$$\frac{\partial\psi_1}{\partial\mu}=-r^2\frac{\partial\phi_1}{\partial r}=-r^2\frac{\partial}{\partial r}\{r^nP_n(\mu)\}=-nr^{n+1}P_n(\mu),$$

and $\dfrac{\partial\psi_2}{\partial\mu}=-r^2\dfrac{\partial\phi_2}{\partial r}=-r^2\dfrac{\partial}{\partial r}\{r^{-n-1}P_n(\mu)\}=(n+1)r^{-n}P_n(\mu)$

Also $\dfrac{\partial\psi_1}{\partial\mu}=(1-\mu^2)\dfrac{\partial\phi_1}{\partial\mu}=(1-\mu^2)r^n\dfrac{\partial P_n}{\partial\mu},$ ...(8)

and $\dfrac{\partial\psi_2}{\partial r}=(1-\mu^2)\dfrac{\partial\phi_2}{\partial\mu}=(1-\mu^2)r^{-n-1}\dfrac{\partial P_n}{\partial\mu}.$ ...(9)

By integrating (8) and (9), we have

$$\psi_1=\frac{1-\mu^2}{n+1}r^{n+1}\frac{\partial P_n}{\partial\mu},\ \psi_2=\frac{1-\mu^2}{n}\frac{1}{r^n}\frac{\partial P_n}{\partial\mu},$$

which are the possible solution for the stream function ψ.

VALUES OF STOKE'S STREAM FUNCTION

I. A simple source on the X-axis

The velocity potential for a simple source of strength + m at the origin is $\phi = m / r$. The outward flux across a sphere of radius r whose centre is at the source, is related to the radial velocity q_r by the relation

$$4\pi m=4\pi r^2q_r \Rightarrow q_r=m/r^2 \text{ and } q_\theta=0 \qquad ...(1)$$

we know that

$$q_\theta = \frac{1}{r\sin\theta}\frac{\partial\psi}{\partial r},\ q_r = -\frac{1}{r^2\sin\theta}\frac{\partial\psi}{\partial\theta}. \quad ...(2)$$

From (1) and (2), it follows that

$$\partial\psi/\partial r = 0 \text{ and } \partial\psi/\partial r = -m\sin\theta$$

$$\Rightarrow \psi(r,\theta) = m\cos\theta = mx/r.$$

II. A uniform line source along the axis

Consider a uniform line source of fluid extending along OX from B to A. Let RR' be a small element of length δx, RP = r and ∠PRA = θ. If + m is the strength per unit length of the line source then the element RR' is effectively a simple source at R of strength + m δx. Hence the contribution to ψ at P (ξ, n) from this source is δψ where

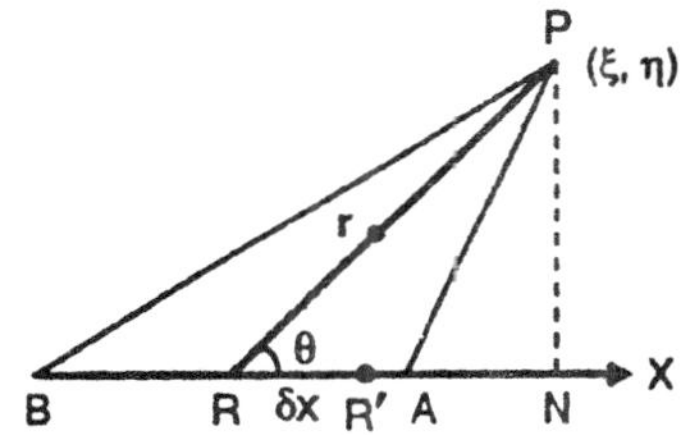

Fig. 3.14

$$\delta\psi = m\ \delta x\cos\theta$$

$$\Rightarrow \psi = \int_0^{BA} m\cos\theta\ dx = \int_0^{BA} \frac{m(\xi - x)}{\sqrt{[(\xi - x)^2 + \eta^2]}}dx$$

$$\Rightarrow \psi = m[(\xi^2 - \eta^2)^{1/2} - \left\{(\xi - BA)^2 + \eta^2\right\}^{1/2}] = m(BP - AP),$$

which shows that the stream surfaces ψ = const., are given by

BP - AP = const.

These are confocal hyperboloids of revolution about BX, having B and A as focii.

III. A doublet along the axis

The velocity potential ϕ for the flow due to a doublet at P is given by

$$\phi = \frac{\mu\cos\theta}{r^2} \quad ...(1)$$

where μ is the strength of the doublet.

The stream function corresponding to equation (1) is obtained as

$$q_r = \frac{\partial \phi}{\partial r} = \frac{2\mu\cos\theta}{r^3} = -\frac{1}{r^2\sin\theta}\frac{\partial \psi}{\partial \theta}$$

$$\Rightarrow \frac{\partial \psi}{\partial \theta} = -\frac{2\mu\sin\theta\cos\theta}{r} \Rightarrow \psi(r,\theta) = -\frac{\mu\sin^2\theta}{r^3} + f(r) \qquad ...(2)$$

Likewise, the two expressions for q_θ give

$$q_\theta = -\frac{1}{r}\frac{\partial \phi}{\partial \theta} = \frac{\mu\sin\theta}{r^3} = \frac{1}{r\sin\theta}\frac{\partial \psi}{\partial r}$$

$$\Rightarrow \frac{\partial \psi}{\partial r} = \frac{\mu\sin^2\theta}{r^2} \Rightarrow \psi(r,\theta) = -\frac{\mu\sin^2\theta}{r} + g(\theta). \qquad ...(3)$$

Comparing the expressions (2) and (3), we have

$$f(r) = g(\theta) = 0,$$

and $$\psi(r,\theta) = \frac{\mu\sin^2\theta}{r} \qquad ...(4)$$

This determines the stream function for a doublet which discharges fluid along the negative portion of the reference axis and attracts fluid along the positive part of the reference axis.

SOLID OF REVOLUTION MOVING ALONG THEIR AXES IN AN INFINITE MASS OF LIQUID

Let the body moves along X-axis with velocity V and dS is an element of the meridian curve, Since the motion is symmetrical about the axis, Stokes's stream function ψ so exists. Thus on the surface, we have

$$-\frac{1}{R}\frac{\partial \psi}{\partial S} = V\frac{\partial R}{\partial S}$$

or $= d\psi = -VR\,dR \Rightarrow \psi = -(1/2)VR^2 + A,$

or $\psi = -\frac{1}{2}Vr^2(1-\mu^2) + A; R = r\sin\theta, \mu = \cos\theta.$ ...(1)

In the case of a sphere, at r = a, we have

$$\psi = -\frac{1}{2}Va^2(1-\mu^2) + A. \qquad ...(2)$$

But the stream function ψ has the solution of the form

$$\frac{1-\mu^2}{n+1}r^{n+1}\frac{\partial P_n}{\partial \mu} \text{ and } \frac{1-\mu^2}{rn^n}\frac{\partial P_n}{\partial \mu}.$$

Assuming $\frac{\partial P_n}{\partial \mu} = 1$

$$\Rightarrow P_n(\mu) = \mu \;\therefore\; n = 1.$$

Hence $\psi = \dfrac{B(1-\mu^2)}{r}$, liquid is at rest at infinity

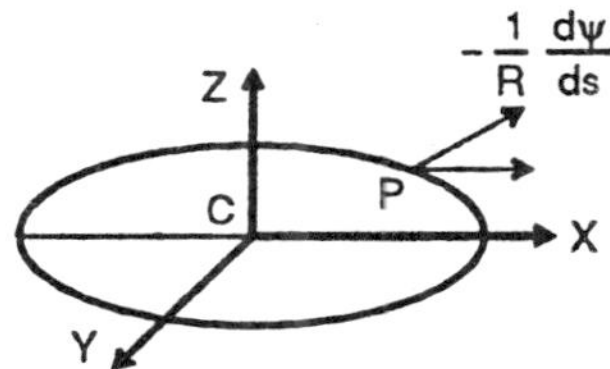

Fig. 3.15

On the surface, we have

$$(B/a)(1-\mu^2) = -\frac{1}{2}Va^2(1-\mu^2) + A.$$

or $B = -\dfrac{1}{2}Va^3,\ A = 0.$

or $\psi = -\dfrac{1}{2}\dfrac{Va^3}{r}(1-\mu^2) = -\dfrac{1}{2}\dfrac{Va^3}{r}\sin^2\theta$

Also $\qquad (1-\mu^2)\dfrac{\partial\phi}{\partial\mu} = \dfrac{\partial\psi}{\partial r} = \dfrac{1}{2}\dfrac{Va^3}{r^2}\sin^2\theta$

or $\dfrac{\partial\phi}{\partial\mu} = \dfrac{1}{2}\dfrac{Va^3}{r^2} \Rightarrow \phi = \dfrac{1}{2}\dfrac{Va^3}{r^2}\cos\theta,$

which is the same velocity potential as obtained, if a sphere is moving in a liquid at rest at infinity.

If the liquid flows past the sphere with velocity V at infinity in the negative X-direction, then

$$\psi = \frac{1}{2}Vr^2\sin^2\theta. \qquad \text{...(5)}$$

Hence the stream function, when the sphere moves in an infinite liquid at rest at infinity, becomes

$$\psi = \frac{1}{2}Vr^2\sin^2\theta - \frac{1}{2}\frac{Va^3}{r^2}\sin^2\theta$$

$$\Rightarrow \psi = \frac{1}{2}Vr^2\left(1-\frac{a^3}{r^3}\right)\sin^2\theta,$$

and $\qquad \phi = Vr\left(1+\dfrac{a^3}{2r^3}\right)\cos\theta.$

SOLVED EXAMPLES

Example 1: *A solid sphere moves through quiescent frictionless liquid whose boundaries are at a distance from it great compared with its radius. Prove that at each instant the motion in the liquid depends only on the position and velocity of the sphere at that instant. Prove that the liquid streams past the sides of the sphere with half the velocity of the sphere.*

Solution: Let O be the centre of the sphere at any instant of time t and U be the velocity of the sphere at that instant along the initial line OX. Since the motion is irrotational, the velocity potential ϕ exists at any point P (r, θ) in the fluid such that $\nabla^2\phi = 0$. The boundary conditions are

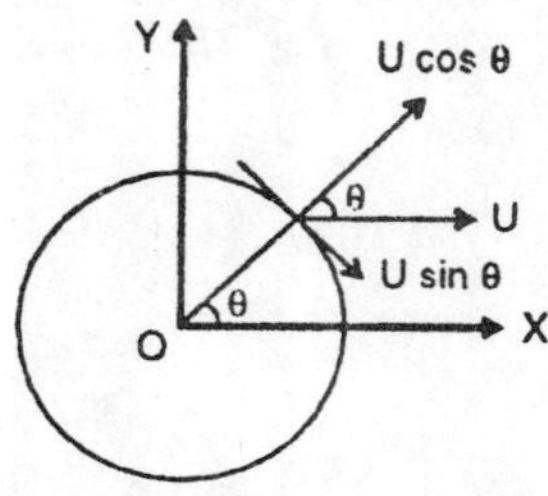

Fig. 3.16

(i) $(-\,\partial\phi/\partial r\,)_{r\,=\,a} = U\cos\theta$ and

(ii) $(-\,\partial\phi/\partial r\,)_{r\,\to\infty} = 0.$

Assuming the suitable form of the velocity potential as

$$\phi = \left(Ar + \frac{B}{r^2}\right)\cos\theta \Rightarrow \frac{\partial\phi}{\partial r} = \left(A - \frac{2B}{r^3}\right)\cos\theta. \qquad \text{...(1, 2)}$$

From the equation (2) and the condition (ii) we have

$$A = \cos\theta = 0 \Rightarrow A = 0,\ \cos\theta \neq 0. \qquad \text{...(3)}$$

Again b using the condition (i), we get

$$-\,(\,2B/a2^2)\cos\theta = -\,U\cos\theta \Rightarrow B = \frac{1}{2}a^3\,U \text{ for all value of } \theta \quad \text{...(4)}$$

Substituting the value of A and B in (1), we have

$$\phi = \frac{1}{2}(a^3U/r^3)\cos\theta. \qquad \text{...(5)}$$

Thus at each instant the motion of the liquid depends only on the position and velocity of the sphere at an instant.

The velocity with which the liquid streams past the sides of the sphere is given by

$$= \left(-\frac{1}{r}\frac{\partial \phi}{\partial \theta}\right)_{r=a} = \left(\frac{1}{2}\frac{a^3 U}{r^3}\sin\theta\right)_{r=a} = \frac{1}{2}U\sin\theta.$$

$\frac{1}{2}\times$ velocity of the sphere along the tangent. **Proved.**

Example 2: *An infinite ocean of an incompressible perfect liquid of density ρ is streaming past a fixed spherical obstacle of radius a. The velocity is uniform and equal to U except in so far as its disturbed by the sphere, and the pressure in the liquid at a great distance from the obstacle is Π. Show that the thrust on that half of the sphere on which the liquid impinges is*

$\pi a^2 \{\Pi - (1/16)\rho U^2\}$.

Solution: The velocity potential of the motion of fluid streaming past the fixed sphere with velocity U in the negative direction of the X-axis (initial line) is given by

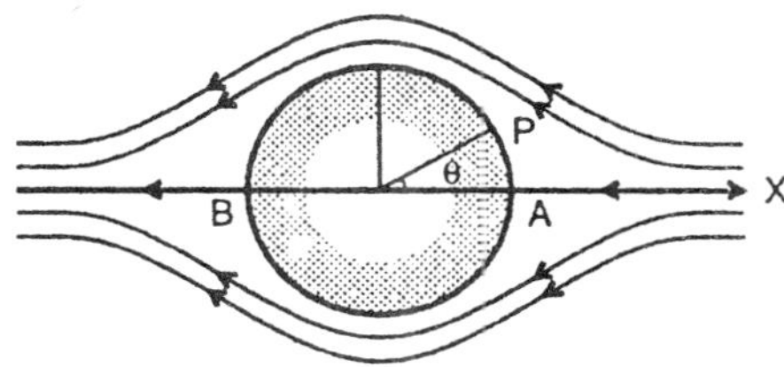

Fig. 3.17

$$\phi = U\left(r + \frac{1}{2}\frac{a^3}{r^2}\right)\cos\theta \qquad ...(1)$$

Let q be the velocity at any point on the boundary of the sphere then

$$q^2 = \left[\left(\frac{\partial \phi}{\partial r}\right)^2 + \left(\frac{1}{r}\frac{\partial \phi}{\partial \theta}\right)^2\right]_{r=a} = \frac{9}{4}U^2\sin\theta. \qquad ...(2)$$

Now for the pressure p at any point, we have

$$p = \Pi, q = U \text{ as } r \to \infty$$

$$\frac{p}{\rho} + \frac{1}{2}q^2 = C = \frac{\Pi}{\rho} + \frac{1}{2}U^2,$$

$$\Rightarrow C = \frac{\Pi}{\rho} + \frac{1}{2}U^2. \qquad ...(3)$$

$$\text{or } \frac{p}{p} = \frac{\Pi}{\rho} + \frac{1}{2}U^2 - \frac{1}{2}q^2$$

or $\dfrac{p}{p}=\dfrac{\Pi}{\rho}+\dfrac{1}{2}U^2-\dfrac{9}{8}U^2\sin\theta,$ ...(4)

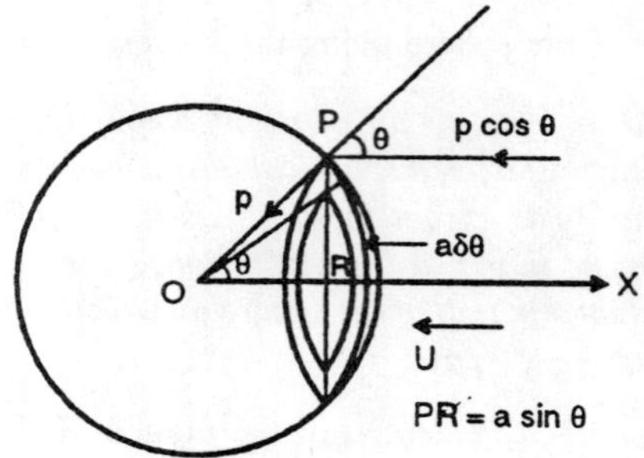

Fig. 3.18

which determines the pressure at any point of the sphere.

The thrust on half of the sphere on which the liquid impinges along the initial line is given by

$$=\int_0^{\pi/2} p\cos\theta\,(2\pi a\sin\theta)a\,d\theta$$

$$=2a^2\pi\rho\int_0^{\pi/2}\left\{(\Pi/\rho)+\frac{1}{2}U^2-(9/8)U^2\sin^2\theta\right\}\sin\theta\cos\theta\,d\theta$$

$$=2a^2\pi\rho\left[\frac{1}{2}\left(\frac{\Pi}{\rho}+\frac{1}{2}U^2\right)-\frac{9U^2}{8}\frac{1}{4}\right]$$

$$=\pi a^2\left[\Pi-(1/16)\rho U^2\right].$$ **proved.**

Example 3: *A stream of water of great depth is flowing with uniform velocity U over a plane level bottom. A hemi-sphere of weight w in water and of radius a, rests with its base on the bottom. Prove that the average pressure between the base of hemisphere and the bottom is less than the fluid pressure at any point of the bottom at a great distance from the hemisphere, if*

U2 > $32w/(11\rho\pi a^2)$.

Solution: Consider the centre O of the hemisphere as origin and (r, θ, ϕ) be the coordinates of a point P referred to the axes through the origin. The velocity potential ϕ of the motion of fluid streaming past the fixed sphere with velocity U is given by

$$\phi=U\left\{r+\frac{1}{2}\frac{a^3}{r^2}\right\}\cos\theta,$$

where x = r sin θ cos ϕ, y = r sin θ sin ϕ, z = r cos θ.

Let q and p be the velocity and pressure at the point P (r, θ, ϕ) on the sphere then

$$q^2 = \frac{9}{4}U^2 \sin^2\theta, \frac{p}{\rho} = \frac{\Pi}{\rho} + \frac{1}{2}U^2 - \frac{9}{8}U^2 \sin^2\theta.$$

The total thrust on the hemisphere due to liquid motion along the X-axis is given by

$$= \int_{\theta=0}^{\pi} \int_{\phi=-\pi/2}^{\pi/2} \left[\rho \left\{ \frac{\Pi}{\rho} + \frac{1}{2}U^2 - \frac{9}{8}U^2 \sin^2\theta \right\} \sin^2\theta \cos\phi \right]$$

$.a \sin\theta \, d\phi \, .ad\theta$

where $p \sin\theta \cos\phi$ is the component of the pressure along X-axis.

$$= a^2 \rho \int_{\theta=0}^{\pi} \int_{\phi=-\pi/2}^{\pi/2} \left(\frac{\Pi}{\rho} + \frac{1}{2}U^2 - \frac{9}{8}U^2 \sin^2\theta \right) . \sin^2\theta \cos\phi \, d\theta \, d\phi$$

$$= 2a^2 \rho \int_{\theta=0}^{\pi} \left(\frac{\Pi}{\rho} + \frac{1}{2}U^2 - \frac{9}{8}U^2 \sin^2\theta \right) \sin^2\theta \, d\theta$$

$$= 2a^2 \rho \left[\left(\frac{\Pi}{\rho} + \frac{1}{2}U^2 \right) \frac{\pi}{2} - \frac{9}{8}U^2 . \frac{3\pi}{8} \right]$$

$$= \pi a^2 [\Pi(11/32)\rho / U^2]$$

Thus the total pressure on the base

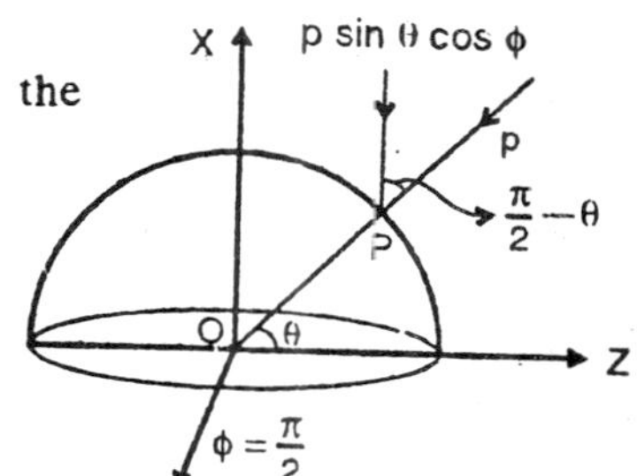

Fig. 3.19

$$= \pi a^2 \left[\Pi - \frac{11}{32} \rho^2 U \right] + w.$$

Average pressure on the base

$$= \frac{\text{Pressure on the base}}{\text{Area if tge base}}$$

$$= \Pi - \frac{11}{32} \rho U^2 + \frac{w}{\pi a^2}.$$

Since Average Pressure < Pressure at great distance

or Π - (11/32) ρU^2 + $(w/\pi a^2)$ < Π

or $U^2 > 32w/(11\rho\pi a^2)$. **proved.**

Example 4: *An infinite homogeneous liquid is flowing steadily past a rigid boundary consisting partly of the horizontal plane y = 0 and partly of a hemispherical boss $x^2 + y^2 + x^2 = a^2$, with irrotational motion which tends, at a great distance from the origin to uniform velocity U parallel to the axis of Z. Find the velocity potential and the surface of equal pressure.*

Solution: The velocity potential of the motion of a liquid steaming past a fixed sphere with velocity U in the negative direction of Z-axis is given by

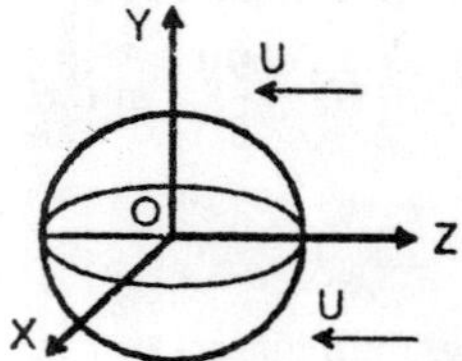

Fig. 3.20

$$\phi = U\left\{r + \frac{a^3}{2r^2}\right\}\cos\theta$$

The velocity perpendicular to the plane y = 0 (*i.e.xz*-plane) vanishes. The plane y = 0 may be taken as a steam surface *i.e.*, the hemisphere above y = 0 is also a steam surface.

The pressure at any point is determined by Bernoulli's theorem

$$\frac{p}{\rho} + \frac{1}{2}q^2 = \text{const.} \qquad ...(1)$$

The surfaces of equal pressure are given by p = const.

$$\Rightarrow q^2 = \text{const.} = \left(-\frac{\partial\phi}{\partial r}\right)^2 + \left(-\frac{1}{r}\frac{\partial\phi}{\partial\theta}\right)^2 = \text{const.}$$

$$\Rightarrow \left[U\left(1 - \frac{a^3}{r^3}\right)\cos\theta\right]^2 + \left[\frac{U}{r}\left(r - \frac{a^3}{2r^2}\right)\sin\theta\right]^2 = \text{const.}$$

$$\Rightarrow \left(1 - \frac{a^3}{r^3}\right)^2 \cos^2\theta + \left(1 + \frac{a^3}{2r^3}\right)\sin^2\theta = \text{const.} \qquad \textbf{Ans.}$$

Example 5: *Liquid of density ρ fills the space between a solid sphere of radius a and density ρ' and a fixed concentric spherical envelope of radius b, prove that the work done by an impulse which starts the solid sphere with velocity U is*

$$\frac{1}{3}\pi a^3 U^2\left\{2\rho'+\frac{2a^3+b^3}{a^3-b^3}\rho\right\}.$$

Solution: We know that the impulse I necessary to start the solid sphere with velocity U is given by

$$I - MU + \frac{1}{2}M'U\frac{2a^3+b^3}{b^3-a^3},$$

where M = mass of the inner sphere $=\frac{4}{3}\pi a^3\rho'$,

and M = mass of the liquid displaced $=\frac{4}{3}\pi a^3\rho$.

Thus we have

$$I=\frac{2\pi a^3 U}{3}\left\{2\rho'+\rho\left(\frac{2a^3+b^3}{a^3-b^3}\right)\right\}.$$

Work done by the impulse

= Impulse × Mean of initial and final velocities.

$$= I\times\frac{0+U}{2}=\frac{1}{2}IU=\frac{1}{3}\pi a^3U^2\left\{2\rho'+\frac{2a^3+b^3}{a^3-b^3}\rho\right\}.$$ **Proved.**

Example 6: *The space between two concentric spherical shells of radii a and b (a > b) is filled with on incompressible fluid of density ρ and the shells suddenly begin to move with velocities U, V in the same direction; prove that the resultant impulsive pressure on the inner shell is*

$$\frac{2\pi\rho b^3}{3(a^3-b^3)}\{3a^3U-(a^3+2b^3)V\}.$$

Solution: The velocity potential at an instant of start is given by

$$\phi=-\frac{1}{a^3-b^3}\left\{(Ua^3-Vb^3)r+\frac{(U-V)a^3b^3}{2r^2}\right\}\cos\theta$$

The resultant impulsive pressure on the surface of the inner spherical shell is given by

$$=\int_0^\pi(-\underline{\omega}\cos\theta).2\pi b\sin\theta.b\,d\theta,$$

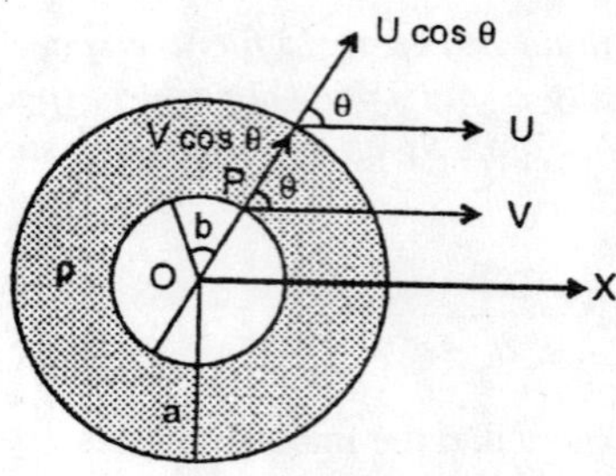

Fig. 3.21

where $\underline{\omega} = -\rho\phi, \phi$ is the velocity potential on the surface of the inner shell (*i.e.*, r = b).

$$= -\frac{2\pi\rho b^3}{a^3 - b^3}\left\{(Ua^3 - Vb^3) + \frac{1}{2}(U - V)a^3\right\}\int_0^{\pi} \cos^2\theta \sin\theta \, d\theta$$

$$= -\frac{2\pi\rho b^3}{3(a^3 - b^3)}\left\{3Ua^3 - (a^3 + 2b^3)V\right\},$$

negative sign is admissible. **Proved.**

Example 7: *Prove that for liquid contained between two instantaneously concentric spheres, when the outer (radius a) is moving parallel to the axis of X with velocity U and the inner (radius b) is moving parallel to the axis of Y with velocity V, the velocity potential is*

$$-\frac{1}{a^3 - b^3}\left\{a^3 Ux\left(1 + \frac{b^3}{2r^3}\right) - b^3 Vy\left(1 + \frac{a^3}{2r^3}\right)\right\}.$$

and find the kinetic energy.

Solution: Since the motion is irrotational consequently the velocity potential ϕ exists such that

$$\nabla^2\phi = 0. \qquad ...(1)$$

The boundary conditions are

(i) $(-\,\partial\phi/\partial r)_{r=a} = U \cos\theta$,

and (ii) $(-\,\partial\phi/\partial r)_{r=b} = V \sin\theta$.

Assuming the velocity potential ϕ be of the form

$$\phi = \left(Ar + \frac{B}{r^2}\right)\cos\theta + \left(Cr + \frac{D}{r^2}\right)\sin\theta, \qquad ...(2)$$

or $$\frac{\partial\phi}{\partial r} = \left(A + \frac{2B}{r^3}\right)\cos\theta + \left(C - \frac{2D}{r^3}\right)\sin\theta. \qquad ...(3)$$

From the boundary conditions (i), (ii) and (3), we obtain

$$-U\cos\theta = \left(A - \frac{2B}{a^3}\right)\cos\theta + \left(C - \frac{2D}{a^3}\right)\sin\theta,$$

and $$-V\cos\theta = \left(A - \frac{2B}{b^3}\right)\cos\theta + \left(C - \frac{2D}{b^3}\right)\sin\theta. \qquad ...(4,\ 5)$$

These relation being true for all values of θ, such that

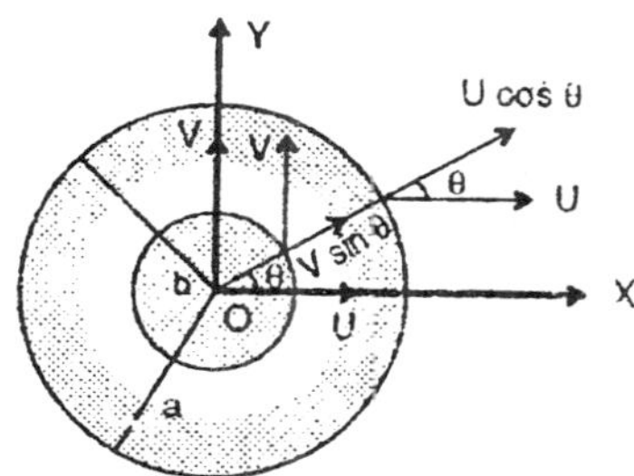

Fig. 3.22

$$A - \frac{2B}{a^3} = -U \text{ and } A - \frac{2B}{b^3} = 0,$$

$$C - \frac{2D}{a^3} = 0 \text{ and } C - \frac{2D}{b^3} = V.$$

$$\Rightarrow A = -\frac{Ua^3}{a^3 - b^3}, \qquad B = -\frac{Ua^3b^3}{2(a^3 - b^3)},$$

and $$C = \frac{Vb^3}{a^3 - b^3}, \qquad D = \frac{Va^3b^3}{2(a^3 - b^3)}.$$

Substituting the value of the constants in (2), the velocity potential ϕ becomes

$$\phi = -\frac{Ua^3}{a^3 - b^3}\left(r + \frac{b^3}{2r^2}\right)\cos\theta + \frac{Vb^3}{a^3 - b^3}\left(r + \frac{a^3}{2r^2}\right)\sin\theta$$

or $$\phi = -\frac{1}{a^3 - b^3}\left\{a^3Ux\left(1 + \frac{b^3}{2r^2}\right) - b^3Vy\left(1 + \frac{a^3}{2r^3}\right)\right\}.$$ **Proved.**

Let T be the kinetic energy of the liquid moving irrotationally at an instant, then

$$T = -\frac{1}{2}\rho \iint \phi(\partial\phi / \partial n)\, dS,$$

where the surface integral is carried over the whole boundary of the liquid and ($-\partial\phi/\partial n$) denotes the outwards normal velocity.

or $T = -\frac{1}{2}\rho\iint_{r=a}\left(\phi\frac{\partial\phi}{\partial n}\right)dS - \frac{1}{2}\rho\iint_{r=b}\left(\phi\frac{\partial\phi}{\partial n}\right)dS$

Since $\quad (-\partial\phi/\partial n)_{r=a} = -U\cos\theta, (-\partial\phi/\partial n)_{r=b} = V\sin\theta$

or $T = \frac{1}{2}\frac{\rho U}{2(a^3-b^3)}\iint_{r=a}\left[a^3U.\left(1+\frac{b^3}{2a^3}\right) - \frac{3}{2}b^3Vy\right]\cos\theta\, dS$

$$-\frac{1}{2}\frac{\rho V}{(a^3-b^3)}\iint_{r=b}\left[\frac{3}{2}a^3Ux - b^3Vx\left(1+\frac{a^3}{2b^3}\right)\right]\sin\theta\, dS$$

or $T = \frac{1}{4}\rho\frac{U^2(2a^3-b^3)}{a(a^3-b^3)}\iint_{r=a}x^2 dS - \frac{3}{4}\rho\frac{UVb^3}{a(a^3-b^3)}\iint_{r=a}xy\, dS$

$$-\frac{3}{4}\rho\frac{UVa^3}{b(a^3-b^3)}\iint_{r=b}xy\, dS + \frac{1}{4}\rho\frac{V^2}{b}\frac{(a^3+2b^3)}{(a^3-b^3)}\iint_{r=b}y^2\, dS$$

Since $\quad \iint_{r=a}x^2 dS = \frac{1}{2}\iint_{r=a}y^2 dS = \frac{1}{2}.(2Ma^2/3) = \frac{4\pi a^4}{3}$

and $\quad \iint_{r=a}xy\, dS = 0$ (Product of inertia)

or $T = \frac{1}{4}\rho\frac{U^2(2a^3+b^3)}{a(a^3-b^3)}.\frac{4}{3}\pi a^4 + \frac{1}{4}\rho\frac{V^2(2a^3+b^3)}{b(a^3-b^3)}.\frac{4}{3}\pi b^4.$

or $T = \frac{1}{3}.\frac{\pi\rho}{(a^3-b^3)}\{2(U^2a^6+V^2b^6)+a^3b^3(U^2+V^2)\}.$ **Ans.**

Example 8: *Incompressible fluid, of density ρ, is contained between two rigid concentric spherical surfaces, the outer one of mass M_1 and radius a; the inner on of mass M_2 and radius b. A normal blow P is given to the outer surface. Prove that the initial velocities of the two containing surfaces (U for the outer and V for the inner) are given by the equation.*

$$\left[M_1 + \frac{2\pi\rho a^3(2a^3+b^3)}{3(a^3-b^3)}\right]U - \frac{2\pi\rho a^3b^3)}{a^3-b^3}V = P,$$

and $$\left[M_2 + \frac{2\pi\rho b^3(2a^3+b^3)}{3(a^3-b^3)}\right]V = \frac{2\pi\rho a^3b^3)}{a^3-b^3}U.$$

Solution: We know that

$$\phi = \frac{1}{a^3-b^3}\left[(Vb^3-Ua^3)r + \frac{(V-U)a^3b^3}{2r^2}\right]\cos\theta. \qquad ...(1)$$

Let P be the normal blow given to the outer sphere

$$M_1U = P - \iint_{r=b}\underline{\omega}\cos\theta\, dS,$$

$$M_2 V = -\iint_{r=b} \underline{\omega} \cos\theta \, dS, \qquad ...(2, 3)$$

where M_1 and M_2 be the masses of outer and inner sphere and ω (= $-\rho\phi$) is an impulsive pressure over an element δs of the surface and ϕ is the velocity potential just before an impulsive action From (1) and (2), we have

$$M_1 U = P - \iint_{r=a} \underline{\rho\phi} \cos\theta \, dS$$

or $$M_1 U = P + \frac{\rho}{a^3 - b^3} \int_0^x \left[(Vb^3 - Ua^3) r + \frac{(V - Ua^3 b^3}{2r^2} \right]_{r=a}$$

$$\cos\theta . \cos\theta . 2\pi a \ \sin\theta \ a d\theta$$

or $$M_1 U = P + \frac{\pi\rho a^3}{(a^3 - b^3)} [3Vb^3 - U(2a^3 + b^3)] \int_0^x \cos^2\theta \sin\theta d\theta$$

or $$\left[M_1 + \frac{2\pi\rho a^3 (2a^3 - b^3)}{3(a^3 - b^3)} \right] U - \frac{2\pi\rho a^3 b^3)}{a^3 - b^3} V = P,$$

which proves the first result.

Again from (1) and (3), we have

$$M_2 V = -\iint_{r=b} \rho\phi \cos\theta \, dS$$

or $$M_2 V - \frac{2\pi b^3 \rho}{a^3 - b^3} \left[(Vb^3 - Ua^3) + \frac{1}{2}(V - U) a^3 \right] \int_0^x \cos^2\theta \sin\theta . d\theta.$$

or $$\left[M_2 + \frac{2\pi\rho b^3 (2b^3 - a^3)}{3(a^3 - b^3)} \right] V = \frac{2\pi\rho a^3 b^3}{a^3 - b^3} U,, \qquad \textbf{Proved.}$$

which proves the second result

Example 8(a): *Prove that at a point on a sphere moving through an infinite liquid the pressure is given by the formula*

$$\frac{p - p_0}{\rho} = \frac{1}{2} af \cos\theta_1 + \frac{1}{8} v^2 (9\cos^2\theta - 5),$$

where v is the velocity, f the acceleration of the sphere, and θ, θ_1 are the angles between the radius and the direction of v, f respectively and p_0 is the hydrostatic pressure.

Solution: We know that

$$q^2 = \frac{a^6}{4r^6} v^2 + \frac{3a^6}{4r^6} v^2 \cos^2\theta,$$

and $$\frac{\partial\phi}{\partial t} = \frac{1}{2} \frac{a^3}{r^2} f \cos\theta_1 - \frac{1}{2} \frac{a^3}{r^3} v^2 + \frac{3a^3}{2r^3} v^2 \cos^2\theta,$$

The pressure equation is given by

$$\frac{p}{\rho}+\frac{1}{2}q^2-\frac{\partial\phi}{\partial t}+V=F(t), \qquad ...(1)$$

where V is the potential due to the external forces.

There is no motion in the liquid such that

$$q=0, \frac{\partial\phi}{\partial t}=0, \text{ and } p=p_0; \text{ so } \frac{p_0}{\rho}+V=F(t) \qquad ...(2)$$

From (1) and (2) we have

$$\frac{p-p_0}{\rho}=\frac{\partial\phi}{\partial t}-\frac{1}{2}q^2=\left(\frac{1}{2}\frac{a^3}{r^2}f\cos\theta_1-\frac{1}{2}\frac{a^3}{r^3}v^2+\frac{3a^3}{2r^3}v^2\cos^2\theta\right)$$

$$-\frac{1}{2}\left(\frac{a^6}{4r^6}v^2+\frac{3a^6}{4r^6}v^2\cos^2\theta\right)$$

There pressure at a point on the surface of the sphere reduces

$$\frac{p-p_0}{\rho}=\frac{1}{2}af\cos\theta_1+\frac{1}{8}v^2(9\cos^2\theta-5). \quad \textbf{proved.}$$

Example 9: *When a sphere of radius a moves in an infinite liquid; show that the pressure at any point exceeds what would be the pressure if the sphere were at a rest by*

$$\frac{a^3}{2r^2}f-\frac{a^3}{8r^6}(4r^3+a^3)q^2+\frac{3}{8}\frac{a^3}{r^6}(4r^3-a^3)q'^2,$$

where q is the velocity of the sphere and q' and f are the resolved parts of its velocity and acceleration in the direction of r and density of the liquid is unity.

Solution: Here v = q, v cosθ = q.

The difference between the pressure at any point of liquid in motion when at rest is given by

$$p-p_0=\frac{\partial\phi}{\partial t}-\frac{1}{2}q^2; \rho=1$$

$$\text{or } p-p_0=\frac{1}{2}\frac{a^3}{r^3}f-\frac{1}{2}\frac{a^3}{r^3}q^2+\frac{3a^3}{2r^3}q'^2-\frac{1}{2}\left(\frac{a^6}{4r^6}q^2+\frac{3a^6}{4r^6}q'^2\right)$$

$$\text{or } p-p_0=\frac{a^3}{2r^2}f-\frac{a^3}{8r^6}(4r^3+a^3)q^2+\frac{3a^3}{8r^6}(4r^3-a^3)q'^2. \quad \textbf{Proved.}$$

Example 10: *A sphere of radius a is in motion in fluid, which is at rest at infinity, the pressure there being* Π: *determine the pressure at any point of the fluid, and show that the pressure on the front hemisphere cut off by* *a plane perpendicular to the direction of motion is the resultant pressure*

$$\pi a^2\left(\Pi-\frac{1}{16}\rho v^2\right) and \frac{1}{3}\pi\rho a^3 f,$$

in the direction respectively opposite to those of the velocity v, and the acceleration f, of the centre of the sphere

Solution: We know that the pressure at any point on the surface of the sphere is given by

$$\frac{p}{\rho}=\frac{\Pi}{\rho}+\frac{1}{2}af\cos\theta_1+\frac{1}{8}v^2(9\cos^2-5),$$

$$\text{or } \frac{p}{\rho}=\frac{1}{\rho}\left\{\Pi+\frac{1}{8}\rho v^2(9\cos^2\theta-5)\right\}+\frac{1}{\rho}\left\{\frac{1}{2}af\cos\theta_1\right\}$$

or $p=p_1+p_\upsilon$

where $\quad p_1=\Pi+\frac{1}{8}\rho v^2(9\cos^2\theta-5)$, and $p_2=\frac{1}{2}af\rho\cos\theta_1$.

Le p_1 and p_2 be the pressures on the hemisphere along the direction opposite to the velocity v and acceleration f respectively. The pressure on the hemisphere along the direction opposite to the velocity v is

$$=\int_0^{\pi/2}(p_1\cos\theta).2\pi a.\sin\theta.ad\theta$$

$$=2\pi a^2\int_0^{\pi/2}\left[\Pi+\frac{1}{8}\rho v^2(9\cos^2\theta-5\right]\sin\theta\cos\theta$$

$$=2\pi a^2\left[\left(\Pi-\frac{5}{8}\rho v^2\right).\frac{1}{2}+\frac{9}{8}\rho V^2.\frac{1}{4}\right]=\pi a^2\left[\Pi-\frac{1}{16}\rho v^2\right] \quad \textbf{proved.}$$

Again pressure on the hemisphere along the direction opposite to the acceleration f is given by

$$=\int_0^{\pi/2}(p_2\cos\theta_1).2\pi a\ \sin\theta_1\ ad\theta_1$$

$$=2\pi a^3\rho f\int_0^{\pi/2}\cos^2\theta_1\sin\theta_1\ d\theta=\frac{1}{3}\pi a^3\rho f \qquad \textbf{proved.}$$

Example 11: *Prove that when the sphere is in motion with uniform velocity U, the pressure at the part of its surface where the radius makes an angle θ with the direction of motion is increased on account of the motion by the amount.*

$$\frac{1}{16}\rho U^2(9\cos 2\theta-1),$$

where ρ is the density of the liquid.

Solution: The velocity potential ϕ of the liquid motion is given by

$$\phi = \frac{1}{2}\frac{Ua^3}{r^2}\cos\theta. \qquad ...(1)$$

Let q be the velocity of the fluid at p_1 then

$$q^2 = \left(\frac{\partial\phi}{\partial r}\right)^2 + \left(\frac{\partial\phi}{r\,\partial\theta}\right)^2 = \frac{U^2a^6}{r^6}\left(\cos^2\theta + \frac{1}{4}\sin^2\theta\right). \qquad ...(2)$$

Let C be the position of the centre of the sphere at an instant of time t and CZ be the direction of the velocity U of the sphere at that time then $U = \dot{z}_0$.

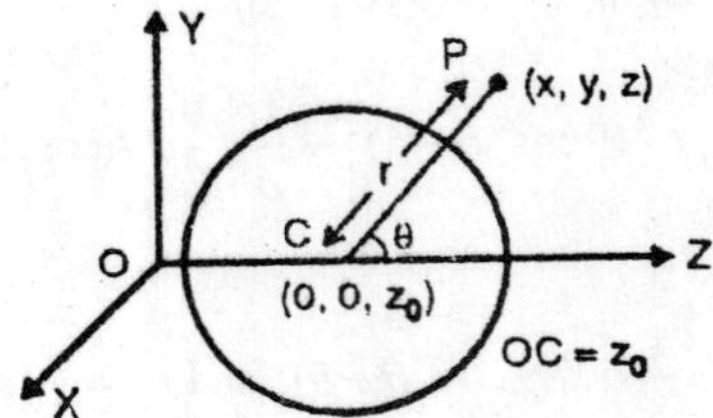

Fig. 3.23

$$r = CP = \sqrt{\{x^2 + y^2 + (z - z_0)^2\}}$$

and $$\cos\theta = \frac{z - z_0}{\sqrt{\{x^2 + y^2 + (z - z_0)^2\}}}.$$

Substituting the value of $\cos\theta$ and r^2 in (1), we have

$$\phi = \frac{1}{2}\frac{Ua^3(z - z_0)}{\{x^2 + y^2 + (z - z_0)^2\}^{3/2}}.$$

or $$\frac{\partial\phi}{\partial t} = \frac{1}{2}\frac{Ua^3(-\dot{z}_0)}{\{x^2 + y^2 + (z - z_0)^2\}^{3/2}} + \frac{3}{2}\frac{Ua^2\dot{z}_0(z - z_0)}{\{x^2 + y^2 + (z - z_0)^2\}^{5/2}}.$$

or $$\frac{\partial\phi}{\partial t} = \frac{1}{2}\frac{U^2a^3}{r^3} + \frac{3}{2}\frac{U^2a^3\cos\theta}{r^3} = \frac{U^2a^3}{2r^3}(3\cos^2\theta - 1). \qquad ...(3)$$

The pressure p at any point (a, θ) on the spherical boundary is to be determined from the equation

$$\frac{p}{\rho} + \frac{1}{2}q^2 - \frac{\partial\phi}{\partial t}F(t),$$

or $$\frac{p}{\rho} + \left\{\frac{1}{2}\frac{U^2a^6}{r^6}\left(\cos^2\theta\frac{1}{4}\sin^2\theta\right) - \frac{U^2a^3}{2r^3}(3\cos^2\theta - 1)\right\}_{r=a} = F(t). \qquad ...(4)$$

$$\text{or } \frac{p}{\rho}+\frac{1}{2}U^2\left(\cos^2\theta+\frac{1}{4}\sin^2\theta\right)-\frac{1}{2}U^2(3\cos^2\theta-1)=F(t). \qquad ...(5)$$

Let p_0 be the pressure on the surface of the sphere when there is no motion, then

$$p = p_0 U = 0 \Rightarrow p_0/\rho = F(t)$$

$$\text{or } p-p_0=-\frac{1}{2}\rho U^2(\cos^2\theta+\frac{1}{4}\sin^2\theta)+\frac{1}{2}\rho U^2(3\cos^2\theta-1),$$

$$\text{or } p-p_0=(1/16)\rho U^2(16\cos^2\theta-2\sin^2\theta-8)$$

$$\text{or } p-p_0=(1/16)\rho U^2\left[9\cos 2\theta-1\right].$$

Therefore the pressure to the surface is increased on account of the motion by the amount

$$(1/16)\rho U^2\left(9\cos 2\theta-1\right). \quad \textbf{Proved.}$$

Example 12: *Find the pressure at any point of a liquid, of infinite extent and at rest at a great distance, through which a sphere is moving under no external forces with constant velocity U, and show that the mean pressure over the sphere is a defect of the pressure* Π *at a great distance* $\frac{1}{4}\rho U^2$, *it being supposed that* Π is sufficiently large for the pressure everywhere to be positive, that is, that

$$\Pi > (5/8)\,\rho U^2.$$

Solution: We know that the pressure p at any point (r, θ) of the liquid motion is given by

$$\frac{p}{\rho}+\frac{U^2a^6}{2r^6}\left(\cos^2\theta+\frac{1}{4}\sin^2\theta\right)-\frac{U^2a^3}{2r^3}(3\cos^2\theta-1)=F(t). \qquad ...(1)$$

Since $p \to \Pi$, $r \to \infty$; $(\Pi/\rho) = F(t)$, then

$$\frac{p-\Pi}{\rho}=\frac{1}{2}\frac{U^2a^3}{r^3}(3\cos^2\theta-1)-\frac{U^2a^6}{2r^6}\left(\cos^2\theta+\frac{1}{4}\sin^2\theta\right)$$

On the surface of the sphere, it becomes

$$p-\Pi=\frac{1}{8}\rho U^2(8\cos^2\theta-\sin^2\theta-1)$$

$$\text{or } p-\Pi+(1/8)\rho U^2(9\cos^2\theta-5)$$

The mean pressure ($= p_m$) over the sphere is given by

$$p_m=\frac{\int p\,dS}{\int dS}=\frac{\int_0^\pi p2\pi a\sin\theta.a\,d\theta}{\int_0^\pi 2\pi a\sin\theta.a\,d\theta}$$

$$\text{or } p_m = \frac{1}{2}\int_0^\pi \left[\Pi + \frac{1}{8}\rho U^2(9\cos 2\theta - 5)\right]\sin\theta\, d\theta$$

$$\text{or } p_m = \frac{1}{2}\left[2\Pi + \frac{1}{8}\rho U^2(-4)\right]$$

$$\text{or } \Pi - p_m = \frac{1}{4}\rho U^2.$$

Thus the defect of the pressure Π at a great distance to the mean pressure over the sphere is $\frac{1}{4}\rho U^2$. **Proved.**

Again the pressure is minimum when $\cos\theta = 0 \Rightarrow \theta = \pi/2$.

Therefore min. pressure $= \Pi - \frac{5}{8}\rho U^2$ which will be positive every where if $\Pi > \frac{5}{8}\rho U^2$. **Proved.**

Example 13: *A sphere of radius a is made to move to incompressible perfect fluid with non-uniform velocity u along the X-axis. If the pressure at infinity is zero. Prove that at a point x in advance of the centre*

$$p = \frac{1}{2}\rho a^3\left[\frac{u}{x^3} + u^2\left(\frac{2}{x^3} - \frac{a^3}{x^6}\right)\right].$$

Solution: A sphere is moving with velocity u along the axis of Z as initial line, the velocity potential of the fluid motion at P(r, θ) is

$$\phi = \frac{1}{2}\frac{ua^3}{r^2}\cos\theta \qquad ...(1)$$

where $r^2 = CP^2 = (X - X_0)^2 + Y^2 + Z^2$

and $\cos\theta = (X - X_0)/\sqrt{\{(X - X_0)^2 + Y^2 + Z^2\}}.$

Substituting the value of r^2 and cos θ in (1), we have

$$\phi = \frac{1}{2}\frac{ua^3(X - X_0)}{\{(X - X_0)^2 + Y^2 + Z^2)\}^{3/2}}$$

$$\text{Now} \quad q^2 = \left(\frac{\partial\phi}{\partial r}\right)^2 + \left(\frac{1}{r}\frac{\partial\phi}{\partial\theta}\right)^2 = \frac{u^2a^6}{r^6}\left(\cos^2\theta + \frac{1}{4}\sin^2\theta\right).$$

$$\text{and} \quad \frac{\partial\phi}{\partial t} = \frac{1}{2}\frac{ua^3(X - X_0)}{\{(X - X_0)^2 + Y^2 + Z^2\}^{3/2}} - \frac{1}{2}\frac{ua^3X_0}{\{(X - X_0)^2 + Y^2 + Z^2\}^{3/2}}$$

$$+ \frac{1}{2}\frac{3ua^3X(X - X_0)^2}{\{(X - X_0)^2 + Y^2 + Z^2\}^{5/2}}$$

$$\text{or } \frac{\partial \phi}{\partial t} = \frac{1}{2}\frac{ua^3}{r^2}\cos\theta - \frac{1}{2}\frac{u^2a^3}{r^3} + \frac{3}{4}\frac{u^2a^3\cos^2\theta}{r^3}. \qquad ...(2)$$

The pressure p at any point is to be determined from the equation

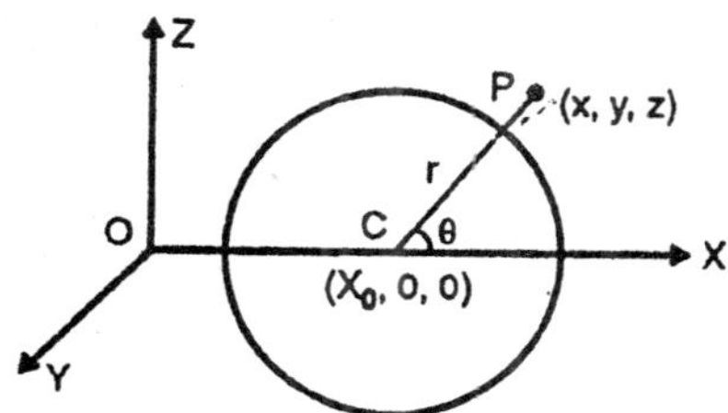

Fig. 3.24

$$\frac{p}{\rho} + \frac{1}{2}q^2 - \frac{\partial \phi}{\partial t} = F(t)$$

$$\text{or } \frac{p}{\rho} + \frac{u^2a^6}{2r^6}\left(\cos^2\theta + \frac{1}{4}\sin^2\theta\right)$$

$$-\left(\frac{1}{2}\frac{ua^3}{r^2}\cos\theta - \frac{1}{2}\frac{u^2a^3}{r^2} + \frac{3}{2}\frac{u^2a^3\cos^2\theta}{r^3}\right) = F(t)$$

As $r \to \infty$, $p = 0$ then $F(t) = 0$.

$$\text{or } \frac{p}{\rho} = -\frac{1}{2}\frac{u^2a^6}{r^6}\left(\cos^2\theta + \frac{1}{4}\sin^2\theta\right) + \frac{1}{2}\frac{ua^3}{r^3}\cos\theta$$

$$-\frac{1}{2}\frac{u^2a^3}{r^3} + \frac{3}{4}\frac{u^2a^3\cos^2\theta}{r^3}. \qquad ...(3)$$

The pressure at a point which is at a distance x in advance of the entre is given by substituting $\theta = 0$, $r = x$ into (3), we have

$$\frac{p}{\rho} = -\frac{1}{2}\frac{u^2a^6}{x^6} + \frac{1}{2}\frac{ua^3}{x^2} - \frac{1}{2}\frac{u^2a^3}{x^3} + \frac{3}{2}\frac{u^2a^3}{x^3}$$

$$\text{or } p = -\frac{1}{2}\rho a^3\left[\frac{u}{x^3} + u^2\left(\frac{2}{x^3} - \frac{a^3}{x^6}\right)\right]. \quad \textbf{Proved.}$$

Example 14: *A rigid surface of radius a is moving in a straight line with velocity U and acceleration f through an infinite incompressible liquid, prove that the resultant fluid pressures over the two hemispheres, into which sphere is divided by a diameteral plane perpendicular to the direction of motion are*

$$\Pi \pi a^2 \pm \frac{1}{4}Mf - (3/64)\frac{MU^2}{a},$$

where Π is the pressure at a great distance, and M is the mass of the fluid displaced by the sphere.

Solution: The pressure p at any point (a, θ) on the spherical boundary is given by

$$\frac{p}{\rho}=\frac{\Pi}{\rho}+(1/8)U^2(9\cos^2\theta-5)+\frac{1}{2}af\cos\theta.$$

Let YOY' be the diameteral plane perpendicular to its direction of motion which divides it into two hemispheres..

The pressure on the first (I) hemisphere becomes

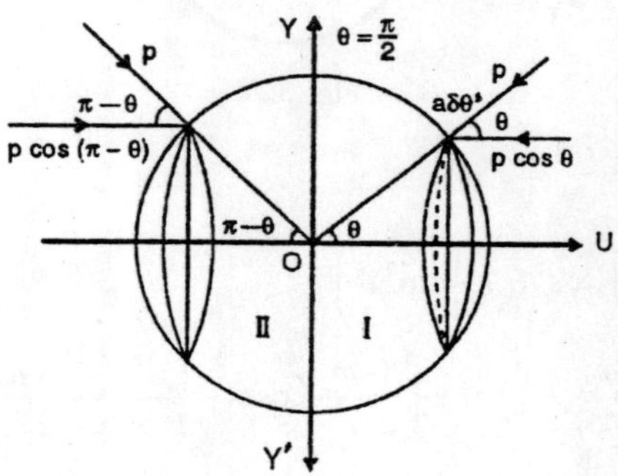

Fig. 3.25

$$=\int_0^{\pi/2} p\cos\theta.2\pi a\sin\theta.ad\theta$$

$$=2\pi a^2\int_0^{\pi/2}\left\{\Pi+\frac{1}{8}\rho U^2(9\cos^2\theta-5)+\frac{1}{2}\rho af\cos\theta\right\}\cos\theta\sin\theta d\theta$$

$$=2\pi a^2\left\{\Pi\frac{\sin^2\theta}{2}-\frac{5}{8}\rho U^2\frac{\sin^2\theta}{2}-\frac{9}{8}\rho U^2\frac{\cos^4\theta}{4}-\frac{1}{2}\rho af\frac{\cos^3\theta}{3}\right\}_0^{\pi/2}$$

$$=2\pi a^2\left\{\frac{\Pi}{2}+\rho\frac{U^2}{8}\left(\frac{9}{4}-\frac{5}{2}\right)+\frac{1}{2}\rho af\frac{1}{3}\right\}$$

$$=\Pi\pi a^2+\frac{1}{4}Mf-\frac{3}{64}\frac{MU^2}{a};M=\frac{4}{3}\pi a^3\rho.$$

Also, the pressure on the second (II) hemisphere becomes

$$=\int_{\pi/2}^{\pi} p\cos(\pi-\theta).2\pi a\sin\theta.ad\theta$$

$$=-2\pi a^2\int_{\pi/2}^{\pi}\left\{\Pi+\frac{1}{8}\rho U^2(9\cos^2\theta-5)+\frac{1}{2}\rho af\cos\theta\right\}\cos\theta\sin\theta\, d\theta$$

$$= 2\pi a^2\left(\Pi\frac{1}{2}+\frac{1}{8}\rho U^2\left(\frac{9}{4}-\frac{5}{2}\right)+\frac{1}{2}\rho af.\frac{1}{3}\right)$$

$$= \prod \pi a^2 - \frac{1}{3}\pi a^2 \rho f - (1/16)\pi a^2 \rho U^2$$

$$= \prod \pi a^2 - \frac{1}{4}Mf - \left(\frac{3}{64}\right)\frac{MU^2}{a}.$$

Hence the resultant pressure over the two hemispheres, respectively, are

$$= \prod \pi a^2 + \frac{1}{4}Mf - \left(\frac{3}{64}\right)\frac{MU^2}{a}.$$ **Proved.**

Example 15: *Show that when a sphere of radius a moves with uniform velocity U through a perfect; incompressible, infinite liquid, the acceleration of particle of liquid at (r, 0) is*

$$3U^2\left(\frac{a^3}{r^4}-\frac{a^6}{r^7}\right).$$

Solution: Assuming the velocity potential □ϕ of the fluid motion at P is

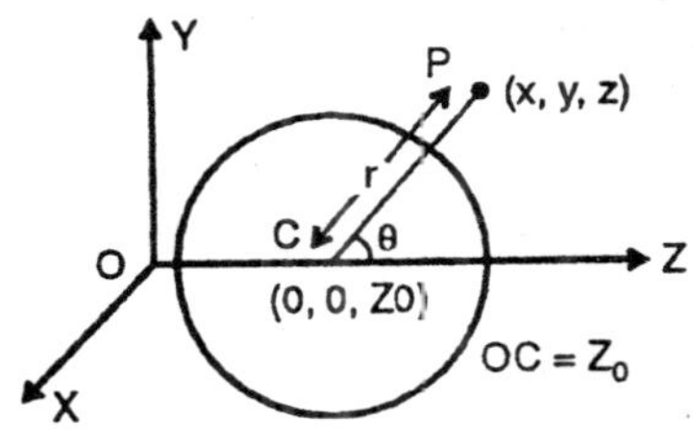

Fig. 3.26

$$\phi = \frac{1}{2}\frac{Ua^3}{r^2}\cos\theta. \qquad ...(1)$$

Let the sphere moves wit velocity U along the Z-axis as initial line, having the centre (0, 0, Z_0) as pole

The relation (1) can be expressed in terms of (x, y, z) as

$$\phi = \frac{1}{2}\frac{Ua^3(z-z_0)}{2\{x^2-y^2+(z-z_0)^2\}^{3/2}}.$$

where $\cos\theta = (z-z_0)/r;\ r^2 = x^2+y^2+(z-z_0)^2.$

$$\text{or } \frac{dz}{dt} = w = -\frac{\partial\phi}{\partial z} = -\frac{1}{2}Ua^3\left[\frac{1}{\{x^2+y^2+(z-z_0)^2\}^{5/2}}\right.$$

$$-\frac{3(z-z_0)}{\{x^2+y^2+(z-z_0)^2\}^{5/2}}\Bigg]$$

or $$\frac{dz}{dt}=-\frac{1}{2}Ua^3\left[\frac{1}{(z-z_0)^3}-\frac{3}{(z-z_0)^2}\right],$$

at the centre of the sphere C (0, 0, z_0).

or $$\frac{dz}{dt}=\frac{Ua^3}{(z-z_0)^3}. \qquad ...(2)$$

Differentiating (2) with regard to t, we have

$$\frac{d^2z}{dt^2}=-\frac{3Ua^3}{(z-z_0)^4}\left\{\frac{dz}{dt}-\frac{dz_0}{dt}\right\};\frac{dz_0}{dt}=U$$

or $$\frac{d^2z}{dt^2}=-\frac{3Ua^3}{(z-z_0)^4}\left\{\frac{3Ua^3}{(z-z_0)^3}-U\right\}$$

or $$\frac{d^2z}{dt^2}=3Ua^3\left\{\frac{1}{(z-z_0)^4}-\frac{a^3}{(z-z_0)^7}\right\};r=z-z_0.$$

or $$\frac{d^2z}{dt^2}=3U^2a^3\left\{\frac{1}{r^4}-\frac{a^6}{r^7}\right\} \text{ or } \frac{d^2z}{dt^2}=2U^2\left\{\frac{a^3}{r^4}-\frac{a^6}{r^7}\right\},$$

which determines the acceleration of a particle of fluid at the point (r, 0). Proved.

Example 17: *A doublet of strength M is placed at the point (0, a, 0) with its axis parallel to the axis of Z. Prove that at points close to the origin the velocity potential of the doublet is approximately $Mz / a^3 + 3\,Myz / a^4$, neglecting terms of the order (r^3 / a^3) and higher powers. Deduce that if a small sphere of radius c placed with its centre at the origin, the velocity potential is then increased by the terms*

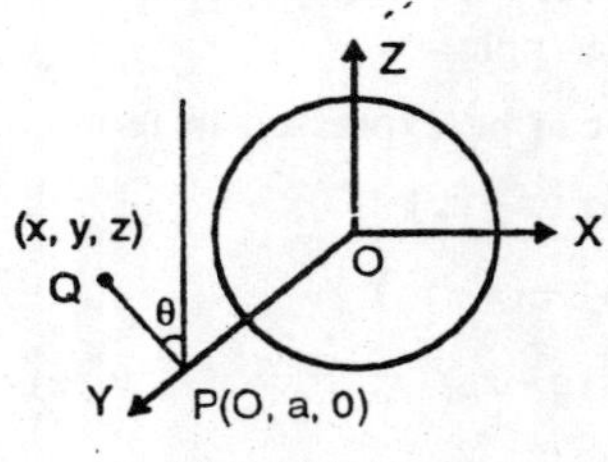

Fig. 3.27

$$\frac{1}{2}\frac{Mc^3}{a^3}.\left(\frac{z}{r^3}\right)+2\frac{Mc^5}{r^4}.\left(\frac{yz}{r^5}\right).$$

Solution: A doublet of strength M is placed at the point P(0, a, 0) with its axis parallel to the axis of Z. Consider a point (x y z) in the liquid. Let PQ makes an angle ϕ due to the doublet is given by

$$\phi=\frac{M\cos\theta}{PQ^2}=\frac{Mz}{\left\{x^2+(y-a)^2+z^2\right\}^{3/2}},$$

$$\text{or } \phi=\frac{Mz}{(r^2-2ay+a^2)^{3/2}},\ r^2=x^2+y^2+z^2$$

$$\text{or } \phi=\frac{Mz}{a^3}\left(1-\frac{2y}{a}+\frac{r^2}{a^2}\right)^{-3/2}=\frac{Mz}{a^3}\left(1+\frac{3}{2}+\frac{2y}{a}+...\right),$$

neglecting the higher order terms.

Thus the velocity potential of the doublet becomes

$$\phi=\frac{Mz}{a^3}+\frac{3Myz}{a^4}.$$ **Proved.**

Again, a small sphere of radius c is placed with its centre at the origin. Let ϕ' be the velocity potential at any point. The velocity however remains unchanged at infinity. Assuming that the increase in the velocity potential be $Az / r^3 + Nyz / r^5$. Then

$$\phi'=\frac{Mz}{a^3}+\frac{3Myz}{a^4}+\frac{Az}{r^3}+\frac{Byz}{r^5}$$

$$\text{or } \phi'=\frac{Mr\cos\theta}{a^3}+\frac{3\,Mr^2\sin\theta\cos\theta\sin\omega}{a^4}+\frac{A\ \cos\theta}{r^3}$$
$$+\frac{B\sin\theta\cos\theta\sin\omega}{r2}$$

$$\text{or } \frac{\partial\phi'}{\partial r}=\frac{Mr\cos\theta}{a^3}+\frac{6\,Mr\sin\theta\cos\theta\sin\omega}{a^4}-\frac{2\,A\ \cos\theta}{r^3}$$
$$-\frac{3B\sin\theta\cos\theta\sin\omega}{r^4}$$

But the velocity potential ϕ is such that $\partial\phi' / \partial r = 0$ at r = c.

$$\text{so } \frac{M\cos\theta}{a^3}-\frac{2A\cos\theta}{c^3}+\frac{6\,Mc\sin\theta\cos\theta\sin\omega}{a^4}$$
$$-\frac{3B\sin\theta\cos\theta\sin\omega}{c^4}=0$$

or $\left(\frac{M}{a^3}+\frac{24}{c^3}\right)\cos\theta+\left(\frac{6Mc}{a^4}-\frac{3B}{c^4}\right)\sin\theta\cos\theta\sin\omega=0$

i.e. $\frac{M}{a^3}+\frac{24}{c^3}=0$ and $\frac{6Mc}{a^4}-\frac{3B}{c^4}=0; A=\frac{Mc^3}{2a^3}$ and $B=\frac{2Mc^5}{a^4}$.

Thus the increase in velocity potential ϕ is

$$=\frac{1}{2}\frac{Mc^3}{a^3}\left(\frac{z}{r^3}\right)+2\frac{Mc^5}{a^4}\left(\frac{yz}{r^5}\right)$$ **Proved.**

Example 18: *An ellipsoidal cavity (semi-axis a, b, c) in a solid initially at rest is filled with an incompressible frictionless fluid initially at rest. Prove that if the solid be moved with velocities y, v, w parallel to the axes of the cavity, and be rotated with angular velocities p, q, r round the semi-axis, the angular momentum of the fluid round the semi-axis at any instant is*

$$\frac{4}{15}\pi\rho bc\frac{(b^2-c^2)^2}{b^3-c^2}p.$$

Solution: The velocity potential is given by

$$\phi=-\frac{b^2-c^2}{b^2+c^2}pyz-\frac{c^2-a^2}{c^2+a^2}qzx-\frac{a^2-b^2}{a^2+b^2}rxy. \quad ...(1)$$

By super-imposing the velocities -u, -v, -w along the axes, the ellipsoid can be reduced to simple rotation about the axes. Thus an expression (-ux-vy-wz) should be added to the velocity potential ϕ that is

$$\phi=-ux-vy-wz-\frac{b^2-c^2}{b^2+c^2}pyz-\frac{c^2-a^2}{c^2+a^2}qzx-\frac{a^2-b^2}{a^2+b^2}rxy. \quad ...(2)$$

Let U, V, W be the velocity components at P (x, y, z) so that

$$U=-(\partial\phi/\partial x),\ V=-(\partial\phi/\partial y),\ W=-(\partial\phi/\partial z) \quad ...(3)$$

The angular momentum of the whole liquid about the axis of X, is given by

$$=\iiint (yW-zV)\rho\,dx\,dy\,dz,\quad \frac{x^2}{a^2}+\frac{y^2}{b^2}+\frac{z^2}{c^2}\le 1$$

$$=-\rho\iiint\left(y\frac{\partial\phi}{\partial z}-z\frac{\partial\phi}{\partial y}\right)dx\,dy\,dz$$

$$=-\rho\iiint y\left\{w+\frac{b^2-c^2}{b^2+c^2}py+\frac{c^2-a^2}{c^2+a^2}qx\right\}$$

$$-z\left[\left\{v+\frac{b^2-c^2}{b^2+c^2}pz+\frac{a^2-b^2}{a^2+b^2}rx\right\}\right]dx\,dy\,dz$$

$$= \rho\left(\frac{b^2 - c^2}{b^2 + c^2}\right) p \iiint (y^2 - z^2)\, dx\, dy\, dz$$

$$= \left(\frac{b^2 - c^2}{b^2 + c^2}\right) p \iiint \rho\{(x^2 + y^2) - (x^2 + z^2)\}\, dx\, dy\, dz$$

$$= \left(\frac{b^2 - c^2}{b^2 + c^2}\right) p \left[M\frac{a^2 + b^2}{5} - M\frac{a^2 + c^2}{5}\right]$$

$$= \frac{4}{15}\pi\rho abc\frac{(b^2 - c^2)^2}{b^2 + c^2} p, M = \frac{4}{3}\pi\rho abc.$$ **Proved.**

Example 19: *A rigid ellipsoidal envelope, without mass, encloses a perfect incompressible fluid of mass M. The equation of the ellipsoid is*

$$x^2 / a^2 + y^2 / b^2 + z^2 (a^2 - 1) = 0.$$

An impulsive couple in the plane of -XY causes the envelope to rotate initially with angular velocity ω. Find the initial velocity potential of the fluid and prove that the moment of the couple is

$$\frac{1}{5} M\omega[(a^2 - b^2)^2 / (a^2 + b^2)].$$

Solution: The component of linear velocities at (x, y, z) of the envelope are ($-y\omega$, $x\omega$, 0). The direction cosines of the normal are proportional to x / a^2, y / b^2 z / c^2. Let ϕ be the velocity potential of the liquid motion, the boundary condition is given by

$$-(x / a^2)\partial\phi / \partial x - (y / b^2)\partial\phi / \partial y - (z / c^2)\partial\phi / \partial z$$

$$= x / a^2(-y\omega) + y / b^2.(x\omega) \qquad ...(1)$$

Assuming ϕ = cxy, which is also a solution δf Laplace's equation.

From (1), we have

$$cxy(a^{-2} + b^{-2}) = xy\omega(a^{-2} - b^{-2})$$

$$\Rightarrow c = -\omega[(a^2 - b^2) / (a^2 + b^2)]$$

Hence the required initial velocity potential is

$$\phi = -\omega xy[(a^2 - b^2) / (a^2 + b^2)].$$

The initial moment of the couple is equal to the total moment of momentum of the fluid, and therefore

$$= \iiint [x - (-\partial\phi / \partial y) - y(-\partial\phi / \partial x)]\rho\, dx\, dy\, dz$$

$$= \rho\omega\frac{a^2 - b^2}{a^2 + b^2}\iiint \{(x^2 + z^2) - (y^2 + z^2)\}\, dxdydz$$

$$= \rho\omega \frac{a^2 - b^2}{a^2 + b^2} \left\{ M \frac{a^2 - c^2}{5} - M \frac{b^2 + c^2}{5} \right\} = \frac{1}{5} M\omega \frac{(a^2 - b^2)^2}{a^2 + b^2}$$ **Proved.**

Example 20: *Prove that if two rigid surfaces of revolution, one of which surrounds the other, are moving along their common axis with velocities U_1, U_2 and space between them is filled with homogeneous liquid, the momentum of the liquid is M_2U_2 - M_1U_1, where M_1, M_2 are the masses of liquid which either surface would contain.*

Solution: Let X-axis be chosen as the axis of revolution. By symmetry the momentary the moment of momentum of the liquid along axes of Y and Z vanishes. The momentum of the liquid along X-axis is

$$\iiint \rho u \, dx \, dy \, dz. \qquad ...(1)$$

Let ϕ is the velocity potential at any point (x y z) of the liquid, then (1) reduces to

$$= -\iiint \rho \, (\partial\phi / \partial x) \, dx \, dy \, dz, \qquad ...(2)$$

the integration extends over the whole volume of liquid.

By using Green's theorem (2) becomes

$$= \rho \iint x \, (\partial\phi / \partial n) \, dS$$

where δx in an element of the normal drawn at the element of the bounding surface δs.

Hence the momentum of the liquid along X-axis is

$$= \rho \iint_{\text{inner}} x \, (\partial\phi / \partial n) \, dS_1 + \rho \iint x \, (\partial\phi / \partial n) \, dS_2,$$

$$= -\rho \iint_{\text{inner}} x l_1 U_1 dS_1 + \rho \iint_{\text{outer}} x l_2 U_2 dS_2,$$

where l_1 and l_2 are cosines of the angles which the outer drawn normals at dS_1, dS_2 make with X-axis,

$$= -U_1 \iint \rho x \, l_1 dS_1 + U_2 \iint \rho x \, l_2 dS_2$$

$$= -U_1 \iint \rho x \, dy \, dx + U_2 \iint \rho x \, dy \, dz$$

$$= -M_2 U_2 - M_1 U_1,$$

where M_1 and M_2 are the masses of the liquid **Proved.**

Example 21: *Prove that the velocity potential at a point P due to a uniform finite line source AB of strength + m per unit length is of the form ϕ = m log f, where*

$$f = \frac{r_2 + x_2}{r_1 + x_1} = \frac{r_1 - x_1}{r_2 - x_2} = \frac{a + l}{a - l},$$

in which AB = 2l, PA = r_1, PB = r_2 NA = x_1, NB = x_2 and $r_1 + r_2$ = 2a, N being the perpendicular drawn from P on the line AB.

Solution: The velocity potential at P but to the entire line distribution is

$$\phi = m\int_0^{2l}\frac{dx}{r} = m\int_0^{2l}\frac{dx}{\left\{(r_2^2 - r_1^2) + (x_1 - x)^2\right\}^{1/2}},$$

or $$\phi = m\int_0^{2l}\frac{dx}{\left\{d^2 - (x_1 - x)^2\right\}^{1/2}}, d^2 = r_1^2 - x_1^2$$

or $$\phi = m\left\{\sinh^{-1}\left[\frac{(x_1 - x)}{d}\right]\right\}_0^{2l}$$

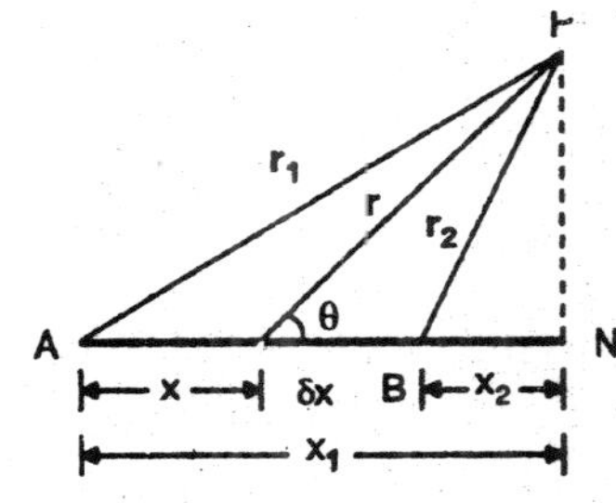

Fig. 3.27(a)

or $\phi = m\left\{\sinh^{-1}(x_1/d) - \sinh^{-1}(x_1 - 2l)/d\right\}$

or $\phi = m \log\left\{\frac{x_1 + \sqrt{(x_1^2 + d^2)}}{x_2 + \sqrt{(x_2^2 + d^2)}}\right\} = m \log\left(\frac{x_1 + r_1)}{x_2 + r_2}\right)$

Since $\quad r_1^2 - x_1^2 = r_2^2 - x_2^2 = d^2, x_1 - x_2 = 2l$

or $\frac{r_1 + x_1}{r_2 + x_2} = \frac{r_2 - x_2}{r_1 - x_1} = \frac{r_1 + r_2 + x_1 - x_2}{r_1 + r_2 + x_2 - x_1} + \left(\frac{x_1 + r_1)}{x_2 + r_2}\right)$

or $\frac{r_1 + x_1}{r_2 + x_2} = \frac{r_2 - x_2}{r_1 - x_1} = \frac{a + l}{a - l} = f(\text{let})$

Thus $\quad \phi = m\log\left(\frac{a + l}{a - l}\right) = m\log f$. **Proved.**

The equi-potentials are given by ϕ = const.

$\frac{a + l}{a - l}$ = const. $\Rightarrow$ a = const. $\Rightarrow$ $r_1 + r_2$ = const., which represents the confocal ellipsoids of revolution about AB with A and B as focii.

Example 22: *Doublets of strengths μ_1, μ_2 are situated at points A_1 (0, 0 c_1) and A_2 (0, 0, c_2) their axes being directed towards and away from the origin respectively. Find the condition that there is no transport of fluid over the surface of the sphere $x_2 + y_2 + z_2 = c_1c_3$.*

Solution: Let OA_2A_1 be the initial line. The axes of the doublets at A_1, A_2 make angles θ_1, θ_2 with A_1P, A_2P. Then the velocity potential at P (r, θ, ϕ) reduces to

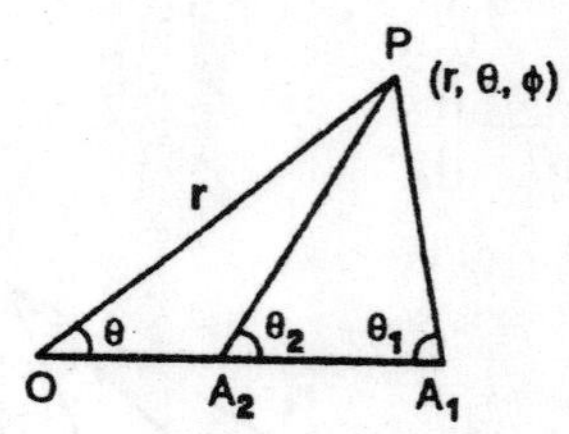

Fig. 3.28

$$\phi = \frac{\mu_2 \cos\theta_2}{(A_2P)^2} + \frac{\mu_1 \cos\theta_1}{(A_1P)^2} \quad ...(1)$$

where

$$(A_1P)^2 = r^2 - 2rc_1 \cos\theta + c_1^2 (A_2P)^2 = r^2 - 2rc_2 \cos\theta + c_2^2,$$

and $$\cos\theta_1 = \frac{c_1 - r\cos\theta}{A_1P}, \cos\theta_2 = \frac{r\cos\theta - c_2}{A_2P}. \quad ...(2)$$

From (1) and (2), we have

$$\phi(r,\theta) = \frac{\mu_2(r\cos\theta - c_2)}{(r^2 + 2rc_2\cos\theta - c_2^2)^{3/2}} + \frac{\mu_1(c_1 - \cos\theta)}{(r^2 - 2rc_1\cos\theta + c_1^2)^{3/2}}$$

We require that $\partial\phi / \partial r = 0$ when $r = (c_1 c_2)^{1/2}$ ...(3)

Since $x^2 + y^2 + z^2 = c_1c_2 \Rightarrow r^2 = c_1c_2 \Rightarrow r = (c_1c_2)^{1/2}$.

$$\therefore \frac{\partial\phi}{\partial r} = \mu_2\left\{\cos\theta(r^2 - 2rc_1\cos\theta + c_2^2)^{-3/2} - 3(r\cos\theta - c_2)\right.$$

$$\left.(r - c_2\cos\theta)(r^2 - 2rc_2\cos\theta + c_2^2)^{-5/2}\right\}$$

$$+\mu_1\left\{-\cos\theta(r^2 - 2rc_1\cos\theta + c_1^2)^{-3/2} - (c_1 - r\cos\theta)\right.$$

$$\left.(r - c_1\cos\theta)(r^2 - 2rc_1\cos\theta + c_1^2)^{-5/2}\right\} \quad ...(4)$$

From (3) and (4), we have

$$\mu_2[\cos\theta(c_1c_1 - 2c_2)(c_1c_1)^{1/2}\cos\theta + c_2^2]^{-3/2}$$

$$-3\left\{(c_1c_1)^{1/2}\cos\theta - c_2\right\}\left\{(c_1c_2)^{1/2} - c_2\cos\theta\right\}$$
$$\times\left\{(c_1c_2 - 2c_2(c_1c_2)^{1/2}\cos\theta + c_1^2\right\}^{-5/2}\Big]$$

$\Rightarrow \mu_2/\mu_1 = (c_2/c_1)^{3/2}$. **Ans.**

Example 23: *A three-dimensional doublet of strength μ whose axis is in the direction OX is distant a from the rigid plane $x = 0$ which is the sole boundary of liquid of density ρ, infinite in extent. Find the pressure at a point on the boundary distant r from the doublet given that the pressure at infinity is p_∞. Show that the pressure on the plane is least at a distance $a\sqrt{5}/2$ from the doublet.*

Solution: Le YY' be a rigid boundary. The image system of a doublet of strength μ placed at A (a, o) consists a doublet of same strength with regard to the rigid boundary at the corresponding point velocity at P reduces to

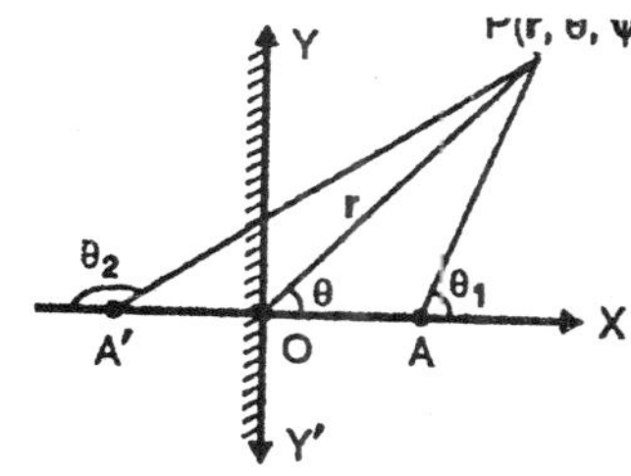

Fig. 3.29

$$\phi = \frac{\mu\cos\theta_1}{(AP)^2} + \frac{\mu\cos\theta_2}{(AP)^2}$$

$$\text{or } \phi = \frac{\mu(r\cos\theta - a)}{(r^2 + a^2 - 2ra\cos\theta)^{3/2}} - \frac{\mu(r\cos\theta + a)}{(r^2 + a^2 + 2ra\cos\theta)^{3/2}} \qquad ...(1)$$

Thus $\quad q_r = -\dfrac{\partial\phi}{\partial r} = -\mu\left\{\cos\theta(r^2 + a^2 - 2ra\cos\theta)^{-3/2}\right.$

$$-3 = (r - a\cos\theta)(r\cos\theta - a)(r^2 + a^2 - 2ra\cos\theta)^{-5/2}$$
$$-\cos\theta(r^2 + a^2 - 2ra\cos\theta)^{-3/2} + 3(r + a\cos\theta)(r\cos\theta + a)$$
$$\left.(r^2 + a^2 + 2ra\cos\theta)^{-5/2}\right\}.$$

$$q_\theta = -\frac{1}{r}\frac{\partial\phi}{\partial\theta} = -\frac{\mu}{r}\{-r\sin\theta(r^2 + a^2 - 2ra\cos\theta)^{-3/2}$$
$$-3r\sin\theta(r\cos\theta - a)(r^2 + a^2 - 2ra\cos\theta)^{-5/2}$$
$$+r\sin\theta(r^2 + a^2 + 2ra\cos\theta)^{-3/2}$$

$$-3ra\sin\theta(r\cos\theta+a)(r^2+a^2+2ra\cos\theta)^{-5/2}\Big\}$$

$$q_\phi=-\frac{1}{r\sin\theta}\frac{\partial\phi}{\partial\theta}=0.$$

At $\theta = \pi / 2$, $q_r = -6\pi ar(r^2+a^2)^{-5/2}$, $q_\theta = 0$, $q_\phi = 0$.

From the Bernoulli's equation, we have

$$p=p_\infty-\frac{1}{2}\rho q^2=p_\infty-18\rho\mu^2a^2r^2(r^2+a^2)^{-5}.$$

$$(dp/dr)=36\rho\mu^2a^2r(2r^2-a^2)(r^2+a^2)^{-6},(dp/dr)_{r=)1/2)_a}=0,$$

Also for $r=\frac{1}{2}a+,\frac{dp}{dr}>0$ and for $r=\frac{1}{2}a-,\frac{dp}{dr}<0.$

Hence p is minimum at $r=\frac{1}{2}a$ on the plane *i.e.*, at a distance $\frac{1}{2}a\sqrt{5}$ from the doublet. **Proved.**

Example 24: *If AB be a uniform line source, and A and B equal sinks of such strength that there is no total gain or loss of fluid. Show that the stream function at any point is*

$$\lambda\{(r_1-r_2)-c_2\}\left(\frac{1}{r_1}-\frac{1}{r_2}\right),$$

where c is the length of AB, r_1 and r_2 are the distance of the point considered from A and B.

Solution: Consider +m be the strength per unit length and c be the length of the line source, then the total strength of the line source is + mc. Let $\angle PAB = \theta_1$ and $\angle PBX = \theta_2$. Let m' be the strength of each sink. Since the total flow is zero, then

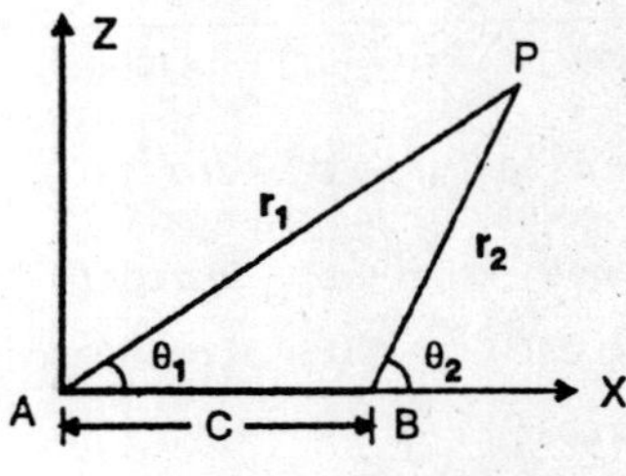

Fig. 3.30

$$-4\pi m' - 4\pi m' + 4\pi c = 0$$

$\Rightarrow m' = (1/2)\ mc.$

Now to neutralize the gain of fluid due to the line source, there must be

two sinks of strength $-\frac{1}{2}mc$ each at A and B. If ψ is the Stoke's stream function at P, then

$$\psi = m(r_1 - r_2) - \frac{1}{2}mc\cos\theta_1 - \frac{1}{2}mc\cos\theta_2$$

$$\text{or } \psi = -mr_1r_2\left(\frac{1}{r_1} - \frac{1}{r_2}\right) - \frac{1}{4}m\left\{\frac{c^2 + r_1^2 - r_2^2}{r_1} + \frac{r_1^2 - r_2^2 - c^2}{r_2}\right\}$$

$$\text{or } \psi = -m\, r_1r_2\left(\frac{1}{r_1} - \frac{1}{r_2}\right) - \frac{1}{4}m\left\{c^2\left(\frac{1}{r_1} - \frac{1}{r_2}\right) + (r_1 - r_2) - \frac{r_2^2}{r_1} + \frac{r_1^2}{r_2}\right\}$$

$$\text{or } \psi = m\left(\frac{1}{r_1} - \frac{1}{r_2}\right)\left\{-r_1r_2 - \frac{1}{4}c^2 + \frac{1}{4}(r_1 - r_2)^2\right\}$$

$$\text{or } \psi = \frac{1}{4}m\left(\frac{1}{r_1} - \frac{1}{r_2}\right)\left\{(r_1 - r_2)^2 - c^2\right\}$$

$$\text{or } \psi = \lambda\left\{(r_1 - r_2)^2 - c^2\right\}\left(\frac{1}{r_1} - \frac{1}{r_2}\right), \lambda = \frac{1}{4}m.$$ **Proved.**

Example 25: *Discuss the motion for which Stokes' steam function is given by*

$$\psi = \frac{1}{2}V(a^4r^{-2}\cos\theta - r^2)\sin^2\theta,$$

where r is the distance from a fixed point and θ is the angle these distance makes with a fixed direction.

Solution:

Here $$\psi = \frac{1}{2}V\frac{a^2}{r^2}\sin^2\theta\cos\theta - \frac{1}{2}Vr^2\sin^2\theta. \quad ...(1)$$

From (1) $\psi = 0$ gives $\theta = 0$ and $(a^4 / r^2)\cos\theta - r^2 = 0$

where $\theta = 0$. is the axis of symmetry for Stoke's stream function and $r^4 = a^4\cos\theta$ is an equation to a meridian curve along which ψ is constant. Hence the motion given by (1) is the motion of a fluid streaming past a solid of revolution, the equation to the curve is $r^4 = a^4\cos\theta$.

Let q_r and q_θ be the velocity components along V and θ increasing respectively. We know that

$$q_r = -\frac{1}{r^2\sin\theta}\frac{\partial\psi}{\partial\theta}$$

$$\text{or } q_r = -V[\{a^4 / r^4)\cos\theta - 1\}\cos\theta - (a^4 / 2r^4)^2\sin^2\theta] \quad ...(2)$$

and $\quad q_\theta = \dfrac{1}{r\sin\theta}\dfrac{\partial\psi}{\partial\theta} = -V\sin\theta[a^4/r^4)\cos\theta + 1]$ ...(3)

Again $\quad (q_r)_{r\to\infty} = V\cos\theta, (q_\theta)_{r\to\infty} = -V\sin\theta,$

which gives V as the resultant velocity along the axis of symmetry.

Let q_n be the normal velocity along the curve, then

$$q_n = q_r\sin\phi - q_\theta\cos\phi - \sin\phi(q_r - q_\theta\cot\phi) \qquad \text{...(4)}$$

where $\quad \cot\phi = (1/r)dr/d\theta = -(1/4)\tan\theta.$ ...(5)

From (2), (3), and (5) the normal velocity, q_n along the boundary of the meridian curve vanishes which shows that the solid of revolution formed by the revolution of $r^4 = a^4 \cos\theta$ about the axis is at rest in the fluid.

Hence the given value of ψ determines the motion due to the streaming of the fluid with velocity V past the solid of revolution which is at rest.

Example 26: *Find the Stokes' stream function ψ where fluid motion is due to a source of strength m (flux $4\pi m$) at a fixed point A, a sink - m at another fixed point B, a translation of the fluid of velocity U in the direction AB being superposed; explain how this solution can be used to deduce the motion of fluid past a certain solid of revolution. If $U = (8m/9a^2)$, where $AB = 2a$, prove that the solid is of axial length 4a, of equatorial radius approximately 1.6a.*

Solution: The stream function, due to a source and a sink of equal strength with a uniform stream U in the negative X-direction, is given by

$$\psi = \frac{1}{2}Ur^2\sin^2\theta + m(\cos\theta_2 - \cos\theta_1) \qquad \text{...(1)}$$

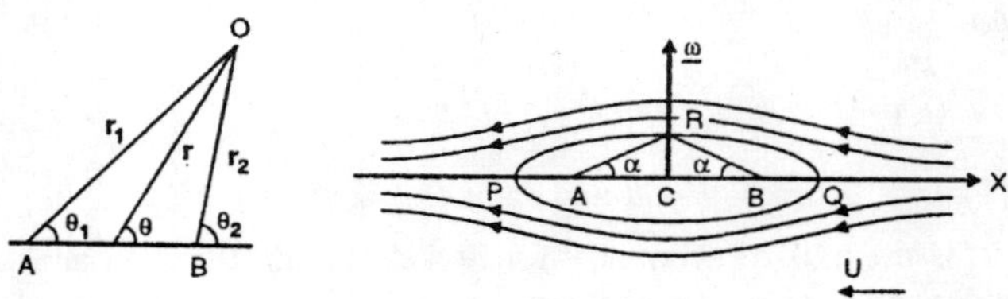

Fig. 3.31

Now $\psi = 0$ for all points on the axis except the portion AB gives

$$\frac{1}{2}Ur^2\sin^2\theta + m(\cos\theta_2 - \cos\theta_1) = 0$$

or $w^2 + b^2(\cos\theta_2 - \cos\theta_1) = 0, b^2 = 2m/U,$ ...(2)

which gives the equation of dividing streamline. Since $\cos\theta_1$, $\cos\theta_2$ are each numerically less than 1, $w^2 > 2b^2$, hence te dividing streamline is closed.

By rotating this streamline about the axis, we obtain a surface of revolution which is taken to be the boundary of a solid

The velocity is zero at the stagnation points P and Q. Let PQ = 2l, CR = h, so that PC = QC = l, then at Q

$$-\frac{m}{(l+a)^2}+\frac{m}{(l-a)^2}=U$$

or $(m/U)\{-(l-a)^2+(l+a)^2=(l^2-a^2)^2$

or $2ab^2l=(l^2-a^2)^2$ ...(3)

Again $h^2+b^2(-\cos\alpha-\cos\alpha)=0 \Rightarrow h^2=2b^2\cos\alpha$ ...(4)

If $U=8m/9a^2$ then $b^2=2m/U=(9/4)a^2$.

Hence the axial length of the solid is $2l$ *i.e.,* $4a$

From (4), we have

$$h^2=2b^2\frac{a}{\sqrt{(h^2+a^2)}} \Rightarrow h=1.6a \text{ approx.}$$ **Proved.**

Example 27: *A solid of revolution is moving along its axis in an infinite liquid, show that the kinetic energy of the liquid is*

$$-\frac{1}{2}\pi\rho\int(\psi/\underline{\omega})(\partial\psi/\partial n)ds,$$

where ψ *is the Stoke's stream function of the time* $\underline{\omega}$ *the distance of a point from the axis and the integral is taken once round a meridian curve of the solid. Hence obtain the kinetic energy of infinite liquid due to the motion of a sphere through it with velocity V.*

Solution: We know that the kinetic energy of the liquid in an irrotational motion is given by

$$T=-\frac{1}{2}\rho\int\phi(\partial\phi/\partial n)ds$$ ...(1)

Let an elementary element ds of the meridian revolves about the axis, so that

$$ds=-2\pi\underline{\omega}\,ds$$ ...(2)

Again $$-\frac{\partial\phi}{\partial n}=\frac{1}{\underline{\omega}}\frac{\partial\phi}{\partial S}.$$ ...(3)

From (1) and (3). We have

$$T=-\frac{1}{2}\rho\int\phi\left(\frac{1}{\underline{\omega}}\frac{\partial\phi}{\partial S}\right).2\pi\underline{\omega}\,dS=\pi\rho\int\phi\,\partial\psi$$

or $T = -\pi\rho \int \psi \dfrac{\partial\phi}{\partial S} dS = -\dfrac{1}{2}\pi\rho \int \dfrac{\psi}{\omega}\dfrac{\partial\psi}{\partial n} dS,$

integration being taken round the whole boundary which is double of the meridian curve. For the K.E of sphere moving with velocity V along the axis, we have

$$\phi = \frac{1}{2}\frac{Va^3\cos\theta}{r^2} \quad \text{and} \quad \psi = -\frac{1}{2}\frac{Va^3\cos^2\theta}{r}$$

Thus $T = -\pi\rho \int \phi d = \dfrac{1}{2}\pi\rho V^2 a^3 \int_0^{\pi} \cos^2\theta \sin\theta \, d\theta$

or $T = \dfrac{1}{3}\pi\rho V^2 a^3 = \dfrac{1}{4} MV^2,$

where $M = \dfrac{3}{4}\pi a^3 \rho$, mass of the liquid displaced by the sphere.

Therefore the kinetic energy of infinite liquid due to the motion of a sphere through it with velocity V is (1/4) MV^2. **Ans.**

Example 28: *Find the Stokes' stream function ψ where fluid motion is due to a source of stream m (flux $4\pi m$) at a fixed point and a translation of the fluid of velocity V. Explain how this solution can be used to deduce the notion of fluid past a blunt-nosed cylindrical body whose diameter is ultimately 4a where $a^2 = m/V$.*

Solution: Here the Stoke's stream function ψ is given as

$$\psi = \frac{1}{2}Vr^2\sin^2\theta + Va^2\cos\theta. \qquad ...(1)$$

The components of velocity vanish at the stagnation point *i.e.*,

$$\partial\psi/\partial r = 0 \text{ and } \partial\psi/\partial\theta = 0.$$

So $Vr\sin^2\theta = 0 \Rightarrow \sin\theta = 0 \Rightarrow \theta = \pi$

and $\sin\theta(Vr^2\sin^2\theta - m) = 0 \Rightarrow V\cos\theta - (m/r^2) = 0$ ...(2)

At θ = π, (2) becomes

$V = m/r^2 \Rightarrow r^2 = m/V = a^2$. (Given)

Therefore the stream lines which pass through the stagnation point is given by

$$-\frac{1}{2}Vr^2\sin^2\theta + m\cos\theta = -Va^2, \quad -\frac{1}{2}Vr^2\sin^2\theta + Va^2\cos\theta = -Va^2,$$

or $r^2\sin^2\theta = 2a^2(1+\cos\theta)$ or $w^2 = 2a^2(1+\cos\theta)$,

which is the dividing stream lines and vies a cylindrical surface when rotated. This is considered as the boundary of the solid. Again, θ → 0, ω → 2*a*. It follows that the diameter of the cylindrical body is 4a. **Proved.**

Example 29: *The space bounded by the paraboloids $x^2 + y^2 = ax$, $x^2 + y^2 = b(z - c)$ (where a, b, c are positive and $b > a$), outside the former and inside the latter, contains liquid at rest. Suddenly the bounding surfaces are made to move with velocities U, V respectively in the direction of the axis of Z. Prove that in the motion, instantaneously set up the surfaces over which the current function is constant are paraboloids of latus-rectum.*

$ab(U - V) / (aU - bV)$.

Solution: The equation of the paraboloids are $x^2 + y^2 = az$,

or $\omega^2 = az \Rightarrow r^2 \sin^2\theta = ar\cos\theta$ ...(1)

and $x^2 + y^2 = b(z - c)$.

or $\omega^2 = b(z - c) \Rightarrow r^2 \sin^2\theta = a(r\cos\theta - c)$ ...(2)

The Stocke's stream function ψ, on the boundaries, is given by

$$\psi = -\frac{1}{2}Ur^2\sin^2\theta + \text{const.} \quad ...(3)$$

we know that $-\dfrac{\partial\phi}{\partial z} = -\dfrac{1}{\omega}\dfrac{\partial\psi}{\partial\omega}$ and $-\dfrac{\partial\phi}{\partial\omega} = \dfrac{1}{\omega}\dfrac{\partial\psi}{\partial z}$.

By eliminating ϕ, we have

$$\frac{\partial^2\psi}{\partial z^2} + \frac{\partial^2\psi}{\partial\omega^2} - \frac{1}{\omega}\frac{\partial\psi}{\partial\omega} = 0.$$

$\Rightarrow \psi = z$ and $\psi = \omega^2$ are the solutions of equation (4). Assuming a solution of the form $\psi = Ar^2\sin^2\theta + Br\cos\theta$. ...(5)

From (1), (3) and (5), we have

$$Aar\cos\theta + Br\cos\theta = -\frac{1}{2}aUr\cos\theta + \text{cosnt.} \quad ...(6)$$

From (2), (3) and (5), we have

$$Ab(r\cos\theta - C) + Br\cos\theta = -\frac{1}{2}bVr\cos\theta + \text{cosnt.} \quad ...(7)$$

From (6) and (7), it follows that

$$Aa + B = -\frac{1}{2}aU, Ab + B = -\frac{1}{2}bV$$

$$\Rightarrow A = \frac{1}{2}\frac{bV - aU}{a - b} \quad \text{and} \quad B = \frac{1}{2}\frac{ab(U - V)}{a - b}$$

Therefore ψ = const. $\Rightarrow Ar^2\sin^2\theta + Br\cos\theta$ = const.

$\Rightarrow A\omega^2 + Bz$ = const. $\Rightarrow A(x^2 + y^2) + Bz$ = const.,

which represents a paraboloid of latus rectum $(-B / A)$. **Proved.**

Example 30: *A model of an aeroplane built to (1/10) th scale is to be rested in a wind tunnel which operates at a pressure of 20 atmospheres. The aeroplane is expected to fly at a speed of 500 km/h. At what speed should the wind tunnel operate to give dynamic similarity model and prototype.*

Solution: For dynamical similarity, we have

$Re_m = Re_p$

or $\dfrac{\rho_m V_m l_m}{\mu_m} = \dfrac{\rho_p v_p l_p}{\mu_p} \Rightarrow V_m = \dfrac{\rho_p \mu_m l_p}{\rho_m \mu_p l_m} V_p$.

Since the coefficient of dynamic viscosity is not significantly altered by pressure changes unless the pressure change is large, so $\mu_m = \mu_p$. Also $\rho_m = 20\ \rho_p$.

Therefore v_m = (1/20) × 10 × v_p = (1/20) × 500 = 250 km/hr. **Ans.**

Example 31: *A fluid flow situation depends upon the velocity v, the density r, several linear dimensions l, l_1, l_2, pressure drop Δp, gravity g, viscosity μ, surface tension s, and bulk modulus of elasticity k. Apply dimensional analysis to these variables to find a set of Π parameters.*

Solution: The relationship between a group of variables is given by

$F(V, \rho, l, l_1, l_2, \Delta p, g, \mu, \sigma, k) = 0$ ...(1)

There are ten variables involving three dimensions, so the number of P group is seven. Consider V, r and l are three repeating variables, so

$\Pi_1 = V^a_1\ \rho^b{}_1\ l^c_1\ \Delta p = (LT^{-1})^a_1\ (ML^{-3})^b_1\ L^c_1(ML^{-1}T^{-2}),$

$\Pi_2 = V^a_2\ \rho^b{}_2\ l^c_{2g} = (LT^{-1})^a_2\ (ML^{-3})^b_2\ L^c{}_2(LT^2),$

$\Pi_3 = V^a_3\ \rho^b_3\ l^c_3\ \mu = (LT^{-1})^a_3\ (ML^{-3})^b_3\ L^c_3(ML^{-1}T^{-1}),$

$\Pi_4 = V^a_4\ \rho^b_4\ l^c_4\ \sigma = (LT^{-1})^a_4\ (ML^{-3})^b_4\ L^c_4(ML^{-2}),$

$\Pi_5 = V^a_5\ \rho^b_5\ l^c_5\ k = (LT^{-1})^a_5\ (ML^{-3})^b_5\ L^c_5(ML^{-1}T^{-2}),$

$\Pi_6 = l/l_1,\ \Pi_7 = l/l_2.$

Equating the indices of the three dimensions and solving, we have

$\Pi_1 = (\Delta p/\rho V^2),\ \Pi_2 = (gl/V^2),\ \Pi_3 = (\mu/Vl\rho),\ \Pi_4 = (\sigma/V^2\rho_l)$

$\Pi_5 = (k/\rho V^2),\ \Pi_6 = (l/l_1),\ \Pi_7 = (l/l_2).$

Thus $f\left(\dfrac{\Delta p}{\rho V^2}, \dfrac{gl}{V^2}, \dfrac{\mu}{Vl_p}, \dfrac{\sigma}{V^2 \rho l}, \dfrac{k}{\rho V^2}, \dfrac{l}{l_1}, \dfrac{l}{l_2}\right) = 0$ ...(2)

Equation (2) may be written as

$$\phi\left(\frac{\Delta p}{\rho V^2}, \frac{V^2}{gl}, \frac{Vl\rho}{\mu}, \frac{V^2 \rho l}{\sigma}, \frac{V}{\sqrt{(k/\rho)}}, \frac{l}{l_1}, \frac{l}{l_2}\right) = 0$$

$$\text{or } \phi\left(\frac{\Delta p}{\rho V^2}, Fr, Re, W, M, \frac{l}{l_1}, \frac{l}{l_2}\right) = 0,$$

where Fr is the Froude number, Re is the Reynolds number, W is the weber number and M is the Mach number.

$$\text{or } \frac{\Delta p}{\rho V^2} = \phi_1\left(Fr, Re, W, M, \frac{l}{l_1}, \frac{l}{l_2}\right),$$

$$\text{or } \Delta p = \rho V^2 \phi_1\left(Fr, Re, W, M, \frac{l}{l_1}, \frac{l}{l_2}\right),$$

where ϕ_1 is to be determined either from analysis or from experiment, **Ans.**

Example 32: *For the flow of a fluid through similar pipes show that* $P = (\rho l V^2/d)\ \phi\ (Vd\rho/\mu)$,

where P is the drop in pressure over the length l of the pipe, d is diameter of the pipe, ρ is the density and μ is viscosity of the fluid and V is the mean velocity of flow through the pipe.

Solution: Let P = f^0 ρ, μ, V, d, l)

$P = Ar^a\ \mu^b\ V^c\ d^{dle}$ ...(1)

Expressing the relation (1) dimensionally, we have

$ML^{-1}T^{-2} = (ML^{-3})^a\ (ML^{-1}T^{-1})^b\ (LT^{-1})^c\ (L)^d\ (L)^e$ ...(2)

Equating powers of M, L, and T, we get

$1 = a + b,\ -1 = -3a - b + c + d + e,\ -2 = -b - c.$

Solving these equations for the powers of ρ, V and d, we get

$a = 1 - b,\ c = 2 - b,\ d = -b - e.$ ...(3)

From (1) and (3), we have

$P = A\rho^{1-b}\ \mu^b V^{2-b}\ d^{-b-ele}$

or $P = A\rho V^2\ (l/d)^e\ (\rho Vd/\mu)^{-b}$

$$\text{or } P = A\ \frac{\rho V^2 l}{d}\left(\frac{i}{d}\right)^{e-1}\left(\frac{\rho Vd}{\mu}\right)^{-b}$$

For geometrically similar pipes (l/d) is a constant and $(l/d)^{e-1}$ can be combined with constant A. Hence

$$P = \left(P = \frac{\rho l V^2}{d}\right)\phi\left(\frac{Vd\rho}{\mu}\right)$$ **Proved.**

Example 33: *Show by means of dimensional analysis that the thrust T of a screw propeller is given by*

$T = \rho d^2 V^2 \phi\, (Vd\rho/\mu,\ dn/V)$,

where r is the fluid density, m its viscosity, d is the diameter of the propeller, V is the speed of advance and n is the revolution per second

Solution: Let $T = f(\rho, \mu, d, V, n)$

$T = A\rho^a \mu^b d^c V^d n^e$. ...(1)

Expressing the relation (1) dimensionally, we have

$MLT^{-2} = (ML^{-3})^a (ML^{-1}T^{-1})^b (L)^c (LT^{-1})^d (T^{-1})^e$ ...(2)

Equating power of M, L and T, we have

$1 = a + b,\ 1 = -3a - b + c + d,\ -2 = -b - d - e.$

Solving these equations for the powers of ρ, V and d, we have

$a = 1-b,\ c = 2 + e - b,\ d = 2 - b - e.$ ...(3)

Substituting (3) into (1), we have

$T = A\rho^{1-b} \mu^b d^{2+e-b} V^{2-b-e} n^e$

$$\text{or } T = A\rho d^2V^2\,\phi\left(\frac{Vd\rho}{\mu}\right)^{-b}\left(\frac{dn}{V}\right)^{e} = A\rho\, d^2V^2\,\phi\left(\frac{Vd\rho}{\mu}, \frac{dn}{V}\right).\ \textbf{Proved.}$$

Example 34(a): *A 1 : 20 model of an airduct is to be tested with water which is 45 times more viscous and 850 times more dense than air. What should be the pressure drop in the prototype if the pressure drop is 3kg/cm² in the model when tested under hydrodynamically similar conditions?*

Solution : The pressure drop is due to viscous effects, for dynamic similarity Reynolds number must be the same in the model and the prototype

i.e., $(Re)_m = (Re)_p$

$$\Rightarrow \left(\frac{VL\rho}{\mu}\right)_m = \left(\frac{VL\rho}{\mu}\right)_p$$

$$\frac{(V)_p}{(V)_m} = \frac{\rho_m L_m \mu_p}{\rho_p L_p \mu_m} = \frac{850}{20 \times 45} = \frac{17}{18}$$

Again, Euler's numbers must be the same for dynamic similarity

$$\frac{(\Delta p)_p}{\rho_p V_p^2} = \frac{(\Delta p)_m}{\rho_m V_m^2}$$

The pressure drop in the prototype becomes

$$\Rightarrow (\Delta p)_p = (\Delta p)_m \frac{\rho_p}{\rho_{m}} \frac{V_p^2}{V_m^2}$$

$$\Rightarrow (\Delta p)_p = 3 \times \frac{1}{850} \times \left(\frac{17}{18}\right)^2$$

$$\Rightarrow (\Delta p)_p = 3.4 \times 10^{-3} \text{ kg/cm}^2$$ **Ans.**

Example 34(b): *The drag of a small submarine hull is desired when it is moving far below the surface of water. A 1 / 10 scale model is to be tested. What dimensionless group should be duplicated between the model and prototype? If the drag of the prototype at 1 knot is desired at what speed should the model be moved to give the drag the drag to be expected by the prototype?*

Solution: A submarine is a wholly submerged body in an infinite mass of fluid at rest. The Reynolds number must be duplicated for the model and the prototype. Equating the Reynolds number for dynamic similarity, we have

$$U_1 l_1 \rho_1/\mu_1 = U_2 l_2 \rho_2/\mu_2 .$$

Here $l_1 = (1/10)\, l_2$, $\rho_1 = \rho_2$ and $\mu_1 = \mu_2$ for the same fluid, (water) for the model and prototype.

$$U_1 = U_2\, (l_2/l_1) = 1 \times 10 = 10 \text{ knots}$$

The drag coefficient $D/\rho U^2 l^2$ should be duplicated to evaluate the drag experienced by the prototype.

$$D_2 = D_1 \frac{\rho_2 U_2^2 l_2^2}{\rho_1 U_1^2 l_1^2} = \frac{1}{10^2} \times 10^2 \times D_1 = D_1 .$$

which shows that if the 1/10 model is moved at 10 knots the drag experienced by it would be the same as that by the prototype at 1 knot speed.

Example 34(c): *The viscous force FD exerted by the fluid on a sphere of diameter D depends on viscosity μ, mass density of fluid ρ and velocity of the sphere U. With dimensional analysis, determine the general form of the equation.*

Solution: The dependence of F_D the expressed as

$$F_D = f\,(D, \mu, \rho, U). \quad ...(1)$$

There are five variables involving three primary units so the number of dimensionless parameters is two. Taking U, D and r as the repeating variables, these two parameters will be

$$\Pi_1 = F_D\, U^a\, D^b\, \rho^c, \; \Pi_2 = \mu\, U^d\, D^e\, \rho^f \quad ...(2)$$

Writing the dimensions of each equantity, we get

$[\Pi_1] = M_0 L_0T_0 = (MLT^{-2}) (LT^{-1})^a (L)^b (ML^{-3})^c$

$[\Pi_2] = M_0L_0T_0 = (ML^{-1}T^{-1}) (LT^{-1})^d (L)^e (ML^{-3})^f$.

Equating the exponents of M, L, and T, we have

$$a = -2,\ b = -2,\ c = -1$$

and $d = -1,\ e = -1,\ f = -1$ ...(3)

Thus $\Pi_1 = F_D/ (\rho D^2U^2)$ and $\Pi_2 = \mu/(UD\rho)$.

Hence $F_D/\rho D^2U^2 = f\ (UD\rho/\mu)$. ...(4)

Thus for spheres, the drag coefficient $C_D = F_D/(\rho U^2D^2)$ is a function of Reynolds number Re.

Example 35: *A mass of fluid is in motion so that the lines of motion lie on the surface of coaxial cylinders. Show that the equation of continuity is*

$$\frac{\partial \rho}{\partial t}+\frac{1}{r^2}\frac{\partial}{\partial \theta}(\rho q_r)+\frac{\partial}{\partial z}(\rho q_z) = 0,$$

where q_r, q_z are the velocities perpendicular and parallel to z.

Solution: Let us concider a point P (r, θ, z) in the fluid. Construct a parallelopiped at P with edges PS (= rδθ), PR (= δr) and PP' (= δz). Let q_r, q_θ and and q_z be the velocity components along PR, PS, PP'. Since the lines of motion of the fluid lie on the surface of coasial cylinders so there is no motion along PR.

Excess of mass of flow-in over flow out along the direction PS

$$= -r\,\delta\theta\frac{\partial}{\partial z}(\rho q_\theta\,\delta r.\delta z)$$

and excess of mass of flow-in over flow out along the direction PP'

$$= -\delta z.\frac{\partial}{\partial z}(\rho q_z\,\delta r.r\delta z).$$

Mass of the fluid element

$$= \rho r\delta\theta.\ \delta r.\ \delta z.$$

Rate of increase in the mass of the fluid element

$$= \frac{\partial}{\partial t}(\rho r\delta\theta.\delta r.\delta z).$$

From the equation of continuity, we have

$$\frac{\partial}{\partial t}(\rho r\delta\theta\,\delta r\,\delta z) = -r\delta\theta.\frac{\partial}{r\partial\theta}(\rho\theta_z\,\delta r\,\delta_z)-\delta_z\frac{\partial}{\partial z}(Pq_z\delta r\delta\theta)$$

$$\Rightarrow \frac{\partial \rho}{\partial t}(r\delta\theta.\delta r\,\delta z) = +\frac{\partial}{r\partial\theta}(r\delta\theta\,\delta r\,\delta z)+\frac{\partial}{\partial z}(\rho q_z)(r\delta\theta\,\delta r\,\delta z) = 0$$

$$\Rightarrow \frac{\partial \rho}{\partial t}+\frac{\partial}{r\partial\theta}(\rho q_\theta)+\frac{\partial}{\partial z}(\rho q_z) = 0.$$

Hence Proved.

Example 36: *If the linesof motion are curves on the surfaces of cones having their vertices at the origin and the axis of z for common axis. Prove that the equation of continuity is*

$$\frac{\partial \rho}{\partial t}+\frac{\partial}{\partial r}(\rho q_r)+\frac{2\rho q_r}{r}+\frac{\cos ec\theta}{r}\frac{\partial}{\partial \omega}(\rho q_\omega) = 0.$$

Solution: Let OZ be the common axis of Z with O as vertex. Let A(r, θ, ω) be a point on the surface of the cone and q_r, q_θ, $q\,\omega$ be the components of the velocity along the edges of the parallelopiped AA' (= δr), AB (= rδθ) and AD (= rsinθ $\delta\,\omega$) respectively. Since the lines of motion are curves on the surfaces of cones so there will be no motion perpendicular to the surface of the cone.

Excess of flow- in over flow out in the direction AA' *i.e.*, from the face ABB'A' and the opposite face in time δt

$$= -\frac{\partial}{\partial r}\{\rho q_w.r\delta\theta.r\sin\theta\delta\omega\}\delta r\delta t$$

and the excess of flow-in over flow out in the direction AD *i.e.*, from the face ABB'A' and the opposite face in time δt.

$$= -\frac{\partial}{r\sin\theta\,\delta\omega}\{\rho q\omega.r\delta\theta\,\delta r\}r\sin\theta\,\delta\omega\,\delta t.$$

Total excess of mass flow-in over flow out in time δt

$$= -\frac{\partial}{\partial r}\{\rho q_r\, r\delta\theta.r\sin\theta\,\delta\omega\}\,\delta r\,\delta t$$

$$-\frac{\partial}{r\sin\theta\,\partial\omega}\{\rho q_\omega.r\delta\theta\,\delta r\}r\sin\theta\,\delta\omega\,\delta r\,\delta t \quad ...(1)$$

Rate of increase in the mass of an element in time δt

$$= \frac{\partial}{\partial t}\{\rho r\,\delta\theta.r\sin\theta\,\delta\omega.\delta r\}\,\delta t \quad ...(2)$$

From the equation of continuity, we have

$$= \frac{\partial \rho}{\partial t}\{r^2\sin\theta\,\delta r\,\delta\theta\,\delta\omega\}\,\delta t = \frac{\partial}{\partial r}\{\rho q_r.r\delta\theta.r\sin\theta\,\delta\omega\}\delta r\,\delta t$$

$$-\frac{\partial}{r\sin\theta\partial\omega}\{\rho q\omega\, r\delta\theta\,\delta r\}r\sin\theta\,\delta\omega\,\delta t$$

$$\Rightarrow \frac{\partial \rho}{\partial t}+\frac{1}{r^2}\frac{\partial}{\partial r}(\rho r^2 q_r)+\frac{1}{r\sin\theta}\frac{\partial}{\partial \omega}(\rho q_\omega) = 0$$

or $$\frac{\partial \rho}{\partial t}+\frac{1}{r^2}\left\{r^2\frac{\partial}{\partial r}(\rho q_r)+(\rho q_r).2r\right\}+\frac{\operatorname{cosec}\theta}{r}\frac{\partial}{\partial \omega}(\rho q_\omega) = 0$$

or $$\frac{\partial \rho}{\partial t}+\frac{\partial}{\partial r}(\rho q_r)+\frac{2\rho q_r}{r}+\frac{\operatorname{cosec}\theta}{r}\frac{\partial}{\partial \omega}(\rho q_\omega) = 0.$$ **Hence Proved.**

Example 37: *If every particle moves on the surface of the sphere, prove that the equation of continuty is*

$$\cos\theta\frac{\partial \rho}{\partial t}+\frac{\partial}{\partial \theta}(\rho\omega\cos\theta)+\frac{\partial}{\partial \phi}(\rho\omega'\cos\theta) = 0.$$

ρ being the density, θ, ϕ the latitude and longitude of any element and ω' the angular velocities of the element in latitude and longitude respectively.

Solution: Consider an elementary parallelopiped PQRS and P'Q'R'S' on the surface of the sphere, whose lenght of the edges are as follows

PP' = δr, PQ = $r\delta\theta$,

PS = $r\cos\theta\ \delta\phi$.

Since ω and ω' are the angular velocities of the element along latitutde and longitude respectively so $r\omega$ and $r\cos\theta\,\omega'$ are the velocities along PQ and PS. Since the particle moves on the surface of the sphere it follows that there will be no velocity normal to the surface of the sphere *i.e.*, the velocity along PP' vanish.

Excess of flow-in over flow out along the faces perpendicular to PQ *i.e.*, from the face QQ' RR' and PP' S'S is

$$= -r\delta\theta.\frac{\partial}{r\partial\theta}(\rho r\omega\delta r.r\cos\theta\delta\phi) \text{ per unit time}$$

Simmilarly excess of flow-in over flow out along the face perpendi-cular to PS is

$$= -r\cos\theta\delta\phi.\frac{\partial}{r\cos\theta\partial\phi}(\rho r\cos\theta\,\omega'\delta r\, r\delta\theta) \text{ per unit time}$$

Total excess of flow-in over flow out through all the faces

$$= -[r\delta\theta\frac{\partial}{r\partial\theta}(\delta r\rho\omega.\rho\cos\theta\phi)$$

$$+r\cos\theta\,\delta\phi\frac{\partial}{r\cos\theta\partial\phi}(\rho r\cos\theta\omega'.\delta r.r\delta\theta)] \text{ per unit time} \qquad ...(1)$$

Rate of increase in the mass per unit time is

$$= \frac{\partial}{\partial t}(\rho\delta r.r\delta\theta.r\cos\theta\,\delta\phi.) \qquad ...(2)$$

The equation of continuity is given by the relation

$$\frac{\partial}{\partial t}(\rho\delta r.r\delta\theta.r\cos\theta\,\delta\phi) = -r\delta\theta.\frac{\partial}{r\partial\theta}(\rho r\omega.\delta r.r\cos\theta\delta\phi)$$

$$-r\cos\theta\delta\phi.\frac{\partial}{r\cos\theta\partial\phi}(\rho r\cos\theta\omega'.\delta r.r\delta\theta)$$

$$\Rightarrow \frac{\partial\rho}{\partial t}+(r^2\cos\theta\,\delta r\,\delta\theta\delta\phi)+r\delta\theta.\frac{\partial}{r\partial\theta}(\rho\omega\cos\theta)r^2\delta r\delta\phi$$

$$+r\cos\theta\delta\phi.\frac{\partial}{r\cos\theta\partial\phi}(\rho\omega'\cos\theta)r^2\delta r\,\delta\theta = 0$$

$$\Rightarrow \frac{\partial\rho}{\partial t}+\frac{1}{\cos\theta}\frac{\partial}{\partial\theta}(\rho\omega\cos\theta)+\frac{1}{\cos\theta}\frac{\partial}{\partial\phi}(\rho\omega'\cos\theta) = 0$$

$$\Rightarrow \cos\theta\frac{\partial\rho}{\partial t}+\frac{\partial}{\partial\theta}(\rho\omega\cos\theta)+\frac{\partial}{\partial\phi}(\rho\omega'\cos\theta) = 0.$$ **Hence proved.**

Example 38: *If the lines of motion are curves on the surfaces of spheres all touching the plane of –XY at the origin O, the equation of continuity is*

$$r\sin\theta\frac{\partial\rho}{\partial t}+\frac{\partial}{\partial\phi}(\rho v)+\sin\theta\frac{\partial}{\partial\theta}(\rho u)+1+2\cos\theta) = 0,$$

where r is the radius CP of one of the spheres, θ the angle PCO, u the velocity in the plane PCO, to a fixed plane through the axis of Z.

Solution: Let C and C' be the centres of the Z-axis of the two consecutive spheres of radii r and r + δr respectively, such that CC' = δr. Join the line CP which meets the second sphere in Q. Consider P be a point on the inner sphere and PQ,PR and PS be the edgesof the elementary parallelopiped of lengths

PR = rδθ,

and PS = rsinθ δϕ,

where ϕ isthe angle which the plane PCO makes with a fixed plane through the Z-axis *i.e.*, with – XOZ plane.

Now we shall determine the lenght of the edge PQ.

CP = r, C' Q = r + δr,

CC' = δr, ∠ PCO = θ.

Let ∠ CC'Q = θ' and ∠ C'QC = λ.

Then $\theta' + \lambda = \theta$

$\Rightarrow \qquad \theta' = \theta - \lambda$

From the Δ CQC', we have

$CQ^2 = C''Q^2 = + CC'^2 - 2C'Q.\ CC' \cos\theta'$

$$\Rightarrow (r+PQ)^2 = (r+\delta r)^2 + (\delta r)^2 - 2(r+\delta r)\delta r \cos(\theta - \lambda)$$

$$\Rightarrow r^2 + 2rPQ + PQ^2 = r^2 + 2r\delta r + \delta r^2 + \delta r^2$$

$$-2(r\delta r + \delta r^2)\cos(\theta - \lambda)$$

Neglecting the squares of the quantities PQ and δr, being small, we have

$$2r.PQ = 2r\,\delta r - 2r\,\delta r.\cos(\theta - \lambda)$$

or $PQ = \delta r\{1 - \cos(\theta - \lambda)\}$

or $PQ = \delta r\{1 - (\cos\theta\cos\lambda + \sin\theta\sin\lambda)\}$

or $PQ = \delta r\{1 - \cos\theta - \lambda\sin\theta)\}$

or $PQ = \delta r\{1 - \cos\theta\}$ (neglecting the small quantities)

Since lines of motion are curves on the surfaces of spheres, so there will be no component of velocity along PQ, u and vbe the velocity components along the edegs PR and PS in the direction of θ and φ increasing.

Excess of flow-in over out in the direction PR

$$= -\frac{\partial}{r\partial\theta}\{\rho u r \sin\theta\,\delta\phi(1 - \cos\theta)\delta r\} r\,\delta\theta \text{ per unit time.}$$

Similarly the excess of flow-in over flow out in the direction PS

$$= -\frac{\partial}{r\sin\theta\partial\phi}\{\rho v r\,\delta\theta\,\delta r(1 - \cos\theta)\delta r\} r\sin\theta\,\delta\phi \text{ per unit time}$$

Total excess of flow-in over flow out through all the faces

$$= [-\frac{\partial}{r\partial\theta}\{\rho u r\sin\theta\,\delta\phi\,(1-\cos\theta)\delta r\} r\,\delta\theta$$

$$-\frac{\partial}{r\sin\theta\partial\phi}\{\rho v r\,\delta\theta\,\delta r\,(1-\cos\theta)\}\, r\sin\theta\,\delta\phi] \qquad ...(1)$$

Volume of the parallelopiped

$= (1 - \cos\theta)\delta r.r\delta\theta\, r\sin\theta\delta\phi.$

Rate of increase in the mass of the parallelopiped.

$$= \frac{\partial}{\partial t}\{\rho(1-\cos\theta)\delta r\,\delta\theta.r\sin\theta\delta\phi\} \text{ per unit time} \quad ...(2)$$

From the equation of continuity, we have

$$\frac{\partial}{\partial t}\{\rho(1-\cos\theta)\delta r\,r\delta\theta.r\sin\theta\delta\phi\}$$

$$= \frac{\partial}{r\partial\theta}\{\rho u\,r\sin\theta\delta\phi(1-\cos\theta)\delta r\}\,r\,\partial\theta$$

$$-\frac{\partial}{r\sin\theta\partial\phi}\{\rho v\,r\,\delta\theta\,(1-\cos\theta)\delta r\}\,r\sin\theta\,\delta\phi$$

$$\Rightarrow \frac{\partial\rho}{r\partial\theta}\{(1-\cos\theta)\delta r\,r\delta\theta\,r\sin\theta\delta\phi\}$$

$$+\frac{\partial}{r\partial\theta}\{\rho u\sin\theta\,(1-\cos\theta)\}r^2\delta r\,\delta\theta\,\delta\phi$$

$$+\frac{\partial}{r\sin\theta\,\partial\phi}\{\rho v\}\,r^2\,(1-\cos\theta)\,\delta r\,\delta\theta\,\delta\phi = 0$$

$$\Rightarrow \frac{\partial\rho}{\partial t}+\frac{1}{r\sin\theta\,(1-\cos\theta)}\frac{\partial}{\partial\theta}\{\rho u\sin\theta\,(1-\cos\theta)\}$$

$$+\frac{1}{r\sin\theta}\frac{\partial}{\partial\phi}(\rho v) = 0$$

$$\Rightarrow \frac{\partial\rho}{\partial t}+\frac{1}{r\sin\theta\,(1-\cos\theta)}[\sin\theta\,(1-\cos\theta)\frac{\partial}{\partial\theta}(\rho u)$$

$$+\rho u\{\cos\theta\,(1-\cos\theta)+\sin^2\theta\}]+\frac{1}{r\sin\theta}\frac{\partial}{\partial\phi}(\rho v) = 0$$

$$\Rightarrow \frac{\partial\rho}{\partial t}+\frac{1}{r}\frac{\partial}{\partial\theta}(\rho u)+\frac{\rho u\,(1+\cos\theta-2\cos^2\theta)}{r\sin\theta\,(1-\cos\theta)}+\frac{1}{r\sin\theta}\frac{\partial}{\partial\phi}(\rho v) = 0$$

$$\Rightarrow \frac{\partial\rho}{\partial t}+\frac{1}{r}\frac{\partial}{\partial\theta}(\rho u)+\frac{\rho u\,(1+2\cos\theta)}{r\sin\theta}+\frac{1}{r\sin\theta}\frac{\partial}{\partial\phi}(\rho v) = 0$$

$$\Rightarrow r\sin\theta\frac{\partial\rho}{\partial t}+\frac{\partial}{\partial\phi}(\rho v)+\sin\theta\frac{\partial}{\partial\theta}(\rho u)(1+2\cos\theta) = 0.$$ **Hence Proved.**

Example 39: *The charge Q through a rotating machine such as a pump, a turbine, or a compressor depends on the shaft work (gH), power supplied P, speed of rotation N, characteristic length D, fluid density ρ and viscosity μ. Find a set of dimensionless parameters.*

Solution: There are seven variables involving three primary units, so the number of dimensionless parameter is four. Hence according to Buckingham's

P theorem there will be four dimensionless group. Let N, ρ and D are the repeating variables, then

$\Pi_1 = QN^a\rho^bD^c$.

Writing the dimensions of each quantity, we have

$M^0L^0T^0 = (L^3T^{-1})\ (T^{-2})^a\ (ML^{-3})^b\ (L)^c$

$\Rightarrow a = -1,\ b = 0,\ c = -3$. Thus $\Pi_1 = Q/ND^3$.

Similarly $\Pi_2 = PN^d\ \rho^e\ D^f$.

$M : e + 1 = 0,$

$L : -3e + f + 2 = 0,$

$T : -3 - d = 0.$

Writing the dimensions of each quantity, we have

$M^0\ L^0\ T^0 = (ML^2T^{-3})\ (T^{-1})^d\ (ML^{-3})^c\ (L)^f$

$\Rightarrow d = -3,\ f = -5,\ e = -1.$

Thus $\Pi_2 = P/PN^3D^5$.

Again $\Pi_3 = (gH)/N^2D^2$ and $\Pi_4 = \rho ND^2/\mu$.

Therefore $\dfrac{Q}{ND^3}\ m = f\left(\dfrac{P}{\rho N^3D^S}, \dfrac{gH}{N^2D^2}, \dfrac{\rho ND^2}{\mu}\right)$ **Ans.**

Example 40: *The loss of pressure Δp for laminar flow in a pipe is a function of pipe length l, its diameter D, mean velocity U, and the dynamic viscosity m. Determine an expression for the pressure loss.*

Solution: There are five variables involving three primary units, so the number of dimensionless parameter is two. Assuming, the pressure loss Δp is of the form

$\Delta p = A\ U^a\ \mu^b l^c D^d,$ ...(1)

where A is any constant.

Writing the dimensions of each variable (1), we have

$(ML^{-1}T^{-2}) = A\ (LT^{-1})^a\ (ML^{-1}T^{-1})^b\ (L)^e\ (L)^d$ ...(2)

Equating the exponents of M, L and T, we have

$b = 1,\ a - b + c + d = -1,\ -a - b = -2,$

$\Rightarrow b = 1,\ a = 1,\ d = -(1 + c).$ ...(3)

From (1) and (3), the pressure loss Dp becomes

$\Delta p = A\mu\ U\ (l^c/D^{1+}) = A\ (\mu U/D)\ (l/D)^c.$ **Ans.**

Example 41: *The velocity distribution of a certain two-dimensional flow is given by*

$$u = Ay + B$$

$$\text{and } v = Ct,$$

where A, B, C are constants. Obtain the equation of the motion of fluid particles in Lagrngian method.

Solution: Let r (x, y) be the position of a given particle at any instant of time t. The path lines for the fluid particle are given by

$$q = \frac{dr}{dt} \Rightarrow \frac{dx}{dt} = u = Ay + B,$$

$$\text{and} \frac{dy}{dt} = v = Ct \qquad ...(1, 2)$$

From (2), we have

$$y = \frac{1}{2}Ct^2 + D,$$

where D is an integration constant.
Initially $y = y_0$, $t = 0$; $D = y_0$

$$\Rightarrow \qquad y = \frac{1}{2}Ct^2 + y_0 \qquad ...(3)$$

Form (1) and (3), we have

$$\frac{dx}{dt} = A\left(\frac{1}{2}Ct^2 + y_0\right) + B.$$

$$\text{or } x = A\left(\frac{1}{6}Ct^2 + y_0 t\right) + Bt + E,$$

where E is in integration constant.

Initially $x = x_0$, $t = {}_0$; $E = x_0$.

$$\text{Therefore } x = A\left(\frac{1}{6}Ct^3 + y_0 t\right) + Bt + x_0. \qquad ...(4)$$

The equations (3) and (4) represent the path lines of fluid etements.

Example 42(a): *Determine the equations of the streamlines at the point in an incompressible fluid having spherical polar coordinates (r, θ, φ). The velocity components are*

$$[2r^{-3} \cos \theta,$$

$$= r^{-3} \sin \theta, 0].$$

Solution: From the definition of the stream lines, we have

$$\frac{dr}{2r^{-3}\cos\theta}=\frac{rd\theta}{r^{-3}\sin\theta}=\frac{r\sin\theta d\phi}{0}.$$

By integrating, the equations of the stream lines are obtained as follows :

ϕ = constant,

and $\dfrac{dr}{2r^{-3}\cos\theta}=\dfrac{rd\theta}{r^{-3}\sin\theta}$,

$\Rightarrow$ r = A $\sin^2\theta$.

The equation ϕ = constant represents that the streamlines lie in the planes which pass through the axis of symmetry.

Example 42(b): *The velocity momponents of a flow in sylindrical polar coordinates are (A $r^2z\cos\theta$, B rz $\sin\theta$, C z^2t). Determine the components of the acceleration of a fluid particle.*

Solution: Let q_r, q_θ and q_z be the components of velocity in cylindrical polar coordinates (r, θ,).

q_r = A $r^2z\cos\theta$,

q_θ = B $r_z\sin\theta$,

q_z = C z^2t ....(1)

Let f_r, f_θ, f_z be the components of acceleration then

$$f_r=\frac{\partial q_r}{\partial t}+q_r\frac{\partial q_r}{\partial r}+\frac{q_\theta}{r}\frac{\partial q_r}{\partial\theta}+q_z\frac{\partial q_r}{\partial z}-\frac{q\theta^2}{r}$$

$$f_\theta=\frac{\partial q_\theta}{\partial t}+q_r\frac{\partial q_\theta}{\partial r}+\frac{q_\theta}{r}\frac{\partial q_\theta}{\partial\theta}+q_z\frac{\partial q_\theta}{\partial z}+\frac{q_1q\theta}{r}$$

$$f_z=\frac{\partial q_z}{\partial t}+q_r\frac{\partial q_z}{\partial r}+\frac{q_\theta}{r}\frac{\partial q_z}{\partial\theta}+q_z\frac{\partial q_z}{\partial z}. \quad ...(2, 3, 4)$$

From (1) and (2, 3, 4), we have

f_r = A $r^2z\cos\theta$ (2A rz $\cos\theta$) + B z $\sin\theta$ (– A $r^2z\sin\theta$) + C z^2t(A r $\cos\theta$) – $B^2rz^2\sin^2\theta$,

or f_r = $rz^2[2A^2r^2\cos^2\theta$ – B(A r + B)$\sin^2\theta$ + AC rt $\cos\theta$].

Similarly

f_θ = ($Ar^2z\cos\theta$) (Bz $\sin\theta$) (Bz $\sin\theta$) (Brz $\cos\theta$) + C z^2t(Br $\sin\theta$) + AB $r^2z^2\sin\theta\cos\theta$,

or f_θ = $rz^2\sin\theta$ [B(2AR + B)$\cos\theta$ + BCt],

and f_z = C z^2 + C z^2t. 2C zt,

or $\qquad f_z = C\ z^2(1 + 2C\ zt^2).$

Example 42(c): *Prove that there are only five independent dimensionless groups in the viscous compressible fluid motion.*

Solution: In the viscous compressible fluid motion, consider the following physical quantities are involved

$L, V, \rho, \mu, K, g, p, C_p, T.$

The fundamental units in which the dimensions of all there quantities may be expressed are length, mass, time and temperature. Consider L, U, ρ and K as base quantities, thus the number of independent dimensionless products will be five.

Consider $\Pi_1 = L^a\ U^b\ \rho^c\ K^d\ \mu$

$\Pi_2 = L^e\ U^f\ \rho^g\ K^h\ g$

$\Pi_3 = L^i\ U^j\ \rho^l\ K^m\ p$

$\Pi_4 = L^n\ U^q\ \rho^r\ K^s\ C_p$

$\Pi_5 = L^u\ U^v\ \rho^w\ K^x\ T \qquad \text{...(1)}$

$\Pi_1 = (L)^a\ (LT^{-1})^b\ (ML^{-3})^c\ (MLT^{-3}\Theta^{-1})^d\ (ML^{-1}T^{-1})$

$\Rightarrow \Pi_1 = L^{a+b-3c+d-1}\ M^{c+d+1}T^{-b-3d-1}\Theta^{-d} = M^0L^0T^0\Theta^0$

If Π_1 is dimensionless then

$a + b - 3c + d = 1,$

$c + d = -1,$

$b + 3d = -1,$

$d = 0.$

$\Rightarrow a = -1,\ b = -1,\ c = -1,\ d = 0$

Then $\Pi_1 = L^{-1}\ U^{-1}\rho^{-1}\mu = \dfrac{\mu}{LU\rho} = \dfrac{v}{LU}, \text{as} \dfrac{\mu}{\rho} = v. \qquad \text{...(2)}$

Similarly $\Pi_2 = (L)^e\ (LT^{-1})^f\ (ML^{-3})^g\ (MLT^{-3}\Theta^{-1})^h\ (LT^{-2})$

$\Rightarrow \Pi_2 = L^{e+f-3g+h+1}\ M^{g+h}\ T^{-f-3h-2}\Theta^{-h} = M^0L^0T^0\Theta^0$

If Π_2 is dimensionless then

$e + f - 3g + h + 1 = 0$

$g + h = 0$

$-f - 3h - 2 = 0$

$-h = 0$

$\Rightarrow e = 1,\ h = 0,\ f = -2,\ g = 0$

$$\Pi_2 = LU^{-2}g = \frac{Lg}{U^2} \qquad ...(3)$$

Similarly $\Pi_3 = (L)^i (LT^{-1})^j (ML^{-3})^l (MLT^{-3}\Theta^{-1})^m (ML^{-1}T^{-2})$

$\Rightarrow \qquad \Pi_3 = L^{i+j-3l+m-1} M^{l+m+1} T^{-j-3m-2} \Theta^{-m} = M^0L^0T^0\Theta*^0$

If Π_3 is dimensionless then

$i + j - 3l + m - 1 = 0$

$l + m + 1 = 0$

$-j - 3m - 2 = 0$

$-m = 0$

$\Rightarrow i = 0, j = -2, l = -1, m = 0$

Thus $\Pi_3 = U^{-2}\rho^{-1}p = \dfrac{p}{\rho U^2}$. ...(4)

Similarly $\Pi_4 = (L)^n (LT^{-1})^q (ML^{-3})^r (MLT^{-3}\Theta^{-1})^s (L^2T^{-2}\Theta^{-1})$

$\Rightarrow \Pi_4 = L^{n+q+3r+s+2} M^{r+s} T^{-q-3s-2} \Theta^{-s-1} = M^0L^0T^0\Theta^0$

If Π_4 is dimensionless then

$n + q - 3r + s + 2 = 0$

$r + s = 0$

$-q - 3s - 2 = 0$

$-s - 1 = 0$

$\Rightarrow n = 1, q = 1, r = 1, s = -1$

Thus $\Pi_4 = LU\rho k^{-1}C_p = \dfrac{LU\rho C_p}{k}$ ...(5)

and $\Pi_5 = (L)^u (LT^{-1})^t (ML^{-3})^w (MLT^{-1}\Theta^{-1})^x \Theta$

$\Rightarrow \Pi_5 = L^{u+v-3w+x} M^{w+x} T^{-v-3x} \Theta^{-x+1} = M^0L^0T^0\Theta^0$

If Π_5 is dimensionless then

$u + v - 3w + x = 0$

$w + x = 0$

$-v - 3x = 0$

$-s + 1 = 0$

$\Rightarrow u = -1, v = -3, w = -1, x = 1$

Thus $\Pi_5 = L^{-1} U^{-3}\rho^{-1}KT = \dfrac{KT}{L\rho U^3}$ ...(6)

Therefore the five dimensionless numbers are

$$Re = \frac{1}{\Pi_1} = \frac{LU}{v}; Fr = \frac{1}{\Pi_2} = \frac{U^2}{Lg};$$

$$Pr = \Pi_1\Pi_4 = \frac{v}{LU}.\frac{LU\rho C_p}{k} = \frac{v\rho C_p}{k};$$

$$\frac{\gamma-1}{\gamma} = \frac{\Pi_3}{\Pi_4\Pi_5} = \frac{p}{\rho U_2}.\frac{k}{LU\rho C_p}.\frac{L\rho U^3}{KT} = \frac{p}{\rho C_p T}$$

and $Ma^2 = \dfrac{1}{\gamma\Pi_3} = \dfrac{\rho U^2}{\gamma p}$.

Example 43: *Check the dimensional homogeneity of the following equations:*

(a) $\dfrac{p}{\rho} + \dfrac{q^2}{2g} + z =$ *const., (b)* $\dfrac{\partial u}{\partial x} + \dfrac{\partial v}{\partial y} = 0.$

Solution: (a) $\dfrac{p}{\rho} + \dfrac{q^2}{2g} + z =$ Const.

The dimensions of each term become

$\dfrac{p}{\rho}$: $(ML^{-1}T^{-2})/(ML^{-3}) = L^2T^{-2}$

$\dfrac{q^2}{2g}$: $(LT^{-1})^{-2} = L$,

$z : L = L$,

which shows that dimensions of second and third terms are L where as the dimensions of the first term is L^2T^{-2}. If we divide the first term by an acceleration term by an acceleration term g i.e., (LT^{-2}), the dimensions reduces to L. Thus the correct form of the equations takes the form

$$\frac{p}{\rho g} + \frac{q^2}{2g} + z = \text{Const.}$$ **Ans.**

(b) $\dfrac{\partial u}{\partial x} + \dfrac{\partial v}{\partial y} = 0.$

The dimensions of each term become

$\dfrac{\partial u}{\partial x}$: $LT^{-1}/L = T^{-1}$

$\frac{\partial v}{\partial y}$: $LT^{-1}/L = T^{-1}$,

⇒ that the equation is dimensionally homogeneous. **Proved.**

Example 44: *The discharge through a horizontal capillary tube depends upon the drop per unit length, the diameter, and the viscosity. Find the form of the equation.*

Solution: The relationship between a group of variables is given by

$F(Q, \Delta p/l, D, \mu) = 0.$

There are our variables involving primary units. The number of dimensionless parameter is one, then

$\Pi = \mu Q^a (\Delta p/l)^b D^c.$ ...(2)

By substituting in the dimensions, we get

$\Pi = (ML^{-1}T^{-1}) (L^3T^{-1})^a (ML^{-2}T^{-2})^b L^0 = M^0L^0T^0$ ...(3)

The exponents of each dimension must be the same on both sides of the equation. Equating the indices of three dimensions, we have

$L : 3a - 2b + c - 1 = 0.$

$M : b + 1 = 0,$

$T : - a - 2b - 1 = 0,$

which gives $a = 1$, $b = -1$, $c = -4$,

$$\Pi = \frac{Q\mu}{(\Delta p/l)D^4}$$

$$\Rightarrow Q = P\frac{\Delta p}{l}\frac{D^4}{\mu}.$$ **Ans.**

Example 45: *A V-notch weir is a vertical plate with a notch of angle ϕ cut into the top of it and placed across an open channel. The liquid in the channel is backed up and forced to flow through the notch. The discharge Q is some function of the elevation H of up stream liquid surface above the bottom of the notch. In addition the discharge depends upon gravity and upon the velocity of approach V_0 to the weir. Determine the form of discharge equation.*

Solution: The relationship between a group of variables is given by

$F(Q, H, g, V_0, \phi) = 0.$ ...(1)

Here ϕ is dimensionless, there are four variables involving three primary units so the number of Π group is one, only two dimensions L, T are used. Let g and H are the repeating variables.

$\Pi_1 = H^a g^b Q = L^a L^b T^{-2b}\ L^3 T^{-1} = M^0 L^0 T^0,$

$\Pi_2 = H^c g^d V_0 = L^c L^d T^{-2d}\ LT^{-1} = M^0 L^0 T^0,$ ...(2)

Then $a + b + 3 = 0,\ c + d + 1 = 0.$

$-2b - 1 = 0,\ -2d - 1 = 0,$

which gives $a = -5/2,\ b = -1/2,\ c = -1/2,\ d = -1/2.$ ...(3)

From (2) and (3), we have

$\Pi_1 = Q/\sqrt{(g)}\ H^{5/2},\ \Pi_2 = V_0/\sqrt{(gH)},\ P_3 = \phi.$

or $f\ \{Q/\sqrt{(g)}\ (H^{5/2}),\ V_0\sqrt{(gH)}, \phi\ \} = 0.$ ...(4)

This may be expressed as

$Q/\sqrt{(g)}\ H^{5/2} = f_1\ (V_0/\sqrt{(gH)},\ \phi\ )$

or $Q = \sqrt{g}\ H^{5/2} f_1\ (V_0/\sqrt{gH},\ \phi),$ ...(5)

where f is an unknown function.

Again, let H and V0 are taken as repeating variables then

$\Pi_1 = H^a V_0^b Q = L^a L^b T^{-b}\ L^3 T^{-1} = M^0 L^0 T^0,$

$\Pi_2 = H^c V_0^d\ g = L^c L^d T^{-b}\ LT^{-2} = M^0 L^0 T^0,$ ...(6)

Equating the indices of the three dimensions, we have

$a + b + 3 = 0,\ c + d + 1 = 0,$

$-b - 1 = 0,\ -d - 2 = 0,$

which gives $a = -2,$

$b = -1,$

$c = 1,$

$d = -2.$

and $\Pi_1 = (Q/H^2 V_0),\ \Pi_2 = (gH/V_0^2),\ \Pi_3 = \phi,$

or $f\ \{(Q/H^2 V_0),\ (gH/V_0^2),\ \phi\ \} = 0$ ...(7)

Equation (7) may be expressed as

$(Q/H^2 V_0) = f_2\ (gH/V_0^2,\ \phi\ \},$

or $Q = H^2 V_0 f_{2m}\ \{V_0\sqrt{(gH)},\ \phi\ \}),$

Since any of the Π parameters may be inverted or raised to any power without affecting their dimensionless nature. **Ans.**

Example 46: *The velocity components for a two- dimensional flow system can be given in the Eulerian system by*

$$u = 2x + 2y + 3t,\ v = x + y + \frac{1}{2}t.$$

Find the displacement of a fluid particle in the Lagrangian system.

Solution: The velocities may be expressed in terms of the displacements as follows

$$\frac{dx}{dt}=2x+2y+3t,$$

$$v=\frac{dy}{dt}x+y+\frac{1}{2}t.$$

$$\frac{dx}{dt}-2x-2y=3t,$$

$$\frac{dy}{dt}-x-y=\frac{1}{2}t.$$

The solution of the simultaneous differential equations can be determined by the operator method as follows:

$$(D - 2)\ x - 2y = 3t,$$

$$-x + (D - 1)\ y = \frac{1}{2}t. \qquad ...(1,2)$$

Eliminating x from (1) and (2), we have

$$D\ (D - 3)\ y = 2t + \frac{1}{2},$$

whose solution is given by

$$y = A + Be^{3t} - \left(\frac{7}{18}\right) t - \left(\frac{1}{3}\right) t_2. \qquad ...(3)$$

Substituting the value of y in the equation (2), we have

$$x = - A + 2Be^{3t} + \left(\frac{1}{3}\right) t_2 - \left(\frac{7}{18}\right). \qquad ...(4)$$

The arbitrary constants A and B are determined by using the initial conditions: $x\ x_0$, $y = y_0$ at $t = t_0 = 0$ in (3) and (4), we have

$$y_0 = A + B,\ x_0 = - A + 2B - \left(\frac{7}{18}\right).$$

Thus $A = -\frac{1}{3}[x_0 - 2y_0 + (7/18)]$,

$$B = \frac{1}{3}[x_0 + y_0 + (7/18)] \qquad ...(5)$$

Using (5) into (3) and (4), we get

$$y = -\frac{1}{3}\left[x_0 - 2y_0 + \left(\frac{7}{18}\right)\right] + \frac{1}{3}[x_0 + y_0 + \frac{7}{18}]e^{3t} - \left(\frac{7}{18}\right)t - \left(\frac{1}{3}\right)t^3,$$

$$x = \frac{1}{3}\left[x_0 - 2y_0 + \left(\frac{7}{18}\right)\right] + \frac{2}{3}\left[x_0 + y_0 + \left(\frac{7}{18}\right)\right]$$

$$e^{3t} - \left(\frac{1}{3}\right)t^2 - \left(\frac{7t}{9}\right) - \left(\frac{7}{18}\right). \qquad ...(6)$$

This determines the displacement of fluid particle in the Lagrangian system.

Example 47(a): *For a two-dimensional flow the velocities at a point in a fluid may be expressed in the Eulerian coordinates by*

$u = x + y + 2t$

and $\quad v = 2y + t.$

Determine the Lagrange coordinates as functions of the initial positions x_0 *and* y_0 *and the time* t.

Solution: Equating the given relations, we have

$$u = \frac{dx}{dt} x + y + 2t$$

$$\text{and } v = \frac{dy}{dt} = 2y + t$$

The solution of the differential equations can be determined by the operator method as follows

$(D - 1)\, x - y = 2t,$

$(D - 2)\, y = t. \qquad ...(1, 2)$

The solution for y, is given as follows:

$$y = Be^{2t} - \frac{1}{4}\,(2t + 1). \qquad ...(3)$$

The solution for x can be determined by substituting the relation (3) into equation (1),

$(D - 1)\, x = Be^{2t} + 1/4\,(6t - 1).$

or, $\quad x = Ae^{t} + Be^{2t} + 1/4(6t + 5), \qquad ...(4)$

Initial $\quad x = x_0,\ y = y_0$ at $t = t_0 = 0.$

Then $\quad B = y_0 + \frac{1}{4},$

and $\quad A = x_0 - y_0 + 1.$

Hence the solution (3) and (4) can be written in the from

$x = F_1(x_0, y_0, t), y = F_2(x_0, y_0, t),$ ...(5)

Where

$$F_1(x_0, y_0, t) = (x_0 - y_0 + 1)\, e^t + (y_0 + \left(y_0 + \frac{1}{4}\right) e^{2t} - (6t + 5),$$

$$F_2(x_0, y_0, t) = \left(y_0 + \frac{1}{4}\right) e^{2t} - \frac{1}{4}(2t + 1).$$

This determines the Lagrange coordinates as a function of the initial positions x_0, y_0 and the time t.

Example 47(b): *The losses Δh/per unit length of pipe in a fluid flow through a smooth pipe depend upon velocity V, diameter D, gravity g, dynamic viscosity μ, and density ρ. With dimensional analysis, determine the general form of the equation.*

Solution: The relationship between a group of variables is given by

$F(\Delta h/l, V, D, \rho, \mu, g) = 0$ ...(1)

Here (Δh/l) is a P parameter. There are five variables and three dimensions are involved so number of P group is two. Let V, D and ρ are repeating variables, so

$\Pi_1 = V^a D^b \rho^c \mu$ and $\Pi_2 = V^d D^e \rho^f g$

$\Rightarrow \Pi_1 = (LT^{-1})^a L^b (ML^{-3})^c ML^{-1} T^{-1} = M^0 L^0 T^0$

and $\Pi_2 = (LT^{-1})^d L^e (ML^{-3})^f LT^{-3} = M^0 L^0 T^0$ ...(2)

Equating the indices of the three dimensions, we have

$a + b - 3c - 1 = 0,\ d + e - 3f + 1 = 0,$

$-a - 1 = 0,\ -d - 2 = 0,$

$c + 1 = 0,\ f = 0,$

which gives $a = -1$, $b = -1$, $c = -1$ and $d = -2$, $e = 1$, $f = 0$,

Thus $\Pi_1 = (\mu/VD\rho)$, $\Pi_2 = (gD/V^2)$, $\Pi_3 = (\Delta h/l)$ ...(3)

or $f\{(V D\rho/ m), (V^2/gD), (\Delta h/l)\} = 0$ ...(4)

Since any of the Π-parameters may be inverted or raised to any power without affecting their dimensionless nature

Equation (4) may be written as

$(\Delta h/l) = \phi(VD\rho/\mu, V^2/gD) \Rightarrow Dh/l = \phi(Re, V^2/Dg),$

where Re is the Reynolds number. The size of the Reynolds number determines the nature of the flow. **Ans.**

Example 48: *The flow through a sluice gate set into a dam is to be investigated by building a model of the dam and sluice at 1/20 scale. Calculate the head at which the model should work to give conditions corresponding to a prototype head of 20 meters. If the discharge from the model under this corresponding head is 0.5 m2/s, estimate the discharge from the prototype dam.*

Solution: The relationship between a group of variable is given by

$$F\ (Q,\ \rho,\ g,\ H,\ D,\ \mu) = 0, \qquad ...(1)$$

where Q is the flow rate, D is the dimension of sluice, H is the head over the sluice, r is the mass density of water, g is the gravity field intensity and m is the viscosity of water. There are six variables and three dimensions are involved so the number of Π group is three, that is

$$\Pi_1 = \rho^a g^b H^c Q = (ML^{-3})^a (LT^{-2})^b L^c (L^3T^{-1}) = M^0L^0T^0 \qquad ...(2)$$

$$\Pi_2 = \rho^d g^e H^f D = (ML^{-3})^d (LT^{-2})^e L^f (L) = M^0L^0T^0 \qquad ...(3)$$

$$\Pi_3 = \rho^l g^m H^n \mu = (ML^{-3})^l (LT^{-3})^m L^n (ML^{-3}T^{-1}) = M^0L^0T^0 \qquad ...(4)$$

Equating the indices of the three dimensions in (2), (3) and (4), we have

M : a = 0, d = 0 l + 1 = 0,

L : – 3a + b + c + 3 = 0, –3d + e + f + 1 = 0, –3l + m + n + 1 = 0,

T : – 2b –1 = 0, –2e = 0, –2m–1 = 0,

which gives a = 0, b = – 1/2, c = – 5/2

d = 0, e = 0, f = – 1

l = – 1, m = – 1/2 n = –3/2

Thus $\Pi_1 = Q/(\sqrt{g}\ H^{5/2})$,

$\Pi_2 = D/H$, $\Pi_3 = \mu/\ (\rho g^{1/2}\ H^{3/2})$

$$\text{or } f\left(\frac{Q}{\sqrt{gH^{5/2}}}, \frac{D}{H}, \frac{\mu}{\rho g^{1/2} H^{3/2}}\right) = 0. \qquad ...(5)$$

Equation (5) maybe written as

$$\frac{Q}{\sqrt{gH^{5/2}}} = f\left(\frac{\rho g^{1/2} H^{3/2}}{\mu}, \frac{D}{H}\right) \qquad ...(6)$$

Multiplying L. H. S. of (6) by (H^2/D^2), we have

$Q = D^2 \sqrt{(gH)}\ \phi\ [(\rho g^{1/2}\ H^{3/2})/\mu,\ D/H]$.

This involves both the Reynolds number and the Froude number.

Since the Reynolds number and the Froude number modelling are not possible to obtain simultaneously because if the Reynolds number for the model is large, the value for the prototype will be even minimum, hence it is essential to consider the Froude number and the value D/H.

$D_m / H_m = D_p/H_p \Rightarrow H_p/H_m = D_p/D_m = 20$

or $H_m = 20/20 = 1m$

or $Q_p/Q_m = (H_l/H_m)^{5/2} = 20^{5/2}$

or $Q_p = (0.5) \times 20^{5/2} = 894 \text{ m}^3/\text{s}$. **Ans.**

Example 49(a): *The velocity q in a three-dimensional flow field for an incompressible flids is given by*

$q = 2xi - yj - zk$

Determine the equations of the streamlines passing through the point (1, 1, 1).

Solution: The equations of stream lines are given by

$$\frac{dx}{u} = \frac{dy}{v} = \frac{dz}{w}$$

$$\Rightarrow \frac{dx}{2x} = \frac{dy}{-y} = \frac{dz}{-z}$$

(i) (ii) (iii)

From, (i) and (ii), we have

$$\frac{dx}{2x} = \frac{dy}{-y} \Rightarrow \frac{dx}{x} + \frac{2dy}{y} = 0$$

By integrating, we obtain

$$\log x + 2 \log y = \log A$$

or $xy^2 = A$, where A is an integration constant.

Form (i) and (iii), we have

$$\frac{dx}{2x} = \frac{dz}{-z} \Rightarrow \frac{dx}{x} + \frac{2dz}{z} = 0$$

By integrating, we have

$xz^2 = B$, where B is an integration constant.

At the point (1, 1, 1) $A = 1 = B$

Hence the required streamlines are

$xy^2 = 1$ and $xz^2 = 1$.

Example 49(b): *Determine the streamlines and the path of the particles*

$$u = \frac{x}{(1+t)}, v = \frac{y}{(1+t)}, w = \frac{z}{(1+t)}.$$

Solution: The equations of the streamlines are given by

$$\frac{dx}{u} = \frac{dy}{v} = \frac{dz}{w}$$

or $$\frac{dx}{x/(1+t)} = \frac{dy}{y/(1+t)} = \frac{dz}{z/(1+t)}$$

or $$\frac{dx}{x} = \frac{dy}{v} = \frac{dz}{w}$$

(i) (ii) (iii)

By integrating (i) and (ii), we have

log x = log A, A is an integration constant.

$\Rightarrow$ x = Ay ...(1)

By integrating (i) and (iii), we have

log x = log z + log B, B is an integration constant.

$\Rightarrow$ x = Bz. ...(2)

Hence the streamlines are given by the intersection of (1) and (2).

$$q = \frac{dr}{dt}$$

or $$\frac{dx}{dt} = \frac{x}{1+t}, \frac{dy}{dt} = \frac{y}{1+t}, \frac{dz}{dt} = \frac{z}{1+t}$$

$$\Rightarrow \frac{dx}{x} = \frac{dt}{1+t}, \frac{dy}{y} = \frac{dt}{1+t}, \frac{dz}{z} = \frac{dt}{1+t}$$

By integrating, we have

log x = log(1 + t) + log A,

or log y = log(1 + t) + log B,

or log z = log(1+ t) +log C

$\Rightarrow$ x = A(1 + t),

y = B(1 + t),

z = C(1 + t),

which gives the required path of the particles.

Example 49(c): *The velocities at a point in a fluid in the Eulerian system are given by u = (x + y + z) + t, v = 2 (x + y + z) + t, w = 3 (x + y + z) +t. Find the displacement of a fluid particle in the Lagrangian system. Also determine the velocity of the fluid particle at* (x_0, y_0, z_0).

Solution: The velocity components may be expressed in terms of the displacement as

$$u=\frac{dx}{dt}=x+y+z+t,$$

$$w=\frac{dz}{dt}=3(x+y+z)+t. \qquad ...(1, 2, 3)$$

$$w=\frac{dz}{dt}=3(x+y+z)+t.$$

The differential equations can be written in form of operator as

(D–1) x – y = z + t

–2x + (D – 2) y = 2z + t

–3t –3y +(D – 3) z = t. ...(4, 5, 6)

Multlplying (4) by (D – 2) and adding to (5), we have

[(D – 1) (D – 2) –2] x = (D – 2) z + 2z +(D – 2)t

+ t(D^2–3D) x = Dz + 1 – t. ...(7)

Multiplying (4) by (2) and (5) by (D – 1) and adding, we have

[(D – 1) (D – 2)–2] y = (D – 1) (2z + t)

+ 2z + 2t(D^2 – 3D)y = 2Dz + 1 + t ...(8)

Multiplying (6) by (D^2 – 3D), we have

(D^2 – 3D) (D – 3)z = 3(D^2 – 3D)x + 3(D^2 – 3D)y + (D^2 – 3D)t

From (7) and (8), we have

(D^2 – 3D) (D – 3)z = 3(Dz + 1 – t) + 3(2Dz + 1 + t)

+ (D^2 – 3D)t D^2(D – 6)z = 3 ...(9)

The solution of the differential equation (9) is given by

$$z = A+Bt+Ce^{6t}-\frac{1}{4}t^2. \qquad ...(10)$$

From the equations (5) and (6), we have

(D – 2)y – 2z = 2x + t,

–3y + (D – 3)z = 3x + t.

Solving the equations, we have

$(D^2 - 5D)y = 2Dx + 1 - t$

$(D^2 - 5D)z = 3Dx + 1 + t.$ ...(11, 12)

From (1), we have

$(D - 1)x = y + z + t$

or $(D - 1)(D^2 - 5D)x = (D^2 - 5D)y + (D^2 - 5D)z + (D^2 - 5D)t$

or $(D - 1)(D^2 - 5D)x = 2Dx + 1 - t + 3Dx + 1 + t - 5$

or $(D^3 - 6D^2)x = - 3$ ...(13)

The solution of the differential equation becomes

$$x = A_1 + B_1 t + C_1 e^{6t} + \frac{1}{4}t^2. \quad ...(14)$$

Proceeding in the same manner, we have

$y = A_2 + B_2 t + C_2 e^{6t}$ ...(15)

Thus the equations (10), (14) and (15) determine the displacement of a fluid particle.

Let $x = x_0, y = y_0, z = z_0$

when $t = t_0 = 0$

The relations (14), (15) and (10) give

$x_0 = A_1 + C_1, y_0 = A_2 + C_2, z_0 = A + C$

Thus $x = x_0 - C_1 + B_1 t + C_1 e^{6t} + \frac{1}{4}t_2,$

$y = y_0 - C_2 + B_2 t + C_2 e^{6t},$

$z = z_0 - C + Bt + Ce^{6t} - \frac{1}{4}t^2$ ...(16, 17, 18)

Substituting thse values in (1), (2) and (3), we obtain the following identities

$$B_1 + 6C_2^{e6}t + \frac{1}{2}t = x_0 + y_0 + z_0 - (C_1 + C_2 + C) + (B_1 + B_2 + B)t + (C_1 + C_2 + C)e^{6t} + t,$$

$$B_2 + 6C_2 e^{6t} = 2(x_0 + y_0 + z_0) - 2(C_1 + C_2 + C) + 2(B_1 + B_2 + B)t + 2(C_1 + C_2 + C)e^{6t} + t,\ B + 6Ce^{6t} - \frac{1}{2}t = 3(x_0 + y_0 + z_0) - 3(C_1 + C_2 + C) + 3(B_1 + B_2 + B)t + 3(C_1 + C_2 + C)e^{6t} + t \quad ...(19, 20, 21)$$

Equating the coeffcients of t, e^{6t} and the constant term, we have

$$\left.\begin{aligned} x_0+y_0+z_0-(C_1+C_2+C)=B_1 \\ C_1+C_2+C=6C_1 \\ B_1+B_2+B+1=\frac{1}{2} \end{aligned}\right\} \quad \text{...(22)}$$

$$\begin{aligned} &2(x_0+y_0+z_0)-2(C_1+C_2+C)=B_2 \\ &2(C_1+C_2+C)=6C_2 \qquad \text{...(23)} \\ &2(B_1+B_2+B)+1=0 \end{aligned}$$

$$\begin{aligned} &3(x_0+y_0+z_0)-3C_1+C_2+C)=B \\ &3(C_1+C_2+C)=6C \qquad \text{...(24)} \\ &3(B_1+B_2+B)+1=-\frac{1}{2} \end{aligned}$$

From these three sets, we obtain

$$C_1=\frac{1}{6}\left(x_0+y_0+z_0+\frac{1}{12}\right),$$

$$C_2=\frac{1}{3}\left(x_0+y_0+z_0+\frac{1}{12}\right),$$

$$C=\frac{1}{2}\left(x_0+y_0+z_0+\frac{1}{12}\right)$$

Also $B_1=-\frac{1}{12}$,

$$B_2=-\frac{1}{6},$$

$$B=-\frac{1}{4}$$

Substituting these values in therelations (16), (17), and (18) an simplifying, we have

$$x=\frac{5}{6}x_0-\frac{1}{6}z_0+\frac{1}{6}\left(x_0+y_0+z_0+\frac{1}{12}\right)e^{6t}-\frac{1}{12}t+\frac{1}{4}t^2-\frac{1}{72},$$

$$y=-\frac{1}{3}x_0+\frac{2}{3}y_0-\frac{1}{3}z_0+\frac{1}{3}\left(x_0+y_0+z_0+\frac{1}{12}\right)e^{6t}-\frac{1}{6}t-\frac{1}{36},$$

$$z=-\frac{1}{2}x_0-\frac{1}{2}y_0+\frac{1}{2}z_0+\frac{1}{2}\left(x_0+y_0+z_0+\frac{1}{12}\right)e^{6t}-\frac{1}{4}t-\frac{1}{4}t^2-\frac{1}{24},$$

Determines the displacement of a flid particle in Largragian description, we have

$$u_1 = \frac{\partial x}{\partial t} = \left(x_0 + y_0 + z_0 + \frac{1}{12}\right)e^{6t} - \frac{1}{12} + \frac{1}{2}t,$$

$$v_1 = \frac{\partial y}{\partial t} = 2\left(x_0 + y_0 + z_0 + \frac{1}{12}\right)e^{6t} - \frac{1}{6},$$

$$w_1 = \frac{\partial z}{\partial t} = 3\left(x_0 + y_0 + z_0 + \frac{1}{12}\right) - \frac{1}{4} - \frac{1}{2}t.$$

Thus the velocity of the flid particle is given by

$$q_1 = u_1 i + v_1 j + w_1 k.$$

Example 50: *Determine the acceleration of a fluid particle of fixed identity for the velocity field,*

$$q = iA\,x^2y + jB\,y^2zt + kCzt^2.$$

Solution: Since $q = iu + jv + kw = iA\,x^2y + jB\,y^2zt + k\,Czt^2$.

$\Rightarrow$ $u = A\,x^2y,$

$v = B\,y^2zt,$

$w = C\,zt^2.$

The acceleration f of the fluid particle is given as

$$f = \frac{\partial q}{\partial t} + u\frac{\partial q}{\partial y} + w\frac{\partial q}{\partial z}$$

$\Rightarrow$ $f = jB\,y^2z + k^2Czt + A\,x^2y(i\ 2A\,xy) + B\,y^2zt(iA\,x^2 + j2B\,yzt)$
$+ C\,zt^2(j\,B\,y^2t + kCt^2)$

$f = A(2A\,x^3y^2 + B\,x^2y^2zt)i + B\,y^2z + (2B\,y^3z^2t^2 + C\,y^2zt^3)\,j$
$+ C(2zt + Czt^4)k.$

which determine the acceleration of a fluid particle.

Example 51: *Find the equation of the streamlines for the flow*

$$q = -\,i(3y^2) - j(6x)$$

at the point (1, 1).

Solution: The equations of streamline are given by

$$\frac{dx}{u} = \frac{dy}{v}$$

Here $q = -\,i(3y^2) - j(6x)$

$\Rightarrow$ $u = -\,3y^2,\ v = -\,6x$

or $\dfrac{dx}{-3y^2} = \dfrac{dy}{-6x}$

$\Rightarrow \dfrac{2dx}{y^2} = \dfrac{dy}{x}$

or $2x\ dx = y^2\ dy$

By integrating, we have

$x^2 = \dfrac{1}{3}y^3 + c$, where c is an integration costant.

At the point (1, 1), $c = \dfrac{2}{3} \Rightarrow 3x^2 = y^3 + 2$,

which determines the equation of the streamlines for the flow field.

Example 52: *An aeroplane is to be built to fly at high altitude where the density and coefficient of viscosity of air is 0.2 and 1/1.25 times the corresponding value at sea-level respectively. Before building the prototype, tests are carried out on a model of liner scale ratio 1/50 in a compressed-air tunnel operated at 20 atmosphere. Estimate the speed of air in the tunnel for dynamic similarity between the model tested in the compressed-air tunnel at sea level and the proposed aeroplane travelling at 640 km/h at high altitude.*

Solution: Since $\rho_a/\rho_s = 0.2$ and $\eta_a/\eta_s = 1/1.25$, where ρ_a, η_a and ρs, η_s represent the density and coefficient of viscosity at the altitude and sea level respectively.

For the prototype flying at altitude $\rho_p = \rho_a = 0.2\ \rho_s$.

For the model tested at 20 atmosphere $\rho_m = 20\ \rho_s$.

Hence $\rho_p/\rho_m = 1/100$ and $\eta_m/\eta_p = 1.25$.

For dynamical similarity, we have

$(Ul\rho/\eta)_m = (Ul\rho/\eta)_p$

$\Rightarrow U_m = \dfrac{\eta_m}{\rho_m l_m} \cdot \dfrac{\rho_p l_p}{\eta_p} U_p = \dfrac{\rho_p}{\rho_m} \dfrac{\eta_m l_p}{\eta_p l_m} U_p$

$\Rightarrow U_m = (1/100) \times 1.25 \times 50 \times 640 = 400$ km./h. **Ans.**

Example 53: *Show that the velocity vector q is everywhere tangent to lines in the-XY plane along which ψ (x, y) = const.*

Solution: We know that

$$d\psi = \frac{\partial \psi}{\partial x} dx + \frac{\partial \psi}{\partial y} dy \quad ...(1)$$

Since ψ (x, y) = const. $\Rightarrow$ $d\psi = 0$ then (1) becomes

$$\frac{\partial \psi}{\partial x} dx + \frac{\partial \psi}{\partial y} dy = 0. \quad ...(2)$$

From the definition of stream function, we have

$$u = -\frac{\partial \psi}{\partial y},$$

$$v = \frac{\partial \psi}{\partial x}$$

or $v\,dx - u\,dy = 0 \Rightarrow \frac{dx}{u} = \frac{dy}{v}.$

It follows that the velocity vector q is tangent to the lines ψ (x, y) = const. **Proved**

Example 54: *A velocity field is given by q = -xi + (y + t) j. Find the stream function and the stream lines for this field at t = 2.*

Solution: Here q = ui + vj = -xi + (y + t) j

$\Rightarrow$ u = -x, v = y = t.

We know that

$$u = -\frac{\partial \psi}{\partial y} = -x \text{ and } v = \frac{\partial \psi}{\partial x} = y + t \quad ...(1, 2)$$

By integrating (1) which regard to y, we have

$$\psi = xy + f(x, t) \quad ...(3)$$

where f(x, t) is an integration constant.

or $$\frac{\partial \psi}{\partial x} = y + \frac{\partial f}{\partial x} \quad ...(4)$$

From (2) and (4), we have

$$y + \frac{\partial f}{\partial x} = y + t$$

$\Rightarrow$ $$f(x, t) = xt + g(t) \quad ...(5)$$

From (3) and (5), we have

$$\psi = xy + xt + g(t)$$

At t = 2, ψ = x (y + 2) + g(2).

The streamlines are given by ψ = const., therefore

$$x (y - 2) = \text{const.},$$

which represent rectangular hyperbolas. **Ans.**

Example 55: *Show that $u = 2Axy$, $v = A(a^2 + x^2 - y^2)$ are the velocity components of a possible fluid motion. Determine the stream function.*

Solution: Here $u = 2Axy$, $v = A(a^2 + x^2 - y^2)$

This will be a possible fluid motion if it satisfies the equation of contnuity *i.e.*,

$$\frac{\partial u}{\partial x}+\frac{\partial v}{\partial y} = 0 \Rightarrow 2Ay - 2Ay=0.$$

which is true. Therefore, the given velocity components constitute a possible fluid motion.

We know that $u = -\dfrac{\partial \psi}{\partial y}$ and $v = \dfrac{\partial \psi}{\partial x}$

So
$$\frac{\partial \psi}{\partial y} = -2Axy,$$

and
$$\frac{\partial \psi}{\partial x} = A(a^2 + x^2 - y^2) \qquad ...(1)$$

By integrating, we have

$$\psi = -Axy^2 + f(x, t). \qquad ...(2)$$

Differentiating (2), we hvae

$$\frac{\partial \psi}{\partial x} = -Ay^2 + \frac{\partial f}{\partial x} \qquad ...(3)$$

From (1) and (3), we have

$$-Ay^2 + \frac{\partial f}{\partial x} = A(a^2 + x^2 - y^2)$$

$$\Rightarrow \frac{\partial f}{\partial x} = A(a^2 + x^2)$$

By integrating, we have

$$f(x,t) = A\left(a^2x + \frac{1}{3}x^3\right) + g(t)$$

Substituting the value of f (x, t) in (2), we have

$$\psi = A\left(a^2x - xy^2 \frac{1}{3}x^3\right) + g(t)$$

which is the required stream fucntion.

Example 56: *If $\phi = A(x^2 - y^2)$ represents a possible flow phenomenon, determine the stream function.*

Solution: Here $\phi = A(x^2 - y^2)$

Since $\frac{\partial \phi}{\partial x} = \frac{\partial \psi}{\partial y} \Rightarrow \frac{\partial \psi}{\partial y} = 2Ax$

or $\psi = Axy + C,$

where C is an integration constant, which is the required stream fnction.

Example 57: *The velocity potentials $\phi_1 = x^2 - y^2$ and $\phi_2 = \sqrt{r} \cos (\theta/2)$ are solutions of the Laplace equation. Prove that the velocity potential $\phi_3 = (x^2 - y^2) + \sqrt{r} \cos (\theta/2)$ satisfies $\nabla^2 \phi_3 = 0$.*

Solution: Here $\phi_1 = x^2 - y^2$ and $\phi_2 = \sqrt{r} \cos (\theta/2)$

The laplace's equation in cartesian coordinates and cylndrical polar coordinates are given as

$$\nabla^2 \phi_1 = \frac{\partial^2 \phi_1}{\partial x^2} + \frac{\partial^2 \phi_1}{\partial y^2} = 2 - 2 = 0$$

and $$\nabla^2 \phi_2 = \frac{\partial^2 \phi_2}{\partial r^2} + \frac{1}{r^2} \frac{\partial^2 \phi_2}{\partial \theta^2} = \frac{1}{r} \frac{\partial \phi_2}{\partial r}$$

or $$\nabla^2 \phi_2 = -\frac{1}{4r^{3/2}} \cos (\theta/2) - \frac{1}{4r^{3/2}} \cos (\theta/2) + \frac{1}{2r^{3/2}} \cos (\theta/2) = 0$$

$\Rightarrow$ that ϕ_1 and ϕ_2 satisfy Laplace's equation.

Again $\phi_3 = (x^2 - y^2) \sqrt{r} \cos \theta/2$

or $\phi_3 = r^2 \cos 2\theta + \sqrt{r} \cos \theta/2$

Since $$\nabla^2 \phi_3 = \frac{\partial^2 f_3}{\partial r^2} + \frac{1}{r^2} \frac{\partial^2 \phi_3}{\partial \theta^2} + \frac{1}{r} \frac{\partial \phi_3}{\partial r}$$

or $$\nabla^{2r} \phi_3 = \left(2\cos 2\theta - \frac{1}{4r^{3/2}} \cos \frac{\theta}{2} \right) - \frac{1}{r^2}$$

$$\left(4r^2 \cos 2\theta + \frac{1}{4r^{3/2}} \cos \frac{\theta}{2} \right) + \frac{1}{r} \left(2r \cos 2\theta \frac{1}{2r^{1/2}} \cos \frac{\theta}{2} \right) = 0$$

$\Rightarrow$ that θ_3 also satisfies Laplace's equation. **Proved.**

Example 58(a): *Prove that if the speed is every where the same, the stream lines are straight.*

Solution: The equation to the streamlines are given by

$$\frac{dx}{u} = \frac{dy}{v} = \frac{dz}{w}$$

or $v\,dx - u\,dy = 0,$

$v\,dz - w\,dy = 0,$

$u\,dz - w\,dx = 0.$

By integrating, we have

$vx - uy = \text{const.},$

$vz - wy = \text{const.},$

$uz - wx = \text{const.}$

This shows that the intersection of these planes represent the straight lines.

Example 58(b): *A one-fourth scale model of a helicopter is to be tested in a wind-tunnel such that kinematic similarity is attained. The helicopter is designed to fly in level flight 40 m/sec in a region where* $v = 1.4 \times 10^{-5}\ m^2/sec$*, further* $v = 2 \times 10^{-5}\ m^2/sec$ *in the wind-tunnel. Determine the velocity of air in the tunnel.*

Solution : Let the subscripts p and m denote the prototype and the model respectively. The kinematical similarity shows that the Reynolds number must be the same. Then

$$\frac{V_p L_p}{v_p} = \frac{V_m L_m}{v_m} \Rightarrow V_m = V_p . \frac{L_p}{L_m} . \frac{v_m}{v_p}$$

$$\Rightarrow \quad V_m = 4 \times \frac{2}{1.4} \times 40 = 228 m/sec$$

This determines the velocity of air over the model in the wind tunnel. For dynamic similarity, the Froude number must be the same for the model and the prototype.

$$\frac{(Fr)_m}{(Fr)_p} = \frac{V_m^2 / gL_m}{V_p^2 / gL_p} = \frac{V_m^2}{V_p^2} \times \frac{L_p}{L_m} = \left(\frac{228}{40}\right) \times 4 = 130$$

$\Rightarrow$ that $(Fr)_m \neq (Fr)_p$

It follows that the dynamic similarity is not attained in the test being conducted.

Example 58(c): *The velocity components in a two-dimensional flow field for an-incompressible fluid are given by*

$u = e^x \cosh y$

and $v = - e^x \sinh y.$

Determine the equation of the streamlines for this flow.

Solution: The equation of the streamlines are given by

$$\text{or} \quad \frac{dx}{u} = \frac{dy}{v} \Rightarrow \frac{dx}{e^x \cosh y} = \frac{dy}{e^x \sinh y}$$

or dx + coth y dy = 0

By integrating, we have

x + log sinhy = log c $\Rightarrow$ sinhy = ce^{-x}

where log c is an integration constant.

Example 58(d): *The velocity field at a point in fluid is given as*

$q = \left(\frac{x}{t}, t, 0\right)$

Obtain path lines and streak lines.

Solution: Here $q = \left(\frac{x}{t}, y, 0\right)$

The differential equations of path lines are given by

$$q = \frac{dr}{dt} = \frac{dx}{dt}i = \frac{dy}{dt}j + \frac{dz}{dt}k = \frac{x}{t}i + yj$$

$$\Rightarrow \qquad \frac{dx}{dt} = \frac{x}{t}, \frac{dy}{dt} = y, \frac{dz}{dt} = 0. \qquad \text{...(1, 2, 3)}$$

By integrating (1), we have

$$\frac{dx}{dt} = \frac{x}{t}$$

$$\Rightarrow \qquad \log x = \log t + \log A$$

$$\Rightarrow \qquad x = At. \qquad \text{...(4)}$$

Let (x_0, y_0, z_0) be the cordinates of the chosen fluid particle at time t = t_0, then

$$x_0 = At_0 \Rightarrow A = \frac{x_0}{t_0}.$$

Form (4), we have $x = \frac{x_0}{t_0}t.$

By integrating (2), we have

$$x = \frac{x_0}{t_0}t.$$

or log y = t + log B

$$\Rightarrow \qquad y = Be^t \qquad \text{...(5)}$$

Aty = y_0, t = t0

$\Rightarrow \quad B = y_0 e{-}t_0$

By integrating (3), we have

$$\frac{dz}{dt} = 0 \Rightarrow z = c \text{ i.e. z isindependent of } t \Rightarrow z = z_0.$$

Hence the path lines are given by

$$x = \left(\frac{x_0}{t_0}\right)t, y = y_0^{et^{-t}0}, z = z_0. \quad ...(6)$$

Let the fluid particle (x_0, y_0, z_0) passes through a fixed point (x_1, y_1, z_1) at an instant of time t = T, where $t_0 \le T \le t$. Then the relation (6) reduces to

$$x_1 = \left(\frac{x_0}{t_0}\right)T, y_1 = y_0^{eT-t_0}, z_1 = z_0$$

$$\text{or } x_0 = \left(\frac{x_1}{T}\right)t_0, y_0 = y_1^{e^{t_0-T}}, z_0 = z_1 \quad ...(7)$$

where T is the parameter. Substituting the relation (7) into (6), we have

$$x = \left(\frac{x_1}{T}\right)t, y = y_1^{e^{t-T}}, z = z_1$$

which gives the equaiton of streaklines passing through the point $(x_1, y_1, z_1,)$.

Example 59: *The velocity components in spherical polar coordinates (r, θ, ϕ) of a flow are*

$$q_r = \left(\frac{r^2}{t^2}\right)\sin\phi, q_\theta = \left(\frac{r}{t}\right)\cot\theta\,\mathrm{cosec}\,\phi,$$

$$q_\phi = \left(\frac{r}{t}\right)\sin\theta\cos\phi.$$

Determine the componets of acceleration of a fluid particle.

Solution: Let q_r, q_θ and θ_ϕ be the components of velocity in spherical polar coordinates (r, θ, ϕ), then

$$q_r = \left(\frac{r^2}{t^2}\right)\sin\phi, q_\theta \left(\frac{r}{t}\right)\cot\theta\,\mathrm{cosec}\,\phi, q_\phi = \left(\frac{r}{t}\right)\sin\theta\cos\phi \quad ...(1)$$

Let f_r, f_θ, f_ϕ be the components of acceleration, then

$$f_r = \frac{\partial q_r}{\partial t} + q_r\frac{\partial q_r}{\partial r} + \frac{q_\theta}{r}\frac{\partial q_r}{\partial \theta} + \frac{q_\phi}{r\sin\theta}\frac{\partial q_r}{\partial \phi} - \frac{q^2\theta + q^2\phi}{r}.$$

$$f_\theta = \frac{\partial q_\theta}{\partial t} + q_r \frac{\partial q_\theta}{\partial r} + \frac{q_\theta}{r}\frac{\partial q_\theta}{\partial \theta} + \frac{q_\phi}{r\sin\theta}\frac{\partial q_\theta}{\partial \phi} - \frac{q^2\phi\cot\theta}{r}.$$

$$f_\theta = \frac{\partial q_\phi}{\partial t} + q_r \frac{\partial q_\phi}{\partial r} + \frac{q_\theta}{r}\frac{\partial q_\phi}{\partial \theta} + \frac{q_\phi}{r\sin\theta}\frac{\partial q_\phi}{\partial \phi} + \frac{q_\theta q_\phi\cot\theta}{r}. \qquad ...(2,3,4)$$

From (1)and (2, 3, 4), we have

$$f_r = -\frac{2r^2}{t^3}\sin\phi + \left(\frac{r^2\sin\phi}{t^2}\right)\left(\frac{2r}{t^2}\sin\phi\right) + \left(\frac{\cos\phi}{t}\right)\left(\frac{r^2\cos\phi}{t^2}\right)$$

$$-\frac{r}{t^2}(\cot^2\theta\cos ec^2\phi + \sin^2\theta\cos^2\phi),$$

$$\text{or } f_\theta = -\frac{r}{t^2}\cot\theta\cos ec\phi + \left(\frac{r^2\sin\phi}{t^2}\right)\left(\frac{\cot\theta\cos ec\phi}{t}\right)$$

$$+\left(\frac{\cot\theta\cos ec\phi}{r}\right)\left(\frac{-r\cos ec^2\theta\cos ec\phi}{t}\right)$$

$$+\left(\frac{\cos\phi}{t}\right)\left(\frac{r\cot\theta\cos ec\phi\cot\phi}{t}\right) + \frac{r^2\cot\theta}{t^3} - \frac{r\sin^2\theta\cos^2\phi\cot\theta}{t^2},$$

$$\text{and } f_\phi = -\frac{r}{t^2}\sin\theta\cos\phi + \left(\frac{r^2\sin\phi}{t^2}\right)\left(\frac{\sin\theta\cos\phi}{t}\right)$$

$$+\left(\frac{\cot\theta\cos ec\phi}{t}\right)\left(\frac{r\cos\theta\cos\phi}{t}\right) + \left(\frac{\cos\phi}{t}\right)\left(-\frac{r\sin\theta\sin\phi}{t}\right)$$

$$+\frac{r^2\sin\theta\sin\phi\cos\phi}{t^3} + \frac{r\sin\theta\cot^2\theta\cot\phi}{t^2}.$$

Example 60: *Determine the acceleration of fluid particle from the flow field*

$q = i(A\,xy^2t) + (B\,xy^2t) + k(C\,xyz)$

Solution: Let f be the acceleration of a fluid particle, then

$$f = \frac{\partial q}{\partial t} + u\frac{\partial q}{\partial x} + v\frac{\partial q}{\partial y} + w\frac{\partial q}{\partial z}$$

or $f = i(A\,xy^2) + j(B\,x^2y) + A\,xy^2t[i(A\,y^2t) + j(2B\,xyt)$
$+ k(C\,yz)] + B\,x^2yt\,[i(2A\,xyt) + j(B\,x^2t)$
$+ k(Cxz)] + Cxyz[k(Cxy)]$

$f = iA[xy^2 + Axy^4t^2 + 2Bx3y^2t^2] + jB[x^2y + 2Ax^2y^3t^2 + Bx^4yt^2]$
$+ kC[Axy^3zt + Bx^3yzt + x2y^2z].$

The components of the acceleration are

$f_x = A[xy^2 + A\ xy^4t^2 + 2B\ x^3y^2t^2]$,

$f_y = B[x^2y + 2A\ x^2y^3t^2 + B\ x^4yt^2]$,

$f_z = C[A\ xy^3zt + Bx^3tzt + x^2y^2z]$.

Example 61: *Determine the acceleration at the point (2, 1, 3) at t = 0.5 sec, if u = yz + t, v = xz – t and w = xy.*

Solution: Let q = iu + jv + kw.

or q = (yz + t)i + (xz – t)j + xyk.

The acceleration f of a fluid particle is given by

$$f = \frac{\partial q}{\partial t} + u\frac{\partial q}{\partial x} + v\frac{\partial q}{\partial y} + w\frac{\partial q}{\partial z}$$

or f = (i – j) + (yz + t)(zj + yk) + (xz – t)(zi + xkk) + xy(yi + xj)

or $f = (1 + xz^2 + xy^2 - tz)i + (-1 + yz^2 + x^2y + zt)j$

$+ (y^2z + x^2z + yt - xt)k$.

At the point (2, 1, 3) and t = 0·5 sec, we have

f = 19·5i + 13·5j + 6·5k.

Thus the comoponents of acceleration of a fluid particle are

$f_x = 19{\cdot}5\text{m/sec}^2$,

$f_y = 13{\cdot}5\text{m/sec}^2$,

$f_z = 6{\cdot}5\text{m/sec}^2$.

Example 62: *The velocity vector q is given by*

q = ix – jy.

Determine the equation of the stream lines.

Solution: From the definition of a stream line, we have

q × dr = 0

or (ix – jy) × (idx + jdy) = 0

or (xdy + ydx) k = 0

or $\frac{dx}{x} = -\frac{dy}{y}$.

By integrating, we obtain

log x + log y = log c

or xy = c,

which represents the rectangular hyperbolas where c is an arbitrary constant.

Example 63(a): *Show that the velocity field $q_r = 0$, $q_\theta = Ar + B/r$, $q_z = 0$, satisfy the equation of motion*

$$\frac{d^2 q_\theta}{dr^2} + \frac{d}{dr}\left(\frac{q_\theta}{r}\right) = 0,$$ *where A and B are arbitrary constants.*

Solution: Here $q_r = 0, q_\theta = Ar + \frac{B}{r}, q_z = 0,$

$$\frac{dq_\theta}{dr} = A - \frac{B}{r^2}, \frac{d^2 q_\theta}{dr^2} = \frac{2B}{r^3}.$$

L.H.S $$\frac{d^2 q_\theta}{dr^2} + \frac{d}{dr}\left(\frac{q_\theta}{r}\right) = \frac{2B}{r^3} + \frac{d}{dr}\left(A + \frac{B}{r^2}\right)$$

$$= \frac{2B}{r^3} - \frac{2B}{r^3} = 0 = R.H.S$$ **Proved.**

Example 63(b): *The particles of a fluid move symmertically in space with regard to a fixed centre; prove that the equation of continuity is*

$$\frac{\partial \rho}{\partial t} + u\frac{\partial \rho}{\partial r} + \frac{\rho}{r^2} \cdot \frac{\partial}{\partial r}(r^2 u) = 0,$$

where u is the velocity at a distance r.

Solution: Let us concider a point P (r, θ, φ) in the fluid. Construct a parallelopiped with its edges PP' (= δr), PQ (= r δθ), and PS (= r sinθ δφ). Let u, v, w be the components of the velocity inthe direction of the respective elements.

Let theorigin O be the fixed centre. Since the fluid particles move symmetrically in space with regard to an origin, it follows that the motion is only along the direction PP' and there is no motion along other edges r δθ and r sinθ δφ.

Excess of mass of flow-in over flow out from the faces PQRS and P'Q'R'S' along PP' is

$$= -\delta r \frac{\partial}{\partial r}\{\rho u . r\delta\theta\, r \sin\theta\, \delta\phi\}$$ per unit time.

The excess of mass of flow-in over flow out along PQ and PS vanish as there is no motion along these directions.

Mass of the fluid inside the element = ρ δr. rδθ. rsinθ δφ.

Rate of increase in the mass of the element

$$= \frac{\partial}{\partial t}(\rho\, \delta r . r\delta\theta . r \sin\theta\, \delta\phi)$$ per unit time.

From the equation of continuity, we have

$$= \frac{\partial}{\partial t}(\rho\,\delta r.r\delta\theta.r\sin\theta\,\delta\phi) = -\delta r\frac{\partial}{\partial r}(\rho u.r\delta\theta.r\sin\theta\,\delta\phi)$$

$$\Rightarrow \frac{\partial\rho}{\partial t}(\delta r.r\delta\theta.r\sin\theta\,\delta\phi) + \frac{\partial}{\partial r}(\rho u r^2)\,\delta r.\delta\theta\sin\theta\,\delta\phi) = 0.$$

$$\Rightarrow \frac{\partial\rho}{\partial t} + \frac{1}{r^2}\frac{\partial}{\partial r}(\rho u r^2) = 0$$

$$\Rightarrow \frac{\partial\rho}{\partial t} + u\frac{\partial\rho}{\partial r} + \frac{\rho}{r^2}\frac{\partial}{\partial r}(r^2 u) = 0.$$ **Hence Proved.**

Example 63(c): *A and B are a simple source and a sink of strengths m and m' respectively, in an infinite liquid. Show that the equation of the stream lines is m cos θ - m' cos θ' = const., where θ and θ' are the angles which AP, BP make with AB, P being any point. Prove also that m > m', the cone defined by the equation*

cosθ = {1- (2m' / m)},

divides the streamlines issuing from A into two sets, one extending to infinity and the other terminating at B.

Solution: For any point P, we have

ψ = m cos θ - m' cos θ'.

The streamline are given by ψ = cosnt., we have

m.cos θ - m' cos θ' = const. ...(1)

For the extreme stream line leaving at A (let an angle α) and leaving at B, we notice that when P is very near to A *i.e.*, $\theta = \alpha$, = $\theta' = \pi$ and when P is very near to B *i.e.*, θ = 0, = θ' = 0. From (1), we have

m cos α + m' = m - m'

$\Rightarrow$m cos α = m - 2m'

$\Rightarrow$cos α = {1 - (2m'/m)}.

This generates the curve cos θ = 1 - {(2m'/m)}. **Proved.**

EXERCISES

1. A source and a sink of equal strengths are placed at the points (0, 0, ± a) inside a sphere of radius r with its centre at origin. Find the velocity at points within the sphere.
2. Prove that the image of a radial doublet in a sphere is another radial doublet and compare the magnitudes. Prove that the velocity at any point of the sphere is proportional to ω r^{-5}, where r is the distance

from the doublet and ω is the perpendicular on the diameter on which it lies.

3. Prove that a uniform stream of velocity U can be obtained as c → ∞ of the field due to a source of strength $2\pi c^2 U$ at (c, 0, o).
4. Prove tha the period of revolution of the liquid relative to the spherical shell is the same as the period of revolution of the shell.
5. Determine the velocity potential φ of the liquid motion if the ellipsoid ahs a velocity components U, V, W parallel to the coordinates axes.
6. Find the equations of motion of a sphere through a liquid extending to infinity in all direction and at rest there, the extraneous force, whose potential is Ω being supposed to act on the sphere and the liquid alike. Prove that the velocity potential at a point P due to a uniform line source OA of strength m per unit length is

$$m\log\frac{OP+AP+OA}{OP+AP-OA}.$$

The liquid is supposed to extend to infinity in all directions and to be at rest there.

7. A sphere of variable radius r moves through an infinite fluid with a variable velocity q in a fixed direction. Find the pressure at any point on its surface and show that the resultant thrust of the fluid on the sphere is

$$\frac{2}{3}\pi\rho a^2(rp+3qr).$$

8. Prove that, when an oblate spheroid of eccentricity sin α moves parallel to its axis with velocity V in infinite fluid, the kinetic energy of the fluid is

$$\frac{1}{3}M'V^2\frac{\tan\alpha-\alpha}{\alpha-\sin\alpha\cos\alpha},$$

where M' denotes the mass of the displaced fluid.

9. Inviscid incompressible fluid is flowing past a fixed circular cylinder of radius a, its undisturbed velocity at a great distance from the cylinder being U parallel to the X-axis. The motion is
10. A sphere of redius a is moving with constant velocity U through an infinite liquid at rest at infinity. If p_0 be the pressure at infinity, shew that the pressure at any point of the surface of the sphere, the radius to which point makes an angle θ with the direction of motion is given by

$$p = p_0 + \frac{1}{2}\rho U^2\left(1 - \frac{9}{4}\sin^2\theta\right).$$

11. Prove that the thrust on that half of the sphere on which the liquid impinges is $\pi a^2\left(\Pi - \frac{1}{16}\rho U^2\right)\pi a^2\left(\Pi - \frac{1}{16}\rho U^2\right)$, where Π is the pressure at infinity, *U* the undisturbed velocity of the liquid and ρ· the density.

12. The space between two concentric spheres of radii a and b is filled with liquid. The psheres have velocities u and v in the same direction. Find the kinetic energy of the liquid.

13. A spherical shell of internal radius a contains a concentric sphere of radius λa and density σ the inervening space being filled with water (unit density) and the whole system is at rest. If the velocity V is suddenly communicated to the shell, prove that the initial velocity U communicated to the sphere is

$U = 3V/[\ 3\ \sigma\ (1 - \lambda^2\ ) + 1 + 2\ \lambda^3].$

14. A hollow spherical shell of inner radius a contains a concentric solid uniform sphere of radius b and density σ and the space between the two is filled with liquid of density ρ. If the shell is suddenly made to move with speed U, prove that a velocity V is imparted to the inner sphere where

$V = 3Ua^3/[\ 2\ (\ \sigma/\rho\ )\ (a^3 - b^3\ ) + (a^3 + 2b^3].$

15. Determine the velocity potential φ of the intial motion when an impulsive force is so applied that the inner sphere starts moving with velocity V and the outer sphere with velocity U in the same direction.

4

Motion in Three Dimensions

We shall now study *irrotational motion in three dimensions.* Particularly the motion of a sphere will be discussed. The treatment is similar to that of the motion of a circular cylinder. However unlike in the case of the circular cylinder the complex potential we will not be of any use in the motion of a sphere, the motion being now in three dimensions in which complex quantities fail to play any role.

The velocity potential ϕ would now satisfy the Laplace's equation in three dimensions, namely

$$\frac{\partial^2\phi}{\partial x^2}+\frac{\partial^2\phi}{\partial y^2}+\frac{\partial^2\phi}{\partial z^2}=0. \qquad ...(1)$$

Therefore study will be much governed by those solutions of (1) which satisfy the boundary conditions in question.

It should be noted that (1) in spherical polar co-ordinates is

$$\frac{\partial^2\phi}{\partial r^2}+\frac{2}{r}\frac{\partial\phi}{\partial r}+\frac{1}{r^2}\frac{\partial^2\phi}{\partial\theta^2}+\frac{\cot\theta}{r^2}\frac{\partial\phi}{\partial\theta}+\frac{1}{r^2\sin^2\theta}\frac{\partial^2\phi}{\partial\omega^2}=0. \qquad ...(2)$$

Also on multiplying by $r_2 \sin\theta$ and regrouping the terms, this equation can be written as

$$\frac{\partial}{\partial r}\left\{r^2\sin\theta\frac{\partial\phi}{\partial\theta}\right\}+\frac{\partial}{\partial r}\left\{\sin\theta\frac{\partial\phi}{\partial\theta}\right\}+\frac{\partial}{\partial\omega}\left\{\frac{1}{\sin\theta}\frac{\partial\phi}{\partial\omega}\right\}=0. \qquad ...(3)$$

In case $\dfrac{\partial^2\phi}{\partial\omega^2}=0$, this equation becomes

$$\frac{\partial}{\partial r}\left\{r^2\frac{\partial\phi}{\partial r}\right\}+\frac{\partial}{\sin\theta\,\partial\theta}\left\{\sin\theta\frac{\partial\phi}{\partial\theta}\right\}=0, \qquad ...(4)$$

whose solutions are of the type $r^n P_n(\cos\theta)$ and $r^{-n-1} P_n(\cos\theta)$ for all integral values of n, where P_n is the Legendre's polynomial of order n.

Thus the general solution of (4) is

$$\phi = \sum_n \left\{ A_n r^n + \frac{B_n}{r^{n+1}} \right\} P_n(\cos\theta).$$

It is also known that

$$P_0(\mu) = 1,\ P_1(\mu) = \mu \text{ and } P_2(\mu) = \frac{1}{2}(3\mu^2 - 1) \text{ etc.}$$

MOTION OF A SPHERE IN A LIQUID AT REST AT INFINITY

A sphere is moving in a liquid at rest at infinity; to calculate the velocity potential and equation of lines of flow.

Let us the origin at the centre of the sphere and the axis of z in the direction of motion, the velocity being now U along the z-axis.

Here the conditions to determine ϕ are

(i) that ϕ satisfies $\nabla^2\phi = 0$ everywhere,

(ii) that space derivative of ϕ vanishes at infinity, *i.e.* $\frac{\partial\phi}{\partial r} = 0$ at infinity,

(iii) that at the surface of the sphere r = a, we must have $-\frac{\partial\phi}{\partial r} = U\cos\theta$

This suggests that ϕ must contain $\cos\theta$ and since $P_1(\cos\theta) = \cos\theta$, there should be only those terms in ϕ which contain P_1.

Thus the suitable form* of ϕ is

$$\phi = \left\{ Ar + \frac{B}{r^2} \right\} \cos\theta \qquad ...(1)$$

so that $\phi = \frac{\partial\phi}{\partial r} = \left\{ A - \frac{2B}{r^3} \right\} \cos\theta$

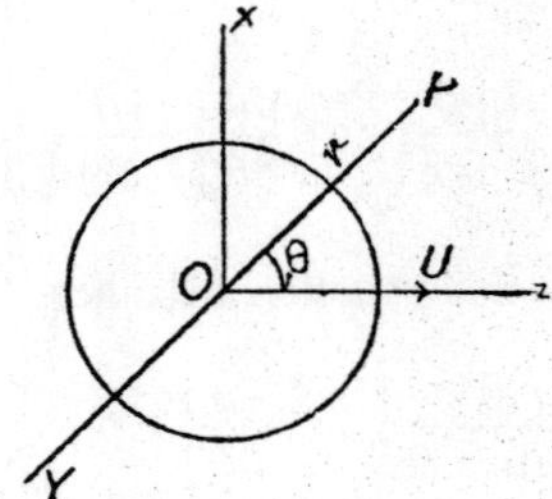

Fig. 4.1

when $r = \infty,\ \frac{\partial\phi}{\partial r} = 0,$

i.e. A cos θ = 0 or A = 0.

when r = a, $-\dfrac{\partial\phi}{\partial r} = U\cos\theta,$

i.e. $B = \dfrac{1}{2}a^3U.$

Thus $\phi = \dfrac{1}{2}\dfrac{a^3U}{r^2}\cos\theta$

This determines velocity potential for the motion.

To determine lines of flow. Referred to centre of the sphere as origin, the differential equation of the lines of flow is

$$\frac{dr}{\partial\phi/\partial r} = \frac{r\,d\theta}{\partial\phi/r\,\partial\theta},$$

i.e.
$$\frac{dr}{a^2U\cos\theta} = \frac{r\,d\theta}{\dfrac{\frac{1}{2}a^3U\sin\theta}{r^3}},$$

i.e. $\dfrac{dr}{r} = \dfrac{2\cos\theta}{\sin\theta}d\theta.$

Integrating, log r = 2 log sinθ + log c

or $r = c\sin^2\theta$,

which is the equation of the lines of flow.

LIQUID STREAMING PAST A FIXED SPHERE

If the sphere be fixed and the liquid be streaming past it with velocity U, the velocity potential can be determined from the foregoing article by imposing a velocity - U on the sphere and the liquid both, thus bringing the sphere to rest. This would amount to adding a term Uz, *i.e.* Ur cos θ to the velocity potential. Thus we have in this case

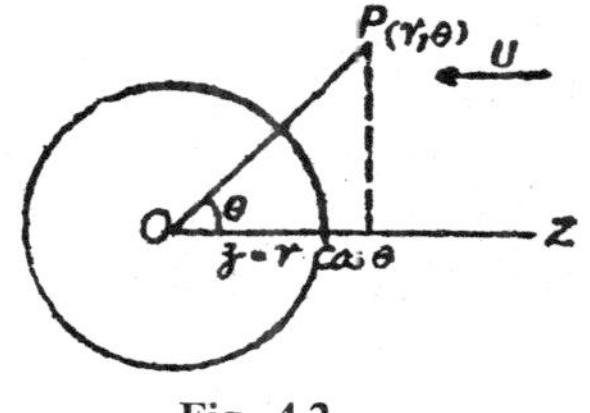

Fig. 4.2

$$\phi = Ur\cos\theta + \frac{1}{2}\frac{Ua^3}{r^2}\cos\theta$$

$$= U\left\{r + \frac{1}{2}\frac{a^3}{r^2}\right\}\cos\theta.$$

And now the stream lines are given by

$$= \frac{dr}{\partial\phi / \partial r} = \frac{r\, d\theta}{\partial\phi / r\, d\theta},$$

i.e.
$$\frac{dr}{U\left\{1 - \frac{a^3}{r^3}\right\}\cos\theta} = \frac{r\, d\theta}{-U\left\{1 + \frac{1}{2}\frac{a^3}{r^3}\right\}\sin\theta},$$

i.e,
$$-2\cot\theta\, d\theta = \frac{2r^3 + a^3}{r(r^3 - a^3)},$$

i.e,
$$-2\cot\theta\, d\theta = \left\{\frac{3r^2}{r^3 - a^3} - \frac{1}{r}\right\}dr.$$

Integrating - 2 log sin θ = log $(r^3 - a^3)$ - log r - log c.

$\sin^2\theta\ (r^3 - a^3) = cr$,

These are the lines of flow relative to the sphere.

CONCENTRIC SPHERES

The liquid is contained in the intervening space between two spheres of radii a and b (a < b). The impulsive forces are so applied that the inner sphere starts moving with velocity U and the outer with velocity V in the same direction; to determine the velocity potential of the initial motion.

The velocity potential ϕ must be such that

(i) it satisfies $\nabla^2\phi = 0$,

(ii) $-\frac{\partial\phi}{\partial r} = U\cos\theta$, when r = a,

(iii) $-\frac{\partial\phi}{\partial r} = V\cos\theta$. when r = b,

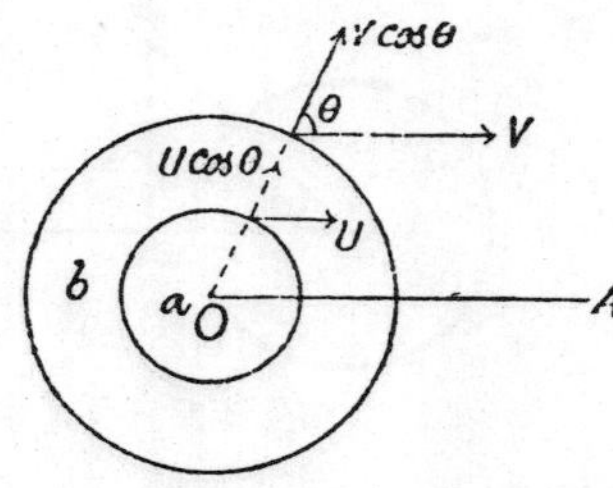

Fig. 4.3

Thus for the motion which is symmetrical about the x-axis, the suitable form of ϕ is

$$\phi = \left\{ Ar + \frac{B}{r^2} \right\} \cos\theta,$$

$$\frac{\partial\phi}{\partial r}\left\{ A - \frac{2B}{r^3} \right\} \cos\theta.$$

$$\therefore\ -\left\{ A - \frac{2B}{a^3} \right\} \cos\theta = U\cos\theta \text{ and } -\left\{ A - \frac{2B}{b^3} \right\} \cos\theta = V\cos\theta$$

i.e.,
$$A - \frac{2B}{a^3} - U \text{ and } A - \frac{2B}{b^3} = -V.$$

$$\therefore\ A = \frac{Ua^3 - Vb^3}{b^3 - a^3} \text{ and } B = \frac{(U-V)a^3b^3}{2(b^3 - a^3)}.$$

$$\text{Thus } \phi = \left\{ \frac{Ua^3 - Vb^3}{b^3 - a^3} \right\} r\cos\theta + \frac{(U-V)a^3b^3}{2(b^3 - a^3)} \frac{\cos\theta}{r^2}.$$

Particular case. *If the outer cylinder is at rest.* Putting V = 0 the velocity potential in this case is given by

$$\phi = \frac{Ua^3}{b^3 - a^3}\left\{ r + \frac{b^3}{2r^3} \right\} \cos\theta.$$

Let I be the impulse necessary to produce the velocity U in the inner sphere of mass M; then by the principle of momentum, we have

$$I - MU = \iint \varpi \cos\theta \, dS.$$

where $\varpi = \rho\phi$ is the impulse pressure (at r = a).

$$\therefore\ I - MU = \rho\frac{Ua^3}{b^3 - a^3}\left\{ a + \frac{b^3}{2a^3} \right\} \int_0^{2\pi} \cos^2\theta (2\pi a \sin\theta) a \, d\theta$$

$$= \frac{\rho U(2a^3 + b^3) 2\pi a^3}{3(b^3 - a^3)}.$$

If M' = mass of the liquid displaced $= \frac{4}{3}\pi a^3 \rho$, then

$$I = MU + \frac{1}{2}\frac{M'U(2a^3 + b^3)}{b^3 - a^3}.$$

Let now the radius b be increased indefinitely, so that

$$\lim_{b\to\infty} \frac{2a^3 + b^3}{b^3 - a^3} = \lim_{b\to\infty} \frac{2(a^3/b^3) + 1}{1 - (b^3/a^3)} = 1.$$

Thus the impulse required to give to a sphere of mass M a velocity U in an infinite liquid is given by

$$I = MU + \frac{1}{2}M'U = (M + \frac{1}{2}M')U$$

effectively increasing the mass of the sphere by an amount $\frac{1}{2}M'$.

EQUATIONS OF MOTION OF A SPHERE

At any time t let $C\ (x_0, y_0, z_0)$ be the centre of the sphere, moving with a velocity v. If U, V, W are the components of v along the axes, then $U = x_0$, $V = y_0$ *and* $W = x_0$.

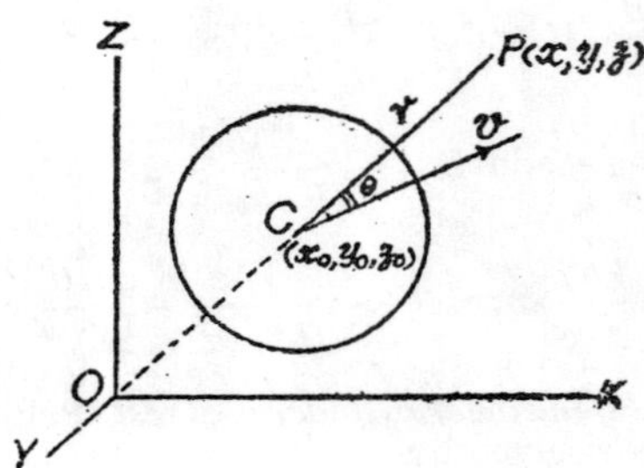

Fig. 4.4

Now at a point $P(x, y, z)$, such that $CP = r$ and $\angle PCv = \theta$, the velocity potential ϕ is given by

$$\phi = \tfrac{1}{2}\frac{va^2}{r^2}\cos\theta$$

$v \cos\theta$ = resolved part of v along CP

= sum of the resolved part of v along CP

$$= \left[U.\frac{x - x_0}{r} + V.\frac{y - y_0}{r} + W.\frac{z - z_0}{r}\right],$$

where $r^2 = (x - x_0)^2 + (y - y_0)^2 + (z - z_0)^2$,

so that $\quad rr = -(x - x_0)x_0 - (y - y_0)y_0 - (z - z_0)z_0.$

$$= -U(x - x_0) - V(y - y_0) - W(z - z_0).$$

$$\therefore\ \phi = \tfrac{1}{2}\frac{a^3}{r^2}\left[U\frac{x - x_0}{r} + V\frac{y - y_0}{r} + W\frac{z - z_0}{r}\right].$$

This gives

$$\frac{\partial\phi}{\partial x} = \frac{a^3 U}{2r^3} - \frac{3a^3}{2r^3}(x - x_0)[U(x - x_0) + ... + ...]\text{etc.}$$

$$\therefore\ q^2 = \left\{\frac{\partial\phi}{\partial x}\right\}^2 + \left\{\frac{\partial\phi}{\partial y}\right\}^2 + \left\{\frac{\partial\phi}{\partial z}\right\}^2$$

$$= \frac{a^6}{4r^6}(U^2 + V^2 + W^2) + \frac{3a^6}{4r^8}\{U^2(x - x_0) + \ldots + \ldots\}^2. \qquad \ldots(1)$$

Also $$\frac{\partial\phi}{\partial t} = \frac{a^3}{2r^3}\{U(x - x_0) + \ldots + \ldots\} - \frac{a^3}{2r^3}\{U^2 + V^2 + W^2\}$$

$$+\frac{3a^3}{2r^5}\{U(x - x_0) + \ldots + \ldots\}^2 \qquad \ldots(2)$$

Omitting the extraneous forces the pressure is given by the equation

$$\frac{p}{\rho} = F(t) + \frac{\partial\phi}{\partial t} - \frac{1}{2}q^2.$$

Putting values of $\dfrac{\partial\phi}{\partial t}$ and q^2 where $r = a$, the pressure on the sphere $r = a$ is given by

$$\frac{p}{\rho} = F(t) + \frac{1}{2}[U(x - x_0) + \ldots + \ldots] - \frac{5}{8}(U^2 + V^2 + W^2)$$

$$+\frac{9}{8a^2}[U(x - x_0) + \ldots + \ldots]^2. \qquad \ldots(3)$$

If $q = \Pi$ at infinity where $r = \infty$ $F(t) = \Pi/\rho$.

Also if f is the acceleration of the sphere, θ, θ_1 the angles that CP makes with the directions of v and f, then

$$v\cos\theta = U\frac{x - x_0}{a} + V\frac{y - y_0}{a} + W\frac{z - z_0}{a} \quad \text{as } r = a \text{ here}$$

and $$f\cos\theta_1 = U\frac{x - x_0}{a} + V\frac{y - y_0}{a} + W\frac{z - z_0}{a}$$

then (3) gives

$$\frac{p}{\rho} = \frac{\Pi}{\rho} + \frac{1}{2}af\cos\theta_1 - \frac{5}{8}v^2 + \frac{9}{8}v^2\cos^2\theta,$$

$$\frac{p}{\rho} = \frac{\Pi}{\rho} + \frac{1}{2}af\cos\theta_1 = \frac{1}{8}v^2(9\cos^2\theta - 5). \qquad \ldots(4)$$

The resultant thrust on the sphere is

$$= -\int p\cos\theta\, ds$$

$$= -\int_0^\pi \rho\left[\frac{\Pi}{\rho} + \frac{1}{8}v^2(9\cos^2\theta - 5)\right]\cos\theta.2\pi a\sin\theta\ a\, d\theta$$

$$-\frac{1}{2}af\rho\int_0^\pi \cos\theta_1 \cos\theta_1 .2\pi a \sin\theta_1 \; a \, d\theta_1$$

$$= 0 - \frac{2}{3}\pi a^3 f\rho$$

$$= -\frac{1}{2}M'f, \text{ where } M' = \frac{4}{3}\pi a^3\rho, \text{ mass of the liquid displaced.}$$

Hence if *X, Y, Z* are the components of resultant thrust, then

$$X = -\frac{2}{3}\pi\rho a^3 U \text{ etc.}$$

If *X', Y', Z'* be the components of the extraneous forces on the sphere and M be the made of the sphere, then the equation of motion parallel to x-axis is

$$MU = -\frac{1}{2}M'U + \frac{\sigma+\rho}{\sigma}X'$$

where σ is the density of the sphere

$$\text{or } MU = \frac{M}{M+\frac{1}{2}M'} + \frac{\sigma+\rho}{\sigma}X'$$

i.e., $$MU = \frac{\sigma-\rho}{\sigma+\frac{1}{2}\rho}X'.$$

Therefore the effect of the presence of the liquid is to reduce the extraneous force in the ratio σ − ρ : σ + ½ ρ.

Kinetic energy. From (1),

$$q^2 = \frac{a^2}{4r^6} + \frac{3a^6}{4r^6}v^2\cos^2\theta$$

as $v^2 = U^2 + V^2 + W^2$

and $$v\cos\theta = \frac{U(x-x_0)}{r} + \frac{V(y-y_0)}{r} + \frac{W(z-z_0)}{r}.$$

The K.E. $= \frac{1}{2}\rho\int q^2 \, dr,$ where *dr* is the elementary volume

$$= \frac{1}{2}\rho\frac{1}{4}a^6v^2\int_{\theta=0}^{\pi}\int_{r=0}^{\infty}\left[\frac{1}{r^6} + \frac{3}{r^6}\cos^2\theta\right] 2\pi r\sin\theta \; r \, d\theta \, dr$$

$$= \frac{1}{4}\pi a\rho^6 v^2\left[\frac{2}{3a^3} + \frac{3}{3a^3}\frac{2}{3}\right]$$

$$= \frac{1}{3}\pi\rho a^3v^2 = \frac{1}{4}M'v^2 \quad \text{where } M' = \frac{4}{3}\pi\rho a^3.$$

STOKE'S STREAM FUNCTION

Motion symmetrical about an axis.

The equation of continuity in cylindrical co-ordinates is

$$\frac{\partial \rho}{\partial t}+\frac{1}{r}\frac{\partial}{\partial r}(\rho r u)+\frac{1}{r}\frac{\partial}{\partial \theta}(\rho v)+\frac{\partial}{\partial z}(\rho w)=0.$$

For incompressible fluids, *i.e.* for liquids, this becomes.

$$\frac{1}{r}\frac{\partial}{\partial r}(ru)+\frac{1}{r}\frac{\partial}{\partial \theta}(v)+\frac{\partial w}{\partial z}=0.$$

If the motion is symmetrical about the axis of z, then due to symmetry about z-axis, v = 0, and the equation of continuity becomes

$$\frac{1}{r}\frac{\partial}{\partial r}(ru)+\frac{\partial w}{\partial z}=0. \qquad ...(1)$$

Instead, if we take axis of x as the axis of symmetry and ϖ the direction perpendicular to x-axis and u and v velocities in these directions, then we get.

$$\frac{1}{\varpi}\frac{\partial}{\partial \varpi}(\varpi v)+\frac{\partial u}{\partial x}=0,$$

[writing ϖ for r, v for u; u for w and x for z in (1)]

or $\varpi\frac{\partial u}{\partial x}+\frac{\partial}{\partial \varpi}(\varpi v)=0$

or $\frac{\partial}{\partial x}(u\varpi)+\frac{\partial}{\partial \varpi}(v\varpi)=0.$

But this is the condition that

$v\varpi\, dx$ - $u\varpi a\varpi$ = an exact differential

$= d\psi$ say

$= \frac{\partial \psi}{\partial x}dx+\frac{\partial \psi}{\partial \varpi}d\varpi$

so that $\quad v\varpi=\frac{\partial \psi}{\partial x}\text{and}-u\varpi\frac{\partial \psi}{\partial \varpi}.$

i.e., $\quad u=\frac{1}{\varpi}\frac{\partial \psi}{\partial \varpi}\text{and } v=\frac{1}{\varpi}\frac{\partial \psi}{\partial x}. \qquad ...(2)$

This function ψ is called the *Stocke' stream function*

The stream lines are given by

$$\frac{dx}{u}=\frac{d\varpi}{v} \text{ or } v\,dx-u\,d\varpi=0$$

or $v\varpi\, dx - u\varpi\, d\varpi = 0$, *i.e.* $d\psi = 0$

or ψ constant of integration.

Thus analogous to two-dimensional stream function, Stokes' stream function ψ = *const. gives stream lines. Stokes' stream function can exist even when* φ *does not, i.e. even if the motion is rotational.*

A PROPERTY OF STOKES' STREAM FUNCTION

2π *times the difference of the values of Stokes' stream function at two points in the same meridian plane is equal to the flow across the angular surface obtained by the revolution round the axis of curve joining the points.*

Let *ds* be an element of the curve and θ its inclination to the axis ; then the outward flow across the surface of revolution

$$= \int_A^B (v\cos\theta - u\sin\theta) 2\pi\varpi\, ds$$

$$= 2\pi \int_A^B \left(\frac{\partial\psi}{\partial x} dx + \frac{\partial\psi}{\partial\varpi} d\varpi \right) \text{ from (2) of the previous article}$$

$$= 2\pi \int_A^B d\psi = 2\pi(\psi_B - \psi_A).$$

This proves the property.

In the light of the above property we may define the value of the Stokes' stream function at any point *P* as $\frac{1}{2\pi}$ times the amount of flow across a surface got by revolving *AP* round the axis, *A* being a fixed point in the meridian plane through *P*. Thus

$$\psi = \frac{1}{2\pi} \int_A^P (v\cos\theta - u\sin\theta) 2\pi\varpi\, ds$$

$$= \int_A^P (v\varpi\, dx - u\varpi\, d\varpi).$$

This also gives

$$u = -\frac{1}{\varpi}\frac{d\psi}{d\varpi} \text{ and } v = \frac{1}{\varpi}\frac{d\psi}{dx}.$$

IRROTATIONAL MOTION

Since the motion is irrotational the velocity potential φ exists and if ψ be the Stokes' stream function, we have

$$u = -\frac{\partial\phi}{\partial x},\ v = -\frac{1}{\varpi}\frac{\partial\phi}{\partial\varpi}.$$

Also $u = -\frac{1}{\varpi}\frac{\partial \phi}{\partial \varpi}$ and $v = \frac{1}{\varpi}\frac{\partial \psi}{\partial x}$,

i.e., $$\frac{\partial \phi}{\partial x} = \frac{1}{\varpi}\frac{\partial \psi}{\partial \varpi} \text{ and } \frac{\partial \phi}{\partial \varpi} = -\frac{1}{\varpi}\frac{\partial \psi}{\partial x}. \quad ...(1)$$

$\frac{\partial^2 \phi}{\partial \varpi\, \partial x} = \frac{\partial^2 \phi}{\partial x\, \partial \varpi}$ gives

$$\frac{\partial}{\partial \varpi}\left(\frac{1}{\varpi}\frac{\partial \psi}{\partial \varpi}\right) = -\frac{\partial}{\partial x}\left(\frac{1}{\varpi}\frac{\partial \psi}{\partial x}\right),$$

i.e., $$\frac{\partial^2 \psi}{\partial x^2} + \frac{\partial^2 \psi}{\partial \varpi^2} - \frac{1}{\varpi}\frac{\partial \psi}{\partial \varpi} = 0. \quad ...(2)$$

Again eliminating ψ from (1), we get

$$\frac{\partial^2 \phi}{\partial x^2} + \frac{\partial^2 \phi}{\partial \varpi^2} + \frac{1}{\varpi}\frac{\partial \phi}{\partial \varpi} = 0. \quad ...(3)$$

Functions ψ and □ϕ satisfy differential equations (2) and (3) and as such these are not interchangeable. This is a fact contrary to the two-dimensional motion.

Note : Equation (3) is Laplace's equation in cylindrical co-ordinates when there is symmetry about x-axis. Thus here also the function ϕ satisfies Laplace's equation while ψ does not.

POLAR CO-ORDINATES

Let (r, θ) be the polar co-ordinates of the point (x, ϖ), so that $x = r\cos\theta$ and $\varpi = r\sin\theta$.

So if q_r and q_θ be the velocities in the directions of r and θ and ψ be the Stokes' stream function,

$$q_r = \frac{1}{\varpi}\frac{\partial \psi}{r\,\partial \theta} = \frac{1}{r^2 \sin\theta}\frac{\partial \psi}{\partial \theta} \quad ...(1)$$

and $$q_\theta = \frac{1}{\varpi}\frac{\partial \psi}{\varpi\,\partial \theta} = \frac{1}{r\sin\theta}\frac{\partial \psi}{\partial r}$$ as $\varpi = r\sin\theta$.

Again if the motion is irrotational, ϕ exists, and we have

$$q_r = -\frac{\partial \phi}{\partial r} \text{ and } q_\theta = -\frac{1}{r}\frac{\partial \phi}{\partial \theta}.$$

Equating the values of q_r and q_θ,

$$\frac{\partial \phi}{\partial r} = \frac{1}{r^2 \sin\theta}\frac{\partial \psi}{\partial \theta}, -\frac{1}{r}\frac{\partial \phi}{\partial \theta} = \frac{1}{r\sin\theta}\frac{\partial \psi}{\partial r}. \quad ...(2)$$

putting ϕ by the relation $\dfrac{\partial^2\phi}{\partial\theta\,\partial r}=\dfrac{\partial^2\phi}{\partial r\,\partial\theta}$, we get

$$\frac{\partial}{\partial\theta}\left(\frac{1}{r^2\sin\theta}\frac{\partial\psi}{\partial\theta}\right)+\frac{\partial}{\partial r}\left\{\frac{1}{\sin\theta}\frac{\partial\psi}{\partial r}\right\}=0,$$

i.e.,
$$r^2\frac{\partial^2\psi}{\partial r^2}+\sin\theta\frac{\partial}{\partial\theta}\left\{\frac{1}{\sin\theta}\frac{\partial\phi}{\partial\theta}\right\}=0.$$

If we take $\mu=\cos\theta$, so that

$$\sin\theta\frac{\partial}{\partial\mu}=\frac{\partial}{\partial\theta},$$

the above equation becomes

$$r^2\frac{\partial^2\psi}{\partial r^2}+\sin^2\theta\frac{\partial}{\partial\mu}\left\{\frac{\partial\psi}{\partial\mu}\right\}=0$$

or
$$r^2\frac{\partial^2\psi}{\partial r^2}+(1-\mu^2)\frac{\partial^2\psi}{\partial\mu^2}=0. \qquad ...(3)$$

Again eliminating ψ from (2), we have

$$\frac{\partial}{\partial r}\left\{r^2\frac{\partial\phi}{\partial r}\right\}+\frac{1}{\sin\theta}\frac{\partial}{\partial\theta}\left\{\sin\theta\frac{\partial\phi}{\partial\theta}\right\}=0,$$

i.e.,
$$\frac{\partial}{\partial r}\left\{r^2\frac{\partial\phi}{\partial r}\right\}+\frac{\partial}{\partial\mu}\left[(1-\mu^2)\frac{\partial\phi}{\partial\mu}\right]=0,$$

where $\mu=\cos\theta$. ...(4)

This is Laplace's equation under the present condition and has solution of the type

$$\phi_1=r^n\,P_n(\mu) \text{ and } \phi_2=r^{-n-1}P_n(\mu).$$

If ψ_1 and ψ_2 are solutions of equation (3), then from (2),

$$\frac{\partial\psi_1}{\partial\mu}=-r^2\frac{\partial\phi_1}{\partial r}=-nr^{n+1}P_n(\mu),$$

$$\frac{\partial\psi_2}{\partial\mu}=-r^2\frac{\partial\phi_2}{\partial r}=(n+1)r^{-n}P_n(\mu).$$

Also
$$\frac{\partial\psi_1}{\partial r}=(1-\mu^2)\frac{\partial\phi_1}{\partial\mu}=(1-\mu^2)r^n\frac{\partial P_n(\mu)}{\partial\mu} \qquad ...(5)$$

and
$$\frac{\partial\psi_2}{\partial r}=(1-\mu^2)\frac{\partial\phi_2}{\partial\mu}=(1-\mu^2)r^{-n-1}\frac{\partial P_n(\mu)}{\partial\mu}. \qquad ...(6)$$

Integrating (5) and (6), we have

$$\psi_1 = \frac{(1-\mu^2)}{n+1} r^{n+1} \frac{\partial P_n(\mu)}{\partial \mu}$$

$$\psi_2 = -\frac{(1-\mu^2)}{n} r^{-n} \frac{\partial P_n(\mu)}{\partial \mu}.$$

These give the possible solution of (3)

MOTION OF A SOLID OF REVOLUTION ALONG ITS AXIS

Let x-axis be the axis of revolution and solid be moving along x-axis with velocity u. The motion being symmetrical about x-axis the Stokes' stream function ψ exists. If ds is an element of the meridian curve, the normal velocity at any point is $u\frac{\partial \varpi}{\partial s}$.

Also normal velocity of the liquid in contact with the surface is $-\frac{\partial \psi}{\varpi\, \partial s}$.

Thus $$-\frac{\partial \phi}{\varpi\, \partial s} = u\frac{\partial \varpi}{\partial s}$$

or $d\psi = -u\varpi\, d\varpi$.

Integrating, $\psi = -\frac{1}{2}u\varpi^2 + \text{const.}$

$= -\frac{1}{2}ur^2 \sin^2\theta + \text{const.}$ as $\varpi = r\sin\theta$

$= -\frac{1}{2}ur^2(1-u^2) + \text{const.}$ where $\mu = \cos\theta$. ...(1)

This is the boundary condition.

But ψ has got to satisfy the equation

$$r^2\frac{\partial^2 \psi}{\partial r^2} + (1-u^2)\frac{\partial^2 \psi}{\partial \mu^2} = 0, \quad \text{where } \mu = \cos\theta,$$

the solutions of this equation being known of the type

$$\frac{1-u^2}{n+1} r^{n+1} \frac{\partial P_n}{\partial \mu} \text{ and } \frac{1-u^2}{nr^n}\frac{\partial P_n}{\partial \mu}. \quad ...(2)$$

Case of a sphere. If the sphere is of radius a, then from (1), on the boundary, $\psi = \frac{1}{2}ua^2(1-u^2) + \text{const.}$

This suggests that those solutions of (2) would be suitable which contain

$(1-u^2)$ and some constant when $r = a$.

Therefore $\frac{\partial P_n}{\partial \mu}$ =const. = 1, so that $P_n(\mu) = \mu$ or $n = 1$.

Also the liquid is at rest at infinity.

$\therefore$ suitable form of ψ is

$$\psi = A\frac{1-u^2}{r}.$$

On boundary,

$$\frac{A(1-u^2)}{a} = -\frac{1}{2}ua^2(1-u^2) + \text{const.}$$

$$\therefore \ A = -\frac{1}{2}ua^3.$$

Thus $$\psi = -\frac{1}{2}\frac{ua^3}{r}(1-u^2) = -\frac{1}{2}\frac{ua^3}{r}\sin^2\theta.$$

Again we know that

$$(1-u^2)\frac{\partial \phi}{\partial \mu} = \frac{\partial \psi}{\partial r} = \frac{1}{2}\frac{ua^3}{r^2}\sin^2\theta.$$

$$\therefore \ \frac{\partial \phi}{\partial \mu} = \frac{1}{2}\frac{ua^3}{r^2}$$

or $\phi = \frac{1}{2}\frac{ua^3}{r^2}\mu = \frac{1}{2}\frac{ua^3}{r^2}\cos\theta.$

This result we have already derived in 6.2, p. 242.

Important Note. In 3.19 page 139, we have derived the velocity potential due to a three-dimensional doublet of strength μ as

$$\phi = \frac{\mu\cos\theta}{r^2}.$$

Also as found in 6.2 p. 242 the velocity potential due to a sphere of radius a, moving with velocity U, is given by

$$\phi = \frac{1}{2}\frac{Ua^3}{r^2}\cos\theta.$$

A comparison of these two results shows that the sphere of radius a moving with velocity U produces the same effect as a doublet of strength $\frac{1}{2}Ua^3$ *at its centre.*

SOLVED EXAMPLES

Example 1: *A solid sphere moves through quiescent frictionless liquid whose boundaries are at a distance from it great compared with its radius. Prove that at each instant the motion in the liquid depends only on the position and velocity of the sphere at that instant. Prove that the liquid streams past the sides of the sphere with half the velocity of the sphere.*

Solution: At any time t let O be the centre of the sphere and V be the velocity with which it moves towards the line OZ. Now ϕ should be such that

(i) it satisfies $\nabla^2\phi = 0$,

(ii) $-\dfrac{\partial\phi}{\partial r} = V\cos\theta$ when $r = a$,

(iii) $\dfrac{\partial\phi}{\partial r} = 0$ when $r \to \infty$.

$\therefore$ The solution of $\nabla^2\phi = 0$ containing $\cos\theta$ in the suitable form (the motion being symmetrical to x-axis) is

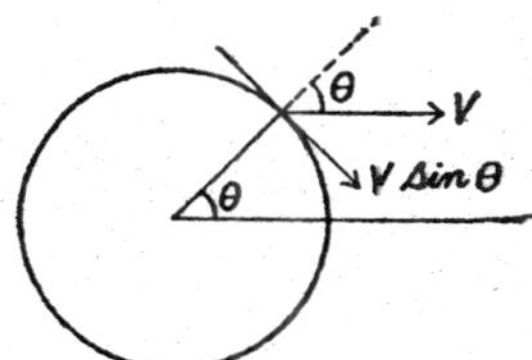

Fig. 4.5

$$\phi = \left\{Ar + \frac{B}{r^2}\right\}\cos\theta,$$

$$\frac{\partial\phi}{\partial r} = \left\{A - \frac{2B}{r^3}\right\}\cos\theta.$$

By condition (ii),

$$\left\{A - \frac{2B}{r^3}\right\}\cos\theta = -V\cos\theta.$$

By condition (iii),

$$A\cos\theta = 0.$$

$$\therefore\ A = 0 \text{ and } \frac{2B}{r^3} = V,\ i.e.\ B = \frac{a^3V}{2}.$$

Thus $$\phi = \frac{1}{2}\frac{a^3V}{r^2}\cos\theta.$$

This depends upon V the velocity of the sphere and the position of the sphere at an instant.

Velocity with which liquid streams past the side of the sphere

$= \text{velocity of slip}$

$$= \left\{-\frac{1}{r}\frac{\partial\phi}{\partial\theta}\right\}_{r=a} = \left\{\frac{1}{2}\frac{a^3V}{r^3}\sin\theta\right\}_{r=a}$$

$= \frac{1}{2}$ the velocity of the sphere along the tangent.

This proves the result.

Example 2: *An infinite ocean of an incompressible perfect liquid of density ρ is streaming past a fixed spherical obstacle of radius a. The velocity is uniform and equal to U except in so far as it is disturbed by the sphere, and the pressure in the liquid at a great distance from the obstacle is II. show that the thrust on that half of the sphere on which the liquid impinges is* $\pi a^2[\text{II} - \frac{1}{16}\rho U^2]$.

Solution: Taking the entre of the sphere as pole, the velocity potential when the liquid is streaming past the fixed sphere with velocity U is given by

$$\phi = Ur\cos\theta + \frac{1}{2}\frac{Ua^3}{r^2}\cos\theta,$$

$$\left\{\frac{\partial\phi}{\partial r}\right\}_{r=a} = Ua\sin\theta - \frac{1}{2}Ua\sin\theta = -\frac{3}{2}aU\sin\theta.$$

If q is the velocity at a point on the sphere, r = a.

$$q = \left[\left\{\frac{\partial\phi}{\partial r}\right\}^2 + \left\{\frac{\partial\phi}{r\,d\theta}\right\}^2\right]_{r=a}$$

In the absence of external forces and steady motion, the pressure p by Bernoulli's equation is given by

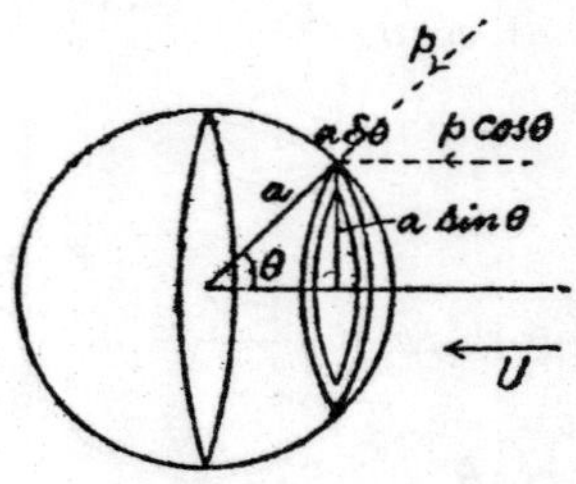

Fig. 4.6

$$\frac{p}{\rho}=\frac{\text{II}}{\rho}+\frac{1}{2}U^2-\frac{1}{2}q^2$$

$$=\frac{\text{II}}{\rho}+\frac{1}{2}U^2-\frac{9}{8}U^2\sin^2\theta. \qquad ...(1)$$

The total thrust on that half on which the liquid impinges

$$=\int_0^{\pi/2}(p\cos\theta)2\pi a\sin\theta.a\,d\theta$$

$$=2\pi a^2\rho\int_0^{\pi/2}\left\{\frac{\text{II}}{\rho}+\frac{1}{2}U^2-\frac{9}{8}U^2\sin^2\theta\right\}\sin\theta\;\cos\theta\,d\theta$$

$$=2\pi a^2\rho\left[\left\{\frac{\text{II}}{\rho}+\frac{1}{2}U^2\right\}\frac{1}{2}-\frac{9}{8}U^2.\frac{1}{4}\right]$$

$$=2\pi a^2(\text{II}-\frac{1}{16}\rho U^2).$$

This proves the result.

Example 3: *A sphere of radius a is moving with constant velocity U through an infinite liquid at rest at infinity. If p_0 be the pressure at infinity, show that the pressure at any point of the surface of the sphere, the radius to which point makes an angle θ with the direction of motion is given by*

$$p=p_0+\frac{1}{2}\rho U^2(1-\frac{9}{4})\sin^2\theta.$$

Solution:

Do yourself

Example 4: *Prove that for liquid contained between two instantaneously concentric spheres, when the outer (radius a) is moving parallel to the axis of x with velocity u and the inner (radius b) is moving parallel to the axis of y with velocity v, the velocity potential is*

$$-\frac{1}{a^3-b^3}\left\{a^3ux\left(I+\frac{b^3}{2r^3}\right)-b^3vy\left(I+\frac{b^3}{2r^3}\right)\right\}$$

and find the kinetic energy.

Solution: The velocity potential ϕ must be such that

(i) $\nabla^2\phi = 0$,

(ii) $-\frac{\partial\phi}{\partial r}=u\cos\theta$ when r = a,

(iii) $-\frac{\partial\phi}{\partial r}=v\sin\theta$ when r = b.

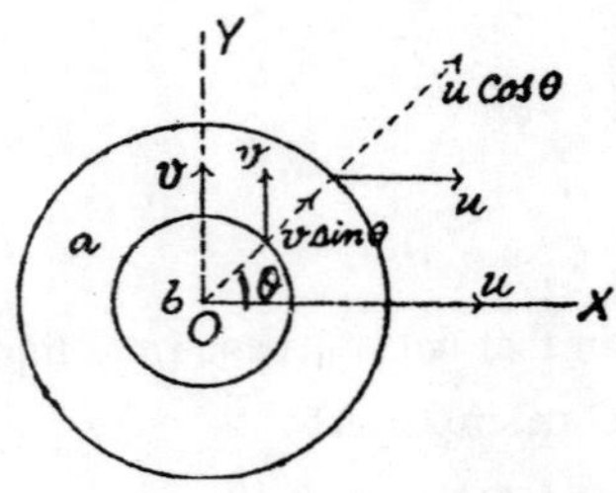

Fig. 4.7

Therefore the suitable form of ϕ is

$$\phi = \left\{ Ar + \frac{B}{r^2} \right\} \cos\theta + \left\{ Cr + \frac{D}{r^2} \right\} \sin\theta.$$

$$\therefore \quad \frac{\partial \phi}{\partial r} = \left\{ A - \frac{2B}{r^3} \right\} \cos\theta + \left\{ C - \frac{2D}{r^3} \right\} \sin\theta.$$

Applying conditions (ii) and (iii),

$$\left\{ A - \frac{2B}{a^3} \right\} \cos\theta + \left\{ C - \frac{2D}{a^3} \right\} \sin\theta = -u\cos\theta$$

and $$\left\{ A - \frac{2B}{b^3} \right\} \cos\theta + \left\{ C - \frac{2D}{b^3} \right\} \sin\theta = -v\sin\theta,$$

giving $$A - \frac{2B}{a^3} = -u, \; C - \frac{2D}{a^3} = 0,$$

$$A - \frac{2B}{b^3} = 0, \; C - \frac{2D}{b^3} = -v.$$

These give $A = -\dfrac{ua^3}{a^3 - b^3}, \; B = -\dfrac{ua^3 b^3}{2(a^3 - b^3)},$

$$C = \frac{vb^3}{a^3 - b^3} \text{ and } D = \frac{va^3 b^3}{2(a^3 - b^3)}.$$

Thus $$\phi = -\frac{ua^3}{a^3 - b^3} \left\{ r + \frac{b^3}{2r^2} \right\} \cos\theta + \frac{vb^3}{a^3 - b^3} \left\{ r + \frac{a^3}{2r^3} \right\} \sin\theta$$

$$= -\frac{1}{a^3 - b^3} \left[a^3 u \left\{ 1 + \frac{b^3}{2r^3} \right\} r\cos\theta - b^3 v \left\{ 1 + \frac{a^3}{2r^3} \right\} r\sin\theta \right]$$

$$= -\frac{1}{a^3 - b^3} \left[a^3 u \left\{ 1 + \frac{b^3}{2r^3} \right\} x - b^3 v \left\{ 1 + \frac{a^3}{2r^3} \right\} y \right].$$

This proves the result.

To calculate K.E., The K.E. of the liquid is given by

$$T = -\frac{1}{2}\rho\iint \phi\frac{\partial\phi}{\partial n}dS$$

$$= \frac{1}{2}\rho\iint\left\{\phi\frac{\partial\phi}{\partial r}\right\}_{r=a} dS - \frac{1}{2}\rho\iint\left\{\phi\frac{\partial\phi}{\partial r}\right\}_{r=b} dS$$

the normal direction being the direction of r

$$= \frac{1}{2}\rho\frac{1}{a^3-b^3}\iint\left[a^3u\left\{1+\frac{b^3}{2a^3}\right\}x - b^3v(1+\frac{1}{2})y\right]\times(-u\cos\theta)\ dS$$

$$= \frac{1}{2}\rho\frac{1}{a^3-b^3}\iint\left[a^3u(1+\frac{1}{2})x - b^3v\left\{1-\frac{a^3}{2b^3}\right\}y\right]\times(-v\sin\theta)\ dS$$

as $\left\{\frac{\partial\phi}{\partial r}\right\}_{r=a} = -u\cos\theta$ and $\left\{\frac{\partial\phi}{\partial r}\right\}_{r=b} = -v\sin\theta$

$$= \frac{1}{2}\rho\frac{u^2(2a^3+b^3)}{(a^3-b^3)a}\iint_{r=a} x^2 dS - \frac{3}{4}\rho\frac{uvb^3}{(a^3-b^3)}\frac{1}{a}\iint_{r=a} xy\ dS$$

$$-\frac{3}{4}\rho\frac{uva^3}{b(a^3-b^3)}\iint_{r=a} xy\ dS + \frac{1}{4}\rho\frac{v^2(a^3+2b^3)}{(a^3-b^3)b}\iint y^2\ dS$$

as when r = a, a cosθ = x and a sinθ = y

and when r = b, b cosθ = x and b sinθ = y

$$\frac{1}{4}\rho\frac{u^2(2a^3+b^3)}{(a^3-b^3)a}.\frac{3}{4}\pi a^4 - 0 - 0 + \frac{1}{4}\rho\frac{v^2(a^3+2b^3)}{(a^3-b^3)b}.\frac{3}{4}\pi b^4$$

as $\iint_{r=a} x^2 dS = \frac{1}{2}\iint_{r=a}(x^2+y^2)dS$

=M.I. of the hollow sphere of radius a about a diameter

$$= \frac{1}{2}\frac{2Ma^2}{3} = \frac{Ma^2}{3}$$

$$= \frac{2\pi a^2.a^2}{3} = \frac{2\pi a^2}{3}$$

and $\iint_{r=b} xy$ =0 being product of inertia.

Thus K.E. $= \frac{1}{3}\frac{\pi\rho}{a^3-b^3}[2(u^2a^6+v^2a^6)+a^3b^3(u^2+v^2)]$.

This gives the required K.E.

Example 5: *Liquid of dencity ρ fills the space between a solid sphere of radius a and density ρ' and a fixed concentric spherical envelope of radius b; prove that the work done by an impulse which starts the solid sphere with velocity V is*

$$\frac{1}{3}\pi a V^2\left\{2\rho'+\frac{2a^3+b^3}{b^3-a^3}\rho\right\}.$$

Solution: Let ϕ be the velocity potential ; then the conditions are

(i) $\nabla^2\phi = 0$,

(ii) $\left\{-\frac{\partial\phi}{\partial r}\right\}_{r=a} = V\cos\theta,$

(iii) $\left\{-\frac{\partial\phi}{\partial r}\right\}_{r=b} = 0.$

Therefore the suitable form for velocity potential is

$$\phi = \left\{Ar+\frac{B}{r^2}\right\}\cos\theta.$$

$$\therefore\quad \frac{\partial\phi}{\partial r} = \left\{A-\frac{2B}{r^3}\right\}\cos\theta.$$

Applying the conditions

$$\left\{A-\frac{2B}{a^3}\right\}\cos\theta = -V\cos\theta, \left\{A-\frac{2B}{b^3}\right\}\cos\theta = 0,$$

i.e. $\qquad A-\frac{2B}{a^3} = -V$ and $A-\frac{2B}{b^3} = 0.$

Solving these, $A = \frac{Va^3}{b^3-a^3}$, $B = \frac{Va^3b^3}{2(b^3-a^3)}$.

Thus $\qquad \phi = \frac{Va^3}{b^3-a^3}\left\{r+\frac{b^3}{2r^2}\right\}\cos\theta.$

Impulsive pressure at a point (r, θ) is given by

$$\varpi = (\rho\phi)_{r=a} = \frac{Va^3\rho}{b^3-a^3}\cdot\left\{a+\frac{b^3}{2r^2}\right\}\cos\theta.$$

$\therefore$ The total impulsive pressure on the sphere

$$= \int_0^\pi \varpi\cos\theta\times(2\pi a\sin\theta)a\,d\theta$$

$$= \frac{Va^3\rho}{(b^3+a^3)}\left\{a+\frac{b^3}{2a^2}\right\}2\pi a^2\times\int_0^\pi \cos^2\theta\,\sin\theta\,d\theta$$

$$=\frac{2\pi Va^3\rho}{3}\left\{\frac{2a^3+b^3}{b^3+a^3}\right\}.$$

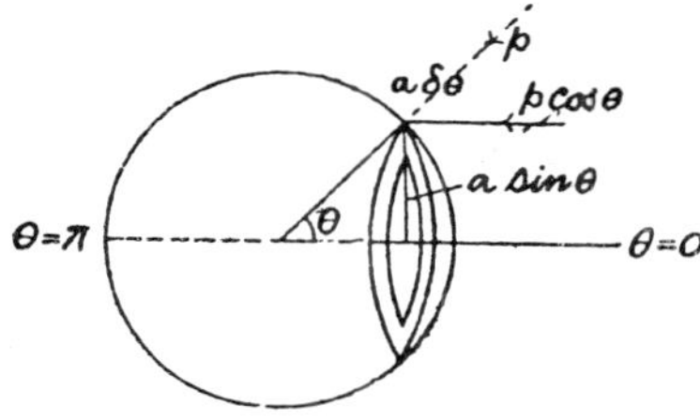

Fig. 4.8

Also the impulse on the sphere because of its mass

$$= mV$$

$$=\frac{3}{4}\pi a^2\rho' V.$$

Therefore total impulse to start the sphere with velocity V

$$=\frac{3}{4}\pi a^3\rho' V+\frac{2\pi Va^3\rho}{3}\left\{\frac{2a^3+b^3}{b^3+a^3}\right\}$$

$$=\frac{2\pi a^3 V}{3}\left[2\rho'+\rho\left(\frac{2a^3+b^3}{b^3+a^3}\right)\right].$$

The work done

= Impulse × (mean of initial and final velocities)

$$=\text{Impulse}\times\frac{0+V}{2}=\text{Impulse}\times\frac{V}{2}$$

$$=\frac{\pi V^2 a^3}{3}\left[2\rho'+\frac{2a^3+b^3}{b^3+a^3}\right].$$

Example 6: *The space between two concentric spherical shells of radii a and b (a > b) is filled with an incompressible fluid of density ρ and the shells suddenly begin to move with velocities U, V in the same direction; prove that resultant impulsive pressure on the inner shell is*

$$=\frac{2\pi\rho b^3}{3(b^3-a^3)}\left\{3a^3U-(a^3+2b^3)V\right\}.$$

Solution:

Here the boundary conditions are

(i) $-\dfrac{\partial \phi}{\partial r} = U\cos\theta$ when $r = a$,

(ii) $-\dfrac{\partial \phi}{\partial r} = V\cos\theta$ when $r = b$.

The suitable form of velocity potential is given by

$$\phi = \left\{Ar + \frac{B}{r^2}\right\}\cos\theta,$$

so that $$\frac{\partial \phi}{\partial r} = \left\{A + \frac{2B}{r^3}\right\}\cos\theta.$$

Applying boundary conditions.

$$\left\{A - \frac{2B}{a^3}\right\}\cos\theta = -U\cos\theta$$

and $$\left\{A - \frac{2B}{b^3}\right\}\cos\theta = -V\cos\theta,$$

i.e., $$A - \frac{2B}{a^3} = -U \text{ and } A - \frac{2B}{b^3} = -V$$

so that $$A = -\frac{Ua^3 - Vb^3}{a^3 - b^3},\ B = -\frac{1}{2}\frac{(U-V)a^3b^3}{a^3 - b^3}.$$

$$\therefore\ \phi = -\frac{1}{a^3 - b^3}\left[(Ua^3 - Vb^3)r + \frac{(U-V)a^3b^3}{2r^2}\right]\cos\theta.$$

If ϖ be the impulsive pressure at any point (b, θ) on the inner sphere,

$$\varpi = \rho(\phi)_{r=b}$$

$$= \frac{-\rho b}{a^3 - b^3}\left\{(Ua^3 - Vb^3) + \frac{1}{2}(U-V)a^3\right\}\cos\theta.$$

The total impulsive pressure on the surface of the inner sphere

$$= \int_0^\pi (\varpi\cos\theta)(2\pi b\sin\theta)b\, d\theta$$

$$= -\frac{2\pi\rho b^3}{(a^3 - b^3)}\left\{(Ua^3 - Vb^3) + \frac{1}{2}(U-V)a^3\right\}\int_0^\pi \cos^2\theta\ \sin\theta\ d\theta$$

$$= -\frac{2\pi\rho b^3}{3(a^3 - b^3)}\left\{(3a^3U - (a^3 + 2b^3)V\right\}.$$

The negative sign may be left for magnitude.

This proves the result.

Example 7: *A spherical shell of internal radius a contains a concentric sphere of radius λa and density σ, the intervening space being filled with water (unit density) and the whole system is at rest. If a velocity V is suddenly communicated to the shell, prove that the initial velocity U communicated to the sphere is*

$$U = \frac{3V}{2\sigma(1-\lambda^3)+1+2\lambda^2}.$$

Solution: The velocity potential ϕ must be such that

(i) $\nabla^2\phi = 0$,

(ii) $-\dfrac{\partial\phi}{\partial r} = V\cos\theta$ when $r = a$,

(iii) $-\dfrac{\partial\phi}{\partial r} = U\cos\theta$ when $r = \lambda a$.

Therefore the suitable form of ϕ is

$$\phi = \left\{Ar + \frac{B}{r^2}\right\}\cos\theta,$$

$$\frac{\partial\phi}{\partial r} = \left\{A - \frac{B}{r^3}\right\}\cos\theta.$$

$\therefore$ conditions (ii) and (iii) give

$$A - \frac{2B}{a^3} = -V \text{ and } A - \frac{2B}{\lambda^3 r^3} = -U,$$

$$B = \frac{a^3\lambda^3(U-V)}{2(1-\lambda^3)}, \quad A = \frac{\lambda^3 U - V}{1-\lambda^3}.$$

$$\therefore \quad \phi = \frac{1}{(1-\lambda^3)}\left[(\lambda^3 U - V)r + \frac{a^3\lambda^3(U-V)}{2r^2}\right]\cos\theta.$$

The impulsive pressure at any point of the sphere r = aλ is given by

$\varpi = (\rho\phi)_{r=a\lambda}$ $\rho = 1$, water being of density 1

$$= \frac{1}{(1-\lambda^3)}\left[(\lambda^3 U - V)a\lambda + \frac{a^3\lambda^3(U-V)}{2\lambda^2 a^2}\right]\cos\theta.$$

And the resultant impulsive pressure on the sphere

$$= \int_0^\pi \varpi\cos\theta . 2\pi(\lambda a)\sin\theta . \lambda a \, d\theta$$

$$= \frac{2\pi\lambda^3 a^3}{(1-\lambda^3)}\left[(\lambda^3 U - V)a\lambda + \frac{a^3\lambda^3(U-V)}{2\lambda^2 a^2}\right]\int_0^\pi \cos^2\theta \sin\theta \, d\theta$$

$$= \frac{2\pi\lambda^3 a^3}{(1-\lambda^3)}\left[\lambda^3 U + \frac{U}{2} - \frac{3V}{2}\right]\frac{3}{2}$$

$$= \frac{2\pi\lambda^3 a^3}{3\,(1-\lambda^3)}\left[2\lambda^3 U + U - 3V\right).$$

Since the sphere of density σ starts moving with velocity U, the equation of motion is

$$\frac{3}{4}\pi(a\lambda)^3\sigma U = -\frac{2\pi\lambda^3 a^3}{3\,(1-\lambda^3)}\left[2\lambda^3 U + U - 3V\right]$$

or $U = \dfrac{3V}{2\sigma\,(1-\lambda^3)+1+2\lambda^3}.$

This proves the result.

Example 8: *A hollow spherical shell of inner radius a contains a concentric solid uniform sphere of radius b and density σ and the space between the two is filled with liquid of density ρ. If the shell is suddenly made to move with speed u, prove that a velocity v is imparted to the inner sphere where*

$$v = \frac{3V}{2(\sigma/\rho)\,(a^3-b^3)+a^3+2b^3}.$$

Solution:

This is just the above example where $\lambda a = b$.

Example 9: *Incompressible fluid, of density ρ, is contained between two rigid concentric spherical surfaces, the outer one of mass M_1 and radius a, the inner one of mass M_2 and radius b. A normal blow P is given to the outer surface. Prove that the initial velocities of the two containing surfaces (U for the outer and V for the inner) are given by the equations.*

$$\left\{M_1 + \frac{2\pi\rho a^3(2a^3+b^3)}{3\,(a^3-b^3)}\right\}U - \frac{2\pi\rho a^3 b^3)}{a^3-b^3}V = P,$$

$$\left\{M_2 + \frac{2\pi\rho b^3(2b^3+a^3)}{3\,(a^3-b^3)}\right\}V = \frac{2\pi\rho a^3 b^3)}{a^3-b^3}U.$$

Solution: If ϕ be the velocity potential, then

(i) $\nabla^2\phi = 0$,

(ii) $-\dfrac{\partial\phi}{\partial r} = U\cos\theta$ when $r = a$,

(iii) $-\dfrac{\partial\phi}{\partial r} = V\cos\theta$ when $r = b$.

$\therefore$ As in 6.4 p. 245,

$$\phi = \frac{1}{a^3 - b^3}\left[(Vb^3 - Ua^3)r + \frac{(V-U)a^3\lambda^3}{2r^2}\right]\cos\theta.$$

The normal blow P on the outer imparts velocity U to the outer and V to the inner. Therefore we have

$$M_1U = P - \iint_{r=a} \varpi\cos\theta\, ds \qquad ...(1)$$

and $$M_2V = -\iint_{r=b} \varpi\cos\theta\, ds, \qquad ...(2)$$

where ϖ is impulsive pressure on an element δs of the boundary surface.

We know that $\varpi = \rho\phi$.

$\therefore$ (1) gives

$$M_2U = P - \int_0^\pi \rho\frac{1}{a^3 - b^3}\left[(Vb^3 - Ua^3)a + \frac{(V-U)a^3b^3}{2a^2}\right]\cos\theta$$

$\times\cos\theta.2\pi a\ \sin\theta.a\ d\theta$, ϕ being taken for r = a

$$= P - \frac{2\pi\rho a^3}{2(a^3 - b^3)}\{3Vab^3aU(2a^3 + b^3)\}\times\int_0^\pi \cos^2\theta\ \sin\theta\ d\theta$$

$$= P + \frac{2\pi\rho a^3}{3\ (a^3 - b^3)}\{3Vb^3 - U(2a^3 + b^3)\},$$

i.e., $$\left\{M_1 + \frac{2\pi\rho a^3(2a^3 + b^3)}{3\ (a^3 - b^3)}\right\}U - \frac{2\pi\rho a^3b^3V}{a^3 - b^3} = P.$$

This gives the first required equation.

Again from (2),

$$M_2V = -\int_0^\pi \rho\frac{1}{a^3 - b^3}\left[(Vb^3 - Ua^3)b + \frac{(V-U)a^3b^3}{2b^2}\right]\cos^2\theta$$

$\times 2\pi b\ \sin\theta.b\ d\theta$

$$= -\frac{2\pi b^3\rho}{a^3 - b^3}\left[(Vb^3 - Ua^3) + \frac{1}{2}(V-U)a^3\right]\int_0^\pi \cos\theta\sin\theta\ d\theta$$

$$= -\frac{2\pi b^3\rho}{3(a^3 - b^3)}\left[V(2b^3 + a^3) - 3Ua^3\right]$$

or $$\left\{M_2 + \frac{2\pi\rho b^3(2b^3 + a^3)}{2\ (a^3 - b^3)}\right\}V = \frac{2\pi\rho b^3a^3}{a^3 - b^3}U.$$

This establishes the second equation.

Example 10: *A stream of water of great depth is flowing with uniform velocity V over a plane level bottom. A hemisphere of weight w in water and of radius a, rests with its base on the bottom. Prove that the average pressure between the base of the hemisphere and the bottom is less than the fluid pressure at any point of the bottom at a great distance from the hemisphere if*

$$V^2 > \frac{32w}{11\pi a^2 \rho}.$$

Solution: Let the liquid be streaming past the fixed hemisphere with velocity V along z-axis and (r, q, w) be the spherical polar co-ordinates of a point referred to centre of the hemisphere as origin.

Then velocity potential ϕ is given by

$$\phi = Vr\cos\theta + \frac{Va^3}{2r^3}\cos\theta,$$

where x = r sin θ, cos ω, y = r sin θ sin ω and z = r cos θ. ...(1)

$$\left\{\frac{\partial\phi}{\partial r}\right\}_{r=a} = V\cos\theta - V\cos\theta = 0.$$

$$\left\{\frac{\partial\phi}{\partial\theta}\right\}_{r=a} = Va\sin\theta - \frac{1}{2}Va\sin\theta = -\frac{3}{2}aV\sin\theta.$$

If q is the velocity at the point P (r, θ, w} (on the sphere), we have

$$q^2 = \left[\left\{\frac{\partial\phi}{\partial r}\right\}^2 + \left\{\frac{\partial\phi}{\partial\theta}\right\}^2\right]_{r=a}$$

$$= \frac{9}{4}V^2\sin^2\theta.$$

Let II be the pressure at infinite distance; then since the motion is steady, we have on a point on the sphere

$$\frac{p}{\rho} = \frac{\text{II}}{\rho} + \frac{1}{2}V^2 - \frac{1}{2}q^2$$

$$= \frac{\text{II}}{\rho} + \frac{1}{2}V^2 - \frac{9}{8}V^2\sin^2\theta.$$

Direction cosines of OP are

$$\left\{\frac{x}{r}, \frac{y}{r}, \frac{z}{r}\right\}$$

where P is (x, y, z),

i.e., (sin θ cos ω, sin θ sin ω, cos θ) by (1).

∴ component of p along x-axis = p sin θ cos ω.

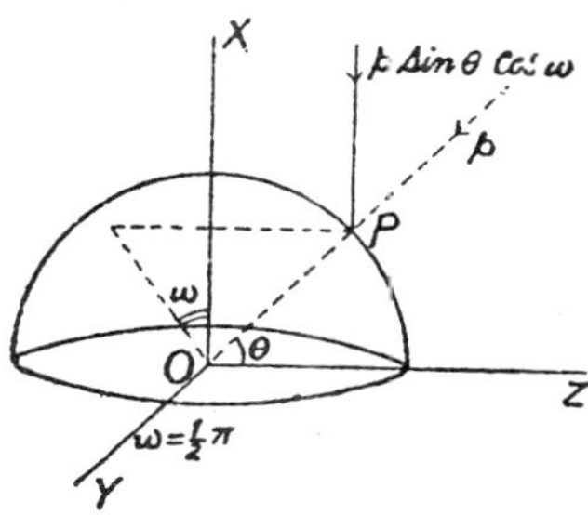

Fig. 4.9

Hence the total thrust on the hemisphere due to liquid along XO

$$= \int_{\theta=0}^{\pi} \int_{\omega=+\pi/2}^{\omega=+\pi/2} (+p \sin\theta \ \cos\omega) a \ \sin\theta \ \cos\omega \ d\theta \ d\omega$$

a sin θ dω a dθ being the element of surface on the sphere at P

$$= a^2 \rho \int_{\theta=0}^{\pi} \int_{\omega=+\pi/2}^{\omega=+\pi/2} \left[\left\{ \frac{\text{II}}{\rho} + \frac{1}{2} V^2 \right\} - \frac{9}{8} V^2 \sin^2\theta \right] \sin^2\theta \ \cos\omega \ d\theta \ d\omega$$

$$= 2a^2 \rho \int_{\theta=0}^{\pi} \left[\left\{ \frac{\text{II}}{\rho} + \frac{1}{2} V^2 \right\} - \frac{9}{8} V^2 \sin^2\theta \right] \sin^2\theta \ d\theta$$

$$= 4a^2 \int_0^{\pi/2} \left[\left\{ \frac{\text{II}}{\rho} + \frac{1}{2} V^2 \right\} \sin^2\theta - \frac{9}{8} V^2 \sin^4\theta \right] d\theta$$

$$= 4a^2 \left[\left\{ \frac{\text{II}}{\rho} + \frac{1}{2} V^2 \right\} \frac{\pi}{4} - \frac{9}{8} V^2 \frac{\Gamma\left(\frac{5}{2}\right)\Gamma\left(\frac{1}{2}\right)}{2\Gamma(3)} \right]$$

$$= \pi a^2 \left[\Pi - \frac{11 \rho V^2}{32} \right].$$

Also there is weight w on the base.

Therefore total pressure on the base

$$= \pi a^2 \left[\Pi - \frac{11 \rho V^2}{32} \right] + w.$$

$$\text{Average pressure on base} = \frac{\text{pressure on base}}{\text{area of the base}}$$

$$= \left[\Pi - \frac{11\rho V^2}{32}\right] + \frac{w}{\pi a^2}.$$

Now the average pressure < pressure at great distance if

$$= \Pi - \frac{11\rho V^2}{32} + \frac{w}{\pi a^2} < \Pi,$$

i.e. if $$\frac{11\rho V^2}{32} > \frac{w}{\pi a^2}$$

or $$V^2 > \frac{32w}{11\rho\pi^2}.$$

This proves the result.

Example 11: *A solid sphere is moving through frictionless liquid. Compare the velocities of slip of the liquid past it at different parts of its surface.*

Prove that when the sphere is in motion with uniform velocity U, the pressure at the part of its surface where the radius makes an angle θ with the direction of motion is increased on account of the motion by the amount.

$$\frac{1}{16}\rho U^2(9\cos 2\theta - 1),$$

where ρ is the density of the liquid.

Solution: At any time *t*, Let C be the centre of the sphere, moving along z=axis with velocity U, so that if C be $(0, 0, Z_0)$,

$$z_0 = U. \qquad ...(1)$$

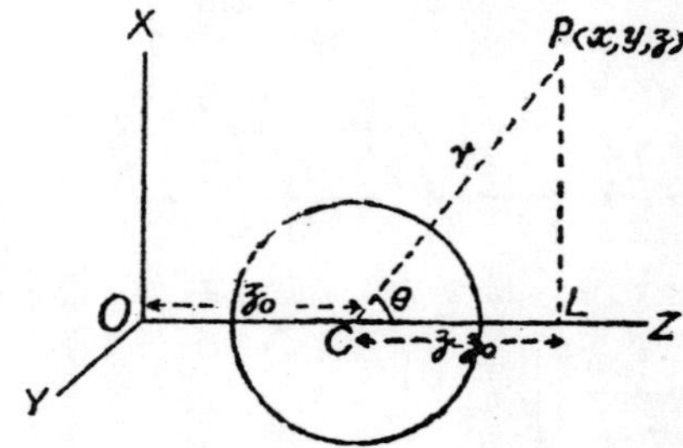

Fig. 4.10

If *P (x, y, z)* referred to O be a point in the liquid and *CP* = r and *PCZ* = *θ*, then, as usual,

$$\phi = \frac{1}{2}\frac{Ua^3}{r^2}\cos\theta. \qquad ...(2)$$

At any point (a, θ), velocity of slip

$$=\left\{-\frac{1}{r}\frac{\partial\phi}{\partial\theta}\right\}_{r=a}\frac{1}{2}U\sin\theta.$$

Also if q be the velocity of fluid at P,

$$q^2=\left\{\frac{\partial\phi}{\partial r}\right\}^2+\left\{\frac{1}{r}\frac{\partial\phi}{\partial r}\right\}^2$$

$$=\left\{\frac{Ua^3}{r^3}\cos\theta\right\}^2+\left\{\frac{Ua^3}{2r^3}\sin\theta\right\}^2$$

$$=\frac{U^2a^6}{r^6}(\cos^2\theta+\frac{1}{4}\sin^2\theta). \qquad ...(3)$$

Again since $r = CP$ = distance between (x, y, z) and $(0, 0, z_0)$

$$= \sqrt{\{x^2 + y^2 + (z - z_0)^2\}},$$

$$\cos\theta=\frac{CL}{CP}=\frac{z-z_0}{\sqrt{\{x^2+y^2+(z-z_0)^2\}}}.$$

$\therefore$ from (2), $\phi=\frac{1}{2}\frac{Ua^3(-z_0)}{[x^2+y^2+(z-z_0)^2]^{3/2}}$

So $$\frac{\partial\phi}{\partial t}=\frac{1}{2}\frac{Ua^3(-z_0)}{[x^2+y^2+(z-z_0)^2]^{3/2}}+\frac{1}{2}\frac{Ua^3\left(-\frac{3}{2}\right)2(z-z_0)^2(-z_0)}{[x^2+y^2+(z-z_0)^2]^{5/2}}$$

$$=-\frac{1}{2}\frac{U^2a^3}{r^3}+\frac{3}{2}\frac{U^2a^3\cos^2\theta}{r^3} \qquad \text{as } z_0 = U \text{ form (1)}$$

again changing to r and θ.

Now if p is the pressure at any point, we have

$$\frac{p}{\rho}+\frac{1}{2}q^2-\frac{\partial\phi}{\partial t}=F(t).$$

Putting $r = a$, the pressure at a point (a, θ) on the sphere is given by

$$\frac{p}{\rho}+\frac{1}{2}U^2(\cos^2\theta+\frac{1}{4}\sin^2)+(\frac{1}{2}U^2-\frac{3}{2}U^2\cos^2\theta)=F(t). \qquad ...(4)$$

Now if p_0 is the pressure on (a, θ) when $U = 0$, *i.e.* when there is no motion, $\frac{p_0}{\rho}=F(t)$.

$$\therefore\ \frac{p-p_0}{\rho}=\frac{1}{2}U^2(3\cos^2\theta-1-\cos^2\theta-\frac{1}{4}\sin^2\theta)$$

$$=\frac{1}{2}U^2(2\cos^2\theta-1-\frac{1}{4}\sin^2\theta)$$

$$=\frac{1}{16}U^2[16\cos^2\theta-8-2\sin^2\theta]$$

$$=\frac{1}{16}U^2[8(1+\cos 2\theta)-8-(1-\cos 2\theta)]$$

$$=\frac{1}{16}U^2[9\cos 2\theta-1).$$

This proves the result.

Example 12: *Find the pressure at any point of a liquid, of infinite extend and at rest at a great distance, through which a sphere is moving under no external forces with constant velocity U, and show that the mean pressure over the sphere is in defect of the pressure* Π *at a great distance by* $\frac{1}{4}\rho U^2$, *it being supposed that* Π *is sufficiently large for the pressure everywhere to be positive, that is that* $\Pi > \frac{5}{8}\rho U^2$.

Solution: Proceeding as in the above example,

$$\phi=\frac{1}{2}\frac{Ua^3}{r^2}\cos\theta,$$

$$q^2=\left\{\frac{\partial\phi}{\partial r}\right\}+\left\{\frac{1}{r}\frac{\partial\phi}{\partial\theta}\right\}^2=\frac{U^2a^6}{r^3}(\cos^2\theta+\frac{1}{4}\sin^2\theta)$$

and $$\frac{\partial\phi}{\partial r}=-\frac{1}{2}\frac{U^2a^3}{r^3}+\frac{3}{2}\frac{U^2a^3\cos^2\theta}{r^3}=-\frac{1}{2}\frac{U^2a^3}{r^5}(1-3\cos^2\theta).$$

$\therefore$ the pressure equation

$$\frac{p}{\rho}+\frac{1}{2}q^2-\frac{\partial\phi}{\partial t}=F(t)$$

gives $$\frac{p}{\rho}+\frac{1}{2}\frac{U^2a^6}{r^6}(\cos^2\theta+\frac{1}{4}\sin^2\theta)+\frac{1}{2}\frac{U^2a^6}{r^3}(1-3\cos^2\theta)=F(t).$$

But $p = \Pi$ when $r = \infty$; $\quad\therefore\ \frac{\Pi}{\rho}=F(t).$

$$\therefore\ \frac{p-\Pi}{\rho}=\frac{1}{2}\frac{U^2a^3}{r^3}\{3\cos^2\theta-1)-\frac{1}{2}\frac{U^2a^3}{r^6}(\cos^2\theta+\frac{1}{4}\sin^2\theta)$$

$$=\frac{1}{2}U^2\{3\cos^2\theta-1-\cos^2\theta-\frac{1}{4}\sin^2\theta\}\quad\text{when } r=a$$

$$=\frac{1}{8}U^2[8\cos^2\theta+\sin^2\theta-4]$$

or $p = \Pi + \frac{1}{8}\rho U^2[9\cos^2\theta - 5].$ ...(1)

Now mean pressure on the sphere

$$= \frac{\int p\, ds}{\int ds} = \frac{\int_0^\pi p\,(2\pi a \sin\theta) a\, d\theta}{2\pi a^2}$$

$$= \frac{1}{2}\int_0^\pi [\Pi + \frac{1}{8}\rho U^2\{9\cos^2\theta - 5\}]\sin\theta\, d\theta$$

$$= \frac{1}{2}[2\Pi + \frac{1}{8}\rho U^2(6-10)] = \Pi - \frac{1}{4}\rho U^2.$$

Clearly $\quad \Pi$ - mean pressure $= \frac{1}{4}\rho U^2,$

which is the devote.

From (1) minimum pressure ρ is given by cos θ = 0.

$$\therefore \; p_{\min.} = \Pi - \frac{5}{8}\rho U^2.$$

This will be positive if $\Pi > \frac{5}{8}\rho U^2.$

Example 13: *A sphere of radius a is made to move in incompressible perfect fluid with non-uniform velocity u along the x-axis. If the pressure at infinity is zero, prove that at a point s in advance of the centre,*

$$p = \frac{1}{2}\rho a^3\left\{\frac{u}{x^2} + u^2\left(\frac{2}{x^3} - \frac{a^3}{x^6}\right)\right\}.$$

Solution: Let the sphere be moving along x-axis with velocity u; then at any point the velocity potential is given by

$$\phi = \frac{1}{2}\frac{ua^3}{r^2}\cos\theta, \qquad ...(1)$$

where $CP = r$ and $\angle PCX = \theta$.

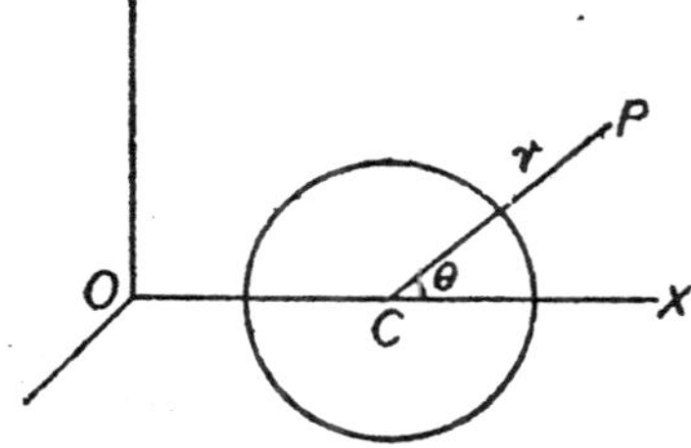

Fig. 4.11

If O be the origin referred to which C is

$(X_0, 0, 0)$

and P is (X, Y, Z), so that

$$r^2 = CP^2 = (X-X_0)^2 + Y^2 + Z^2$$

and $$\cos\theta = \frac{X - X_0}{\sqrt{\{(X-X_0)^2 + Y^2 + Z^2\}}},$$

$$\therefore \ \phi = \frac{1}{2}\frac{ua^3(X - X_0)}{\{(X-X_0)^2 + Y^2 + Z^2\}^{3/2}}. \qquad ...(2)$$

Here u is not uniform.

$$q^2 = \left\{\frac{\partial\phi}{\partial r}\right\}^2 + \left\{\frac{1}{r}\frac{\partial\phi}{\partial\theta}\right\}^2 \text{ from (1)}$$

$$= \frac{u^2a^6}{r^6}(\cos^2\theta + \frac{1}{4}\sin^2\theta). \qquad ...(3)$$

and from (2),

$$= \frac{\partial\phi}{\partial t} = \frac{1}{2}\frac{ua^3(X - X_0)}{[(X-X_0)^2 + Y^2 + Z^2]^{3/2}} + \frac{1}{2}\frac{ua^3(-X_0)}{[(X-X_0)^2 + Y^2 + Z^2]^{3/2}}$$

$$+ \frac{1}{2}\frac{ua^3(-\frac{3}{2})2(X - X_0)^2(-X_0)}{[(X-X_0)^2 + Y^2 + Z^2]^{5/2}}$$

$$= \frac{1}{2}\frac{ua^3}{r^2}\cos\theta - \frac{1}{2}\frac{u^2a^3}{r^3} + \frac{3}{2}\frac{u^2a^3\cos^2\theta}{r^3} \qquad \text{as } X = \text{u}.$$

The pressure p at any point is given by

$$\frac{p}{\rho} + \frac{1}{2}q^2 - \frac{\partial\phi}{\partial t} = F(t),$$

i.e., $$\frac{p}{\rho} = F(t) - \frac{1}{2}\frac{u^2a^6}{r^6}(\cos^2\theta + \frac{1}{4}\sin^2\theta)$$

$$+ \frac{1}{2}\frac{ua^3}{r^2}\cos\theta - \frac{1}{2}\frac{u^2a^3}{r^2} + \frac{3}{2}\frac{u^2a^3\cos^2\theta}{r^3}.$$

But when $r = \infty$, $p = 0$; $\quad \therefore \quad F(t) = 0$.

$$\therefore \ \frac{p}{\rho} = -\frac{1}{2}\frac{u^2a^6}{r^6}(\cos^2\theta + \frac{1}{4}\sin^2\theta)$$

$$+ \frac{1}{2}\frac{ua^3}{r^2}\cos\theta - \frac{1}{2}\frac{u^2a^3}{r^3} + \frac{3}{2}\frac{u^2a^3\cos^2\theta}{r^3}.$$

The point at a distance x in advance to the entre is given by $\theta = 0$ and $r = x$ and there we have

$$\frac{p}{\rho} = -\frac{1}{2}\frac{u^2a^6}{r^6} + \frac{1}{2}\frac{ua^3}{x^2} - \frac{1}{2}\frac{u^2a^3}{x^3} + \frac{3}{2}\frac{u^2a^3}{x^3},$$

i.e., $$p = \frac{1}{2}\rho a^3\left[\frac{u}{x^2} + u^2\left\{\frac{2}{x^3} - \frac{a^3}{x^6}\right\}\right].$$

This proves the result.

Example 14: *Prove that at a point on the sphere moving through an infinite liquid the pressure is given by the formula*

$$\frac{p - p_0}{\rho} = \frac{1}{2}af\cos\theta_1 + \frac{1}{8}v^2(9\cos^2\theta - 5),$$

where v is the velocity, f the acceleration of the sphere, and θ, θ_1 are the angles between the radius and the directions of v, f respectively, and p_0 is the hydrostatic pressure.

Solution: At any time t, let (x_0, y_0, z_0) be the centre C of the moving sphere and U, V, W the components of velocity v of the centre so that

$$x = U, y_0 = V, x = W. \qquad ...(1)$$

Let there be a point P (x, y, z) in the fluid, so that

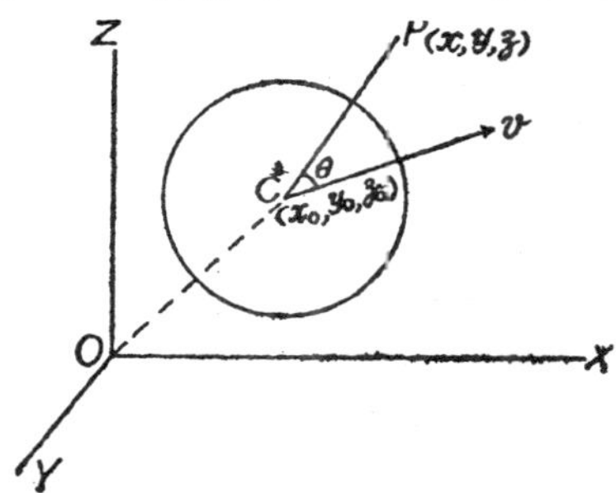

Fig. 4.12

$CP = r$ and $\angle PCv = \theta$.

Then $$\phi = \frac{1}{2}\frac{a^3v}{r^2}\cos\theta$$

Now $v\cos\theta$ = resolved part of v along CP

= sum of the components of v along CP

$$= U\frac{x - x_0}{r} + V\frac{y - y_0}{r} + W\frac{z - z_0}{r}$$

where $r2 = CP^2 = (x - x_0)^2 + (y - y_0)^2 + (z - z_0)^2$,

$\frac{x-x_0}{r}, \frac{y-y_0}{r}, \frac{z-z_0}{r}$ being the d.c.'s of CP

$$\therefore \phi = \frac{1}{2}\frac{a^3}{r^2}\left[U\frac{x-x_0}{r} + V\frac{y-y_0}{r} + W\frac{z-z_0}{r}\right]$$

$$= \frac{1}{2}\frac{a^3}{r^3}[U(x-x_0)+V(y-y_0)+W(z-z_0)]. \qquad ...(2)$$

Also since $r2 = (x - x_0)^2 + (y - y_0)^2 + (z - z_0)^2$,

$$\therefore\ rr = -(x - x_0)\, x_0 - (y - y_0)\ y_0 - (z - z_0)\, z_0$$

$$= -[(x-x_0)\ U + (y-y_0)\ V + (z-z_0)\ W]. \qquad ...(3)$$

$$\frac{\partial\phi}{\partial x} = \frac{1}{2}\frac{a^3U}{r^3} - \frac{3a^3}{4r^4}[U(x-x_0)+V(y-y_0)+W(z-z_0)]\frac{\partial r}{\partial x}$$

$$= \frac{1}{2}\frac{a^3U}{r^3} - \frac{3a^3(x-x_0)}{2r^5}[U(x-x_0)+V(y-y_0)+W(z-z_0)]$$

$$\text{as } \frac{\partial r}{\partial x} = \frac{x-x_0}{r}.$$

Similarly expressions for $\frac{\partial\phi}{\partial y}$ and $\frac{\partial\phi}{\partial z}$ can be written.

$$\therefore\ q^2 = \left\{\frac{\partial\phi}{\partial x}\right\}^2 + \left\{\frac{\partial\phi}{\partial y}\right\}^2 + \left\{\frac{\partial\phi}{\partial z}\right\}^2$$

$$= \frac{a^6}{4r^6}(U^2+V^2+W^2) - \frac{3a^6}{2r^8}[U(x-x_0)+V(y-y_0)$$

$$+W(z-z_0)]^2 + \frac{9a^6}{4r^8}[U(x-x_0)+V(y-y_0)+W(z-z_0)]^2$$

$$= \frac{a^6}{4r^6}(U^2+V^2+W^2) + \frac{3a^6}{4r^8}[U(x-x_0)+V(y-y_0)$$

$$+W(z-z_0)]^2$$

$$\frac{a^6}{4r^6}v^2\ \frac{3a^6}{4r^6}v^2\cos^2\theta. \qquad ...(4$$

Again $\frac{\partial\phi}{\partial t} = \frac{1}{2}\frac{a^3}{r^3}[U(x-x_0)+V(y-y_0)+W(z-z_0)]$

$$+\frac{1}{2}\frac{a^3}{r^3}[U(-x_0)+V(-y_0)+W(-z_0)$$

$$-\frac{3a^3}{2r^4}r[(x-x_0)+V(y-y_0)+W(z-z_0)]$$

$$=\frac{1}{2}\frac{a^3}{r^2}f\cos\theta_1-\frac{1}{2}\frac{a^3}{r^3}[U^2+V^2+W^2]$$

$$+\frac{3a^6}{2r^5}[U(x-x_0)+V(y-y_0)+W(z-z_0)]^2$$

as $\dfrac{U(x-x_0)}{r}+\dfrac{V(y-y_0)}{r}+\dfrac{W(z-z_0)}{r}=f\cos\theta_1$

and applying (1) and (3).

$$\therefore\ \phi=\frac{1}{2}\frac{a^3}{r^2}f\cos\theta_1-\frac{1}{2}\frac{a^3}{r^3}v^2+\frac{2a^3}{2r^3}v^2\cos^2\theta.$$

If V is the potential due to external forces, then pressure equation is

$$\frac{p}{\rho}+\frac{1}{2}q^2-\frac{\partial\phi}{\partial t}+V=F(t).$$

If p_0 is hydrostatic pressure, *i.e.* when there is no motion *i.e.,* when $q = 0$ and $\dfrac{\partial\phi}{\partial t}=0$. $\dfrac{p_0}{\rho}+V=F(t)$.

Subtracting.

$\dfrac{p-p_0}{\rho}\dfrac{\partial\phi}{\partial t}-\dfrac{1}{2}q^2$ Putting values of q^2 and $\dfrac{\partial\phi}{\partial t}$

$$=\left(\frac{1}{2}\frac{a^3}{r^2}f\cos\theta_1-\frac{1}{2}\frac{a^3}{r^3}v^2+\frac{3a^3}{2r^3}v^2\cos^2\theta\right)$$

$$-\frac{1}{2}\left(\frac{a^6}{4r^6}v^2+\frac{3a^6}{4r^6}v^2\cos^2\theta\right).$$

If the point is on the sphere, putting $r = a$,

$$\frac{p-p_0}{\rho}=\frac{1}{2}af\cos\theta_1-\frac{1}{2}v^2+\frac{3}{2}v^2\cos^2\theta-\frac{1}{8}v^2-\frac{3}{8}v^2\cos^2\theta$$

$$=\frac{1}{2}af\cos\theta_1+\frac{1}{8}v^2(9\cos^2\theta-5).$$

This proves the result.

Example 15: *When a sphere of radius a moves in an infinite liquid, show that the pressure at any point exceeds what would be the pressure of the sphere were at rest by*

$$\frac{a^3}{2r^2}f\,\frac{a^3}{8r^6}(4r^3+a^3)q^2+\frac{3}{8}\frac{a^3}{r^6}(4r^3-a^3)q'^2,$$

where q is the velocity of the sphere and q' and f are the resolved parts of its velocity and acceleration in the direction of r and the density of the liquid is unity.

Solution: If at any time t, (x_0, y_0, z_0) be the centre of the sphere, then we proceed as in the previous example. In this example, we have

$$q^2 = U^2 + V^2 + W^2,$$

$$q' = U\frac{x-x_0}{r} + V\frac{y-y_0}{r} + W\frac{z-z_0}{r}$$

and $f = U\dfrac{x-x_0}{r} + V\dfrac{y-y_0}{r} + W\dfrac{z-z_0}{r}$,

$$\phi = \frac{1}{2}\frac{a^3}{r^3}[U(x-x_0)+V(y-y_0)+W(z-z_0)].$$

Now as in previous example,

$$q^2 = \left(\frac{\partial\phi}{\partial x}\right)^2 + \left(\frac{\partial\phi}{\partial y}\right)^2 + \left(\frac{\partial\phi}{\partial z}\right)^2$$

$$= \frac{a^6}{4r^6}(U^2+V^2+W^2) + \frac{3a^6}{4r^8}[U(x-x_0)$$

$$+V(y-y_0)+W(z-z_0)]^2$$

$$= \frac{a^6}{4r^6}q^2 + = \frac{3a^6}{4r^6}q'^2$$

and $\dfrac{\partial\phi}{\partial t} = \dfrac{1}{2}\dfrac{a^3}{r^3}[U(x-x_0)+V(y-y_0)+W(z-z_0)]$

$$+\frac{1}{2}\frac{a^3}{r^3}[U(-x_0)+V(-y_0)+W(-z_0)]$$

$$+\frac{3a^3}{2r^5}[U(x-x_0)+V(y-y_0)+W(z-z_0)]^2$$

putting value of rr in the last term

$$= \frac{1}{2}\frac{a^3}{r^2}f - \frac{1}{2}\frac{a^3}{r^2}q^2 + \frac{3a^3}{2r^3}q'^2.$$

If p be the pressure when the sphere is moving, then

$$p + \frac{1}{2}q^2 - \frac{\partial\phi}{\partial t} + V = F(t) \quad \text{as } \rho = 1.$$

If ρ_0 be the pressure when the sphere is not moving, then

Putting $q^2 =$ and $\dfrac{\partial\phi}{\partial t} = 0$,

$$\therefore\ p - \rho_0 = \frac{\partial \phi}{\partial t} - \frac{1}{2}q^2$$

$$= \left(\frac{1}{2}\frac{a^3}{r^2}f - \frac{1}{2}\frac{a^3}{r^2}q^2 + \frac{3a^3}{r^3}q'^2\right) + \frac{1}{2}\left(\frac{a^6}{4r^6}q^2 + \frac{3a^6}{4r^6}q'^2\right)$$

$$= \frac{a^3}{2r^2}f - \frac{a^3}{8r^6}(4r^6 + a^3)q^2 + \frac{3a^3}{8a^6}(4r^3 - a^3)q'^2.$$

This proves the result.

Example 16: *An infinite homogeneous liquid is flowing steadily past a rigid boundary consisting partly of the horizontal plane $y = 0$, and partly of a hemispherical boss $x2 + y^2 + z^2 = a^2$, with irrotational motion which tends, at a great distance from the origin, to uniform velocity V parallel to the axis of z. Find the velocity potential and the surfaces of equal pressure.*

Solution: For the liquid streaming past a fixed sphere of radius a with velocity V parallel to z-axis,

$$\phi = Vr\cos\theta + \frac{1}{2}\frac{Va^3}{r^2}\cos\theta.$$

Clearly, if y-axis e taken vertically, the motion is such that velocity perpendicular to $y = 0$ plane is zero.

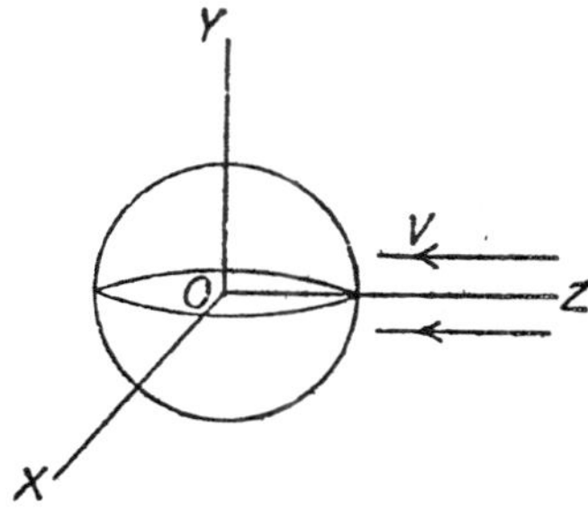

Fig. 4.13

This means that $y = 0$ is a stream surface and the hemisphere above it is also a stream surface.

Therefore for the hemispherical boss $x2 + y^2 + z^2 = a^2$ on the horizontal base $y = 0$, the velocity potential is given by

$$\phi = Vr\cos\theta + \frac{1}{2}\frac{Va^3}{r^2}\cos\theta.$$

Now to determine surface of equal pressure.

Since V is constant and the hemisphere is at rest, the motion is steady. The pressure p is therefore given by

$$\frac{p}{\rho}+\frac{1}{2}q^2=C.$$

This surface of equipressure are given by p = const.

i.e., q^2 = const.,

i.e., $$\left\{\frac{\partial\phi}{\partial r}\right\}^2+\left\{\frac{1}{r}\frac{\partial\phi}{\partial\theta}\right\}^2=\text{const.},$$

$$\left[V\left\{1-\frac{a^3}{r^3}\right\}\cos\theta\right]^2+\left[V\frac{1}{r}\left\{r+\frac{a^3}{2r^3}\right\}\sin\theta\right]=\text{const.},$$

i.e., $$\left\{1-\frac{a^3}{r^3}\cos^2\theta+\left\{1+\frac{a^3}{2r^3}\right\}\sin^2\theta=\text{const.}\right.$$

as V is constant.

Example 17: *Show that when a sphere of radius a moves with uniform velocity U through a perfect, incompressible, infinite fluid, the acceleration of a particle of fluid at (r, 0) is*

$$3U^2\left\{\frac{a^3}{r^4}-\frac{a^6}{r^7}\right\}.$$

Solution: Let the sphere be moving along z-axis with velocity U. Also let $(0, 0, z_0)$ be the co-ordinates of the centre C at any instant of time t. The velocity potential at a point P (x, y, z) is given by

$$\phi=\frac{1}{2}\frac{Ua^3}{r^2}\cos\theta,$$

where $CP = r$ and $\angle PCZ = \theta$.

The point (r, 0) is a point on z-axis.

Now for a point on z-axis, the components of acceleration along the axes of x and y are zero and the along z-axis is $\dfrac{d^2z}{dt^2}$.

The velocity potential in terms of x, y, z can be written as

$$\phi=\frac{1}{2}Ua^3\frac{z-z_0}{[x^2+y^2+(z-z_0)^2]^{3/2}}$$

as $r^2=x^2+y^2+(z-z_0)^2$ and $\cos\theta=\dfrac{z-z_0}{r}$.

Clearly $$\frac{dz}{dt}=w=-\frac{\partial\phi}{\partial z}=-\frac{1}{2}Ua^3\left[\frac{1}{\{x^2+y^2+(z-z_0)^2\}^{3/2}}\right.$$

$$-\frac{3(z-z_0)^2}{\left\{x^2+y^2+(z-z_0)^2\right\}^{5/2}}\Bigg]$$

$$=\frac{1}{2}Ua^3\left[\frac{1}{(z-z_0)^3}-\frac{3}{(z-z_0)^3}\right] \text{ at a point}(0, 0, z)$$

$$=\frac{Ua^3}{(z-z_0)^3},$$

so that $\dfrac{d^2z}{dt^2}=-\dfrac{3Ua^3}{(z-z_0)^4}\left[\dfrac{dz}{dt}-\dfrac{dz_0}{dt}\right]$ at $(0,0,z)$

$$=-\frac{3Ua^3}{(z-z_0)^3}\left[\frac{Ua^3}{(z-z_0)^3}-U\right] \qquad \text{as } \frac{dz_0}{dt}=U$$

$$=3U^2a^3=\left[\frac{1}{(z-z_0)^4}-\frac{a^3}{(z-z_0)^7}\right]$$

$$=3U^2a^3=\left[\frac{1}{r^4}-\frac{a^3}{r^7}\right] \qquad \text{as for point } (r,\ 0),\ \ r = z - z_0.$$

This proves the result.

Alternative Method. Bringing the sphere of rest by superimposing velocities - U to sphere and liquid, the velocity potential is given by

$$\phi=Ur\cos\theta+\frac{Ua^3}{2r^2}\cos\theta=U\left\{r+\frac{a^2}{2r^2}\right\}\cos\theta,$$

so that $r=\dfrac{\partial\phi}{\partial r}=-U\left\{1+\dfrac{a^3}{r^3}\right\}\cos\theta,$ ...(1)

$$r\theta=-\frac{1}{r}\frac{\partial\phi}{\partial r}=U\left\{1+\frac{a^3}{2r^3}\right\}\cos\theta. \qquad ...(2)$$

also $\qquad r=+U\left\{1-\dfrac{a^3}{r^3}\right\}\sin\theta\ \theta-U\left\{\dfrac{3a^3}{r^4}\right\}r\cos\theta$

$$=U\left\{1-\frac{a^3}{r^3}\right\}\sin\theta\ \theta+\frac{3a^3}{r^4}U^2\left\{1-\frac{a^3}{r^3}\right\}\cos^2\theta, \qquad ...(3)$$

putting value of r.

Now for a point (r, 0), the velocity is purely along the direction of r and also the acceleration purely along r, so that

$\theta = 0,$

and the acc. = r only {(r, 0)}

$$= \frac{3a^3}{r^4} U^2 \left\{1 - \frac{a^3}{r^3}\right\} \text{ from (3) putting } \theta = \theta = 0$$

$$= 3U^2 \left\{\frac{a^3}{r^4} - \frac{a^6}{r^7}\right\}.$$

This proves the result.

Example 18: *A rigid sphere of radius a is moving in a straight line with velocity u and acceleration f through an infinite incompressible liquid; prove that the resultant fluid pressure cover the two hemispheres into which the sphere is divided by a diametral plane perpendicular to its direction of motion are*

$$\Pi \pi a^2 \pm \frac{1}{4} Mf - \frac{3}{64} \frac{Mu^2}{a},$$

where Π *is the pressure at a great distance and M is the mass of the fluid displaced by the sphere.*

Solution: Let the sphere move along z-axis with velocity u and C, (0, 0, z_0) be the centre of the moving sphere at any time t, so that

$z_0 = v.$

and $u = f,$

the motion being in a straight line.

Now $\phi = \frac{1}{2} \frac{ua^3}{r^2} \cos\theta,$

where $r^2 = CP^2 = x^2 + y^2 + (z - z_0)^2,$

$$\cos\theta = \cos \angle PCZ = \frac{z - z_0}{r}$$

$$q^2 = \left\{\frac{\partial \phi}{\partial r}\right\}^2 + \left\{\frac{\partial \phi}{r\, \partial \theta}\right\}^2$$

$$= \frac{u^2 a^6}{r^6} (\cos^2\theta + \frac{1}{4} \sin^2\theta).$$

(1) can also be written as

$$\phi = \frac{ua^3 (z - z^0)}{2r^3},$$

so that $\frac{\partial \phi}{\partial t} = \frac{ua^3}{2r^3}(-z_0) + \frac{ua^3}{2r^3}(z - z_0) + \frac{3u^2 a^3}{2r^5}(z - z_0)^2$

$$= \frac{u^2a^3}{2r^3} + \frac{fa^3}{2r^2}\cos\theta + \frac{3u^2a^3}{2r^5}\cos^2\theta$$

$$= \frac{u^2a^3}{2r^3}(3\cos^2\theta - 1) + \frac{fa^3}{2r^2}\cos\theta.$$

The pressure equation is (as $p = \Pi$ when $r = \infty$)

$$\frac{p}{\rho} = \frac{\Pi}{\rho} - \frac{1}{2}q^2 + \frac{\partial\phi}{\partial t}$$

$$= \frac{\Pi}{\rho} - \frac{1}{2}u^2(\cos^2\theta + \frac{1}{4}\sin^2\theta) + \frac{1}{2}af\cos\theta + \frac{1}{2}u^2(3\cos^2\theta - 1)$$

when $r = a$

$$= \frac{\Pi}{\rho} + \frac{u^2}{8}(9\cos^2\theta - 5) + \frac{1}{2}af\cos\theta.$$

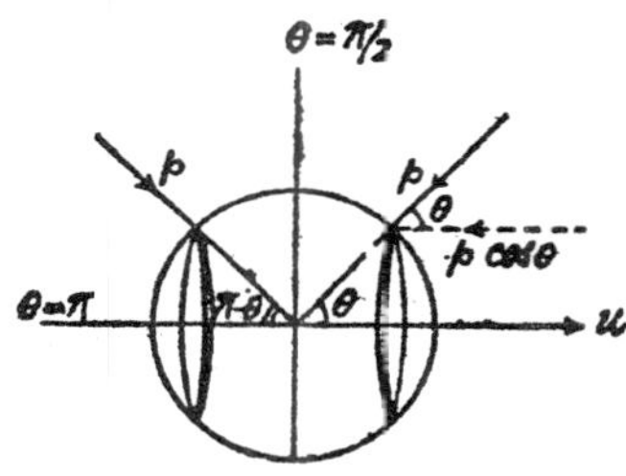

Fig. 4.14

Now on the front hemisphere the pressure

$$= \int_0^{\pi/2} (p\cos\theta)2\pi a\ \sin\theta.a\ d\theta$$

$$= 2\pi a^2\rho\int_0^{\pi/2}\left[\frac{\Pi}{\rho} + \frac{u^2}{8}(9\cos^2\theta - 5) + \frac{1}{2}af\cos\theta\right]$$

$$\cos\theta\sin\theta\ d\theta$$

$$= 2\pi a^2\rho\left[\frac{\Pi}{2\rho} + \frac{u^2}{8}(9.\frac{1}{4} - \frac{5}{2}) + \frac{1}{2}af.\frac{1}{3}\right]$$

$$= \Pi\pi a^2\frac{1}{4}Mf - \frac{3}{64}\frac{Mu^2}{a}, \quad \text{where } M = \frac{3}{4}\pi a^2\rho.$$

Again pressure on the other hemisphere

$$= \int_{\pi/2}^{\pi} p\cos(\pi - \theta)2\pi a\sin\theta.a\ d\theta$$

$$= 2\pi a^2\rho\int_{\pi/2}^{\pi}\left[\frac{\Pi}{\rho} + \frac{u^2}{8}(9\cos^2\theta - 5) + \frac{1}{2}af\cos\theta\right]$$

$$\times \cos\theta \sin\theta \, d\theta$$

$$= 2\pi a^2 \rho \left[\frac{\Pi}{\rho}\frac{1}{2} + \frac{u^2}{8}(9.\frac{1}{2} - \frac{5}{2}) - \frac{1}{2}af.\frac{1}{3} \right]$$

$$= 2\pi a^2 - \frac{1}{3}\pi a^3 \rho f - \frac{1}{16}\pi a^2 \rho u^2$$

$$= 2\pi a^2 - \frac{1}{4}Mf - \frac{3}{64}\frac{Mu^2}{a}.$$

This proves the result.

Example 19: *A sphere of radius a is in motion in fluid, which is at rest at infinity, the pressure being* Π *; determine the pressure at any point of the fluid, and show that the pressure on the front hemisphere cut off by a plane perpendicular to the direction of motion is the resultant of pressure*

$$\pi a^2 (\Pi - \frac{1}{16}\rho V^2) \text{ and } \frac{1}{3}\pi \rho a^3 f,$$

in the directions respectively opposite to those of the velocity V, and the acceleration f of the centre of the sphere.

Solution: Proceeding exactly as in 5.5, we have

$$\frac{p}{\rho} - \frac{\Pi}{\rho} + \frac{1}{4}af \cos\theta_1 + \frac{1}{8}u^2(9\cos^2\theta - 5)$$

$$= \frac{1}{\rho}[\Pi + \frac{1}{8}\rho V^2(9\cos^2\theta - 5)] + \frac{1}{2}af \cos\theta_1$$

$$= \frac{p_1}{\rho} + \frac{p_2}{\rho}, \quad \text{where } p_1 = \Pi + \frac{1}{8}\rho V^2 (9\cos^2\theta - 5)$$

$$\text{and } p_2 = \frac{1}{2}a\rho f \cos\theta_1.$$

The pressure on the hemisphere along the direction opposite to V is

$$= \int_0^{\pi/2} (p_1 \cos\theta) 2\pi a \sin\theta \, a \, d\theta$$

$$= 2\pi a^2 \int_0^{\pi/2} [\Pi + \frac{1}{8}\rho V^2 (9\cos^2\theta - 5)] \sin\theta \cos\theta \, d\theta$$

$$= \pi a^2 \Pi - \frac{1}{16}\pi \rho a^2 V^2$$

$$= \pi a^2 (\Pi - \frac{1}{16}\rho V^2).$$

and the pressure on the hemisphere in the direction opposite to f

$$= \int_0^{\pi/2} p_2 \cos\theta_1 . 2\pi a \sin\theta_1 . a \, d\theta_1$$

$$= 2\pi a^2 \int_0^{\pi/2} \frac{1}{2} a\rho f \cos\theta_1 \cos\theta_1 \sin\theta_1 \, d\theta_1$$

$$= \frac{1}{3}\pi\rho a^3 f.$$

Thus the pressure on the hemisphere is the resultant of

$$\pi a^2 (\Pi - \frac{1}{16}\rho V^2) and \frac{1}{3}\pi\rho a^3 f,$$

which are in directions opposite to V and f.

Example 20: *If AB be a uniform line-source, and A and B equal sinks of such strength that there is no total gain or loss of fluid, show that the stream function at any point is*

$$c\left\{(r_1 - r_2)^2 - c^2\right\}\left(\frac{1}{r_1} - \frac{1}{r_2}\right),$$

where c is the strength of AB, and r_1, r_2 are the distances of the point considered from A and B.

Solution: C is the length of the line source ; so if m is strength per unit length, the strength of the line-source is mc. To neutralize the gain of fluid due to this line source there must be sinks of strengths $-\frac{1}{2}mc$ each at A and B.

If $\angle PAB = \theta_1$ and $\angle PBX = \theta_2$, then

$$\cos\theta_1 = \frac{c^2 + r_1^2 - r_2^2}{2cr_1} \text{ and } \cos\theta_2 = -\frac{r_2^2 + c^2 - r_1^2}{2cr_1}.$$

If ψ is the Stokes' stream function at P,

$$\psi = m(r_1 - r_2) - \frac{1}{2}mc\cos\theta_1 - \frac{1}{2}mc\cos\theta_2$$

$$= -mr_1r_2\left\{\frac{1}{r_1} - \frac{1}{r_2}\right\} - \frac{m}{4}\left[\frac{c^2 + r_1^2 - r_2^2}{r_1} + \frac{r_1^2 - r_2^2 - c^2}{r_2}\right]$$

putting values of $\cos\theta_1$ and $\cos\theta_2$

$$= -mr_1r_2\left\{\frac{1}{r_1} - \frac{1}{r_2}\right\} - \frac{m}{4}\left[c^2\left\{\frac{1}{r_1} - \frac{1}{r_2}\right\} + (r_1 - r_2) - \frac{r_2^2}{r_1} + \frac{r_1^2}{r_2}\right]$$

$$= -m\left\{\frac{1}{r_1}-\frac{1}{r_2}\right\}(r_1r_2+\frac{1}{4}c^2)-\frac{m}{4}\left[(r_1-r_2)-\left\{\frac{r_2^3-r_1^3}{r_1r_2}\right\}\right]$$

$$= -m\left(\frac{1}{r_1}-\frac{1}{r_2}\right)(r_1r_2+\frac{1}{4}c^2)-\frac{m}{4}(r_1-r_2)\left[1+\frac{r_1^2+r_1r_2+r_2^2}{r_1r_2}\right]$$

$$= -m\left\{\frac{1}{r_1}-\frac{1}{r_2}\right\}(r_1r_2+\frac{1}{4}c^2)-\frac{m}{4}\left\{\frac{r_1-r_2}{r_1r_2}\right\}(r_1-r_2)^2$$

$$= -\frac{m}{4}\left\{\frac{1}{r_1}-\frac{1}{r_2}\right\}[4r_1r_2+c^2-(r_1-r_2)^2]$$

$$= C\left\{\frac{1}{r_1}-\frac{1}{r_2}\right\}[(r_1-r_2)^2-c^2], \qquad \text{when } C=\frac{m}{4}.$$

This proves the result.

Example 21: *Discuss the motion for which Stokes' stream function is given by*

$$\psi = \frac{1}{2}V\left\{a^4r^{-2}\cos\theta - r^2\right\}\sin^2\theta,$$

where r is the distance from a fixed point and θ *is the angle this distance makes with the fixed direction.*

Solution: We are given

$$\psi = \frac{1}{2}\frac{Va^4}{r^2}\sin^2\theta\cos\theta - \frac{1}{2}Vr^2\sin^2\theta. \qquad ...(1)$$

Clearly this is the sum of two different terms. Let us consider the liquid streaming past a fixed surface of revolution with velocity V along the positive direction of x-axis (axis of revolution) ; then for this motion, ψ is such that.

$$-\frac{1}{\varpi}\frac{\partial\psi}{\partial\varpi} = +V \text{ or } \psi = -\frac{1}{2}V\varpi^2 = -\frac{1}{2}Vr^2\sin^2\theta$$

as $\varpi = r\sin\theta$.

Thus the second term of (1) is due to liquid streaming past a surface of revolution along +ve direction, which would have given rise to

$$\psi = \frac{1}{2}V\frac{a^4}{r^2}\sin^2\theta\cos\theta \qquad ...(2)$$

had the surface of revolution been moving in an infinite liquid at rest at infinity.

But if a solid of revolution moves in an infinite liquid at rest at infinity, with velocity V parallel to x-axis in negative direction, we have on the surface

$$\psi = \frac{1}{4}Vr^2 \sin^2\theta + \text{const.} \qquad ...(3)$$

Write (2) as $\psi = \frac{1}{2}V\left(\frac{a^4}{r^4}\cos\theta\right)r^2\sin^2\theta.$

Thus at boundary surface [comparing from (3)],

$$\frac{a^4}{r^4}\cos\theta = \text{const.}=1,$$

Then the solid of revolution is $r^4 = a^4 \cos\theta$,

Thus the motion is due to liquid streaming past the solid of revolution $r^4 = a^4 \cos\theta$, with velocity V along the positive direction of x-axis.

Example 22: *A solid of revolution is moving along its axis in an infinite liquid; show that the kinetic energy of the liquid is*

$$-\frac{1}{2}\pi\rho\int\frac{\psi}{\varpi}\frac{\partial\psi}{\partial n}ds,$$

where ψ is the Stokes' stream function of the motion, ϖ the distance of a point from the axis and the integral is taken once round a meridian curve of the solid. Hence obtain the kinetic energy of infinite liquid due to the motion of a sphere through it with velocity V.

Solution: We know that when the motion is irrotational, the K.E. is given by

$$T = -\frac{1}{2}\rho\int\phi\frac{\partial\phi}{\partial n}dS \qquad ...(1)$$

For a solid of revolution,

$$dS = 2\pi\varpi\ ds,$$

where ds is an elementary are of the meridian curve of the boundary.

$$-\frac{\partial\phi}{\partial n} = \text{ outward normal velocity} = \frac{1}{\varpi}\frac{\partial\psi}{\partial s}.$$

$\therefore$ (1) gives $T = \pi\rho\int\phi\ d\psi$

$= [\pi\rho\phi\psi] - \pi\rho\int\psi\ d\phi$ integrating by parts

$$= -\pi\rho\int\psi\ d\phi. \qquad ...(2)$$

The first term is zero because ψ = constant along the curve in the meridian plane along which the integration is taken.

Again $-\frac{\partial\phi}{\partial s} = -\frac{1}{\varpi}\frac{\partial\psi}{\partial n}.$

$\therefore$ (2) gives

$$T=-\pi\rho-\int\frac{\psi}{\varpi}\frac{\partial\psi}{\partial n}ds$$

$$=-\frac{1}{2}\pi\rho-\int\frac{\psi}{\varpi}\frac{\partial\psi}{\partial n}ds,$$

the integration being now taken round the whole boundary which is double of the meridian curve.

To calculate K.E of the moving sphere. For the moving sphere with velocity V along the axis of rotation, we have

$$\phi=\frac{1}{2}\frac{Va^3\cos\theta}{r^2}\text{ and }\psi=-\frac{1}{2}\frac{Va^3\cos\theta}{r}.$$

$$\therefore\ T=-\pi\rho\int\phi\,d\psi$$

$$=\frac{1}{2}\pi\rho V^2a^3\int_0^\pi\cos^2\theta\sin\theta\,d\theta$$

putting values of ϕ and ψ when $r=a$

$$=\frac{1}{2}\pi\rho V^2a^3$$

$$=(\frac{4}{3}\pi\rho a^3).\frac{1}{4}V^2=\frac{1}{4}M.V^2,$$

where M is the mass of the liquid displaced by the sphere.

Example 23: *A doublet of strength M is placed at the point (0, a, 0) with its axis parallel to the axis of z; prove that at points close to the origin the velocity potential of the doublet is approximately* $\frac{Mz}{a^3}+\frac{3Myz}{a^4}$ *neglecting terms of the order* $\frac{r^3}{a^5}$ *and higher powers.*

Deduce that if a small sphere of radius c is placed with its centre at the origin, the velocity potential is then increased by the terms

$$\frac{1}{2}\frac{Mc^3z}{a^3}+2\frac{Mc^5yz}{a^4r^5}.$$

Solution: Let doublet be at B *(0, a, 0)* with its axis parallel to the axis of z. Take a point P (x, y, z) in the liquid. If PB makes angle θ with z-axis, then

$$\cos\theta\frac{z}{\sqrt{\{x^2+(y-z)^2+z^2\}}}.$$

Also the velocity potential of the doublet is given by

$$\phi = \frac{M\cos\theta}{BP^2}$$

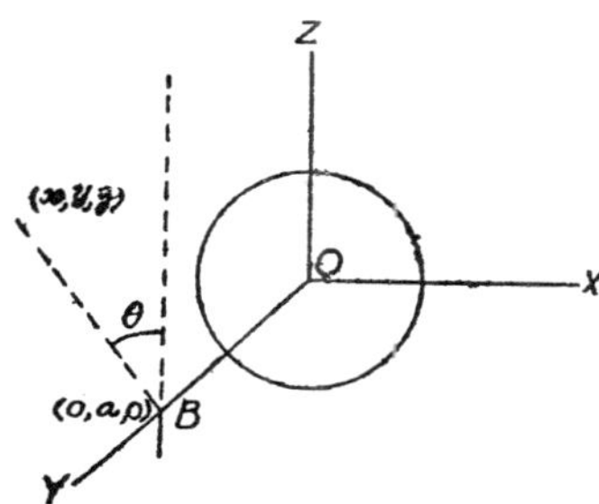

Fig. 4.15

$$= \frac{Mz}{[x^2+(y-a)^2+z^2]^{3/2}}$$

$$= \frac{Mz}{(r^2-2ay+a^2)^{3/2}}, \text{ where } r^2 = x^2+y^2+z^2$$

$$= \frac{Mz}{a^2}\left\{1-\frac{2y}{a}+\frac{r^2}{a^2}\right\}^{-3/2}$$

$= \dfrac{Mz}{a^3}\left\{1+\dfrac{3y}{a}\right\}$ neglecting other terms, r being small for point close to the origin

$$= \frac{Mz}{a^3}+\frac{3Myz}{a^4}.$$

This proves the first part of the result.

Now when the sphere of radius c is inserted, let ϕ_1 be the velocity potential at any point. Then ϕ_1 is such that $\left(\dfrac{\partial\phi_1}{\partial r}\right) = 0,$ where $r = c$.

The velocity at infinity however remains unchanged. Let us assume that the increase in velocity potential be

$$= \frac{Az}{r^3}+\frac{Byz}{r^5},$$

so that $\quad \phi_1 = \dfrac{Mz}{a^3}+\dfrac{3Myz}{a^4}+\dfrac{Az}{r^3}+\dfrac{Byz}{r^5}$

$$= \frac{Mr\cos\theta}{a^3}+\frac{3Mr^2\sin\theta\cos\theta\sin\omega}{a^4}+\frac{A\cos\theta}{r^2}$$

$$+\frac{3B\sin\theta\sin\omega\cos\theta}{r^4}.$$

But $\left(\dfrac{\partial\phi_1}{\partial r}\right)=0,$ where $r=c$,

i.e., $$\left(\frac{M}{a^3}+\frac{2A}{c^3}\right)\cos\theta+\left(\frac{6Mc}{a^4}+\frac{3B}{c^4}\right)\sin\theta\cos\theta\sin\omega=0,$$

i.e., $$\frac{M}{a^3}+\frac{2A}{c^3}=0 \text{ and } \frac{6Mc}{a^4}+\frac{3B}{c^4}=0,$$

i.e., $$A=\frac{Mc^3}{2a^3},\ B=\frac{2Mc^5}{a^4}.$$

$\therefore$ increase in velocity potential

$$=\frac{Az}{r^3}+\frac{Byz}{r^5}=\frac{Mc^3z}{2a^3r^3}+\frac{2Mc^5}{a^4}\frac{yz}{r^5}.$$

This proves the result.

Example 24: *A solid is bounded by the exterior portions of two equal spheres (of radius a) which cut one another orthogonally ; and is surrounded by an infinite mass of liquid. If the solid is set in motion with velocity u in the direction of the line of centres, show that the velocity potential of the resulting motion is*

$$\frac{1}{2}a^3u\left\{\frac{\cos\theta}{r^2}+\frac{\cos\theta'}{r'^2}-\frac{\cos\Theta}{2\sqrt{2}R^2}\right\},$$

where r, r', R are the radii vectors of a point, measured respectively from the centres of the two spheres and from the point midway between them, and θ, θ', Θ *are the angles which these radii vectors make with the direction of motion of the solid.*

Solution: Let A and B be the centres of two orthogonal spheres of radius a and O be the mid-point of AB.

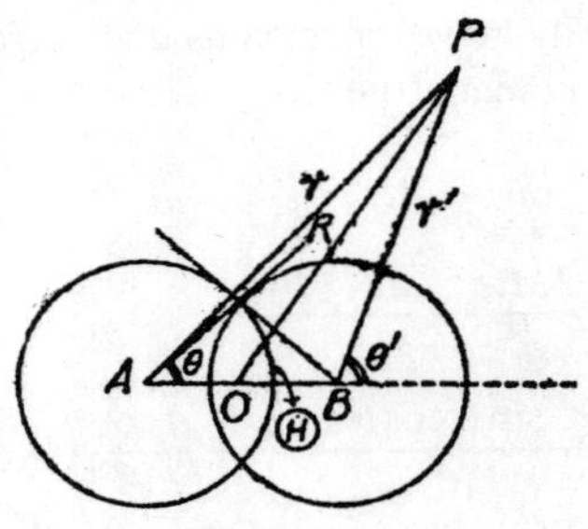

Fig. 4.16

Than since $AB = a\sqrt{2}$,

$$AO = BO = \frac{a\sqrt{2}}{2} + \frac{a}{\sqrt{2}}.$$

$$\therefore AB.OA = BA.BO$$

$$= a\sqrt{2}.\frac{a}{\sqrt{2}} = a^2$$

This shows that O is the inverse point of B with respect to the first circle; also O is the inverse point of A with respect to the second circle.

We know that the effect of the motion of a sphere of radius a with velocity u is the same as that of a doublet of strength $\frac{1}{2}ua^3$ placed at the centre of the sphere

Thus we can replace spheres with doublets of strengths $\frac{1}{2}ua^3$ each at A and B. The image of doublet at B is a doublet at its inverse point O, the strength of this image doublet $= (\frac{1}{2}ua^3).\frac{a^3}{AB^3} = \frac{1}{2}ua^3\frac{a^3}{(a\sqrt{2})^3} = \frac{1}{4\sqrt{2}}ua^3$ with sense opposite to that at B.

The same doublet is image of A with respect to other sphere. Thus the image system for the given boundaries is

(i) a doublet $\frac{1}{2}ua^3$ at A,

(ii) a doublet $\frac{1}{2}ua^3$ at B,

(iii) a doublet $-\frac{1}{4\sqrt{2}}ua^3$ at O.

Therefore the velocity potential ϕ is given by

$$\phi = \frac{1}{2}ua^3\frac{\cos\theta}{r^2} + \frac{1}{2}ua^3\frac{\cos\theta'}{r'^2} - \frac{1}{4\sqrt{2}}ua^3.\frac{\cos\Theta}{R^2}$$

$$= \frac{1}{2}a^3u\left\{\frac{\cos\theta}{r^2} + \frac{\cos\theta'}{r'^2} - \frac{\cos\Theta}{2\sqrt{2}R^2}\right\}.$$

This proves the result.

Example 25: *A thin stratum of incompressible fluid is contained between two concentric spheres; show that the velocity at any point is equivalent to the components.*

$$-\frac{1}{\sin\theta}\frac{\partial\psi}{\partial\omega}, \frac{\partial\psi}{\partial\theta}$$

along the meridian and parallel respectively. Also if the fluid be homogeneous and the motion irrotational, prove that

$$\frac{\partial\phi}{\partial\theta}=\frac{1}{\sin\theta}\frac{\partial\psi}{\partial\omega}, -\frac{1}{\sin\theta}\frac{\partial\phi}{\partial\omega}=\frac{\partial\psi}{\partial\theta}$$

and deduce that $\phi+i\psi=F(e^{i\omega}\tan\frac{1}{2}\theta)$.

Solution: Here ρ is constant and there being no motion along the radius vector the equation of continuity becomes

$$\frac{\partial}{\partial\theta}(v\sin\theta)+\frac{\partial w}{\partial\omega}=0.$$

But this is the condition that

$v\sin\theta\, d\omega - w\, d\theta = 0$

is an exact differential say $-d\psi$, so that

$$v\sin\theta\, d\omega - w\, d\theta = -\left\{\frac{\partial\psi}{\partial\omega}d\omega+\frac{\partial\psi}{\partial\theta}d\theta\right\},$$

i.e., $$v=-\frac{1}{\sin\theta}\frac{\partial\psi}{\partial\omega} \text{ and } w=\frac{\partial\psi}{\partial\theta}. \qquad ...(1)$$

This proves the first part of the result provided ψ is the stream function. To show that ψ is the stream function, the differential equation of stream lines is

$$\frac{a\,d\theta}{v}=\frac{a\sin\theta\, d\omega}{w}, i.e., \frac{d\theta}{v}=\frac{\sin\theta\, d\omega}{\omega},$$

i.e., $$v\sin\theta\, d\omega-\omega\, d\theta=0,$$

i.e., $$d\psi=0$$

or ψ = const. gives stream lines.

Thus ψ is the stream function for the motion.

2nd part. The motion being irrotational, velocity potential ϕ exists and we have

$$v=-\frac{\partial\phi}{\partial\theta} \text{ and } w=-\frac{\partial\phi}{\sin\theta\,\partial\omega}. \qquad ...(2)$$

From (1) and (2),

$$\left.\begin{aligned} -\frac{\partial\phi}{\partial\theta}&=-\frac{1}{\sin\theta}\frac{\partial\phi}{\partial\omega}, i.e., \frac{\partial\phi}{\partial\theta}=\frac{1}{\sin\theta}\frac{\partial\psi}{\partial\omega} \\ \text{and} \qquad & -\frac{\partial\phi}{\sin\theta\,\partial\omega}=\frac{\partial\phi}{\partial\theta}. \end{aligned}\right\} \qquad ...(3)$$

This proves the second part of the result.

Again from (3), we have

$$\frac{\partial}{\partial\theta}(\phi+i\psi)=\frac{1}{\sin\theta}\frac{\partial}{\partial\omega}(\psi-i\phi)$$

$$=-\frac{i}{\sin\theta}\frac{\partial}{\partial\omega}(\phi+i\psi),$$

So that if $\psi+i\psi=\zeta$, we have

$$\frac{\partial\zeta}{\partial\theta}=-\frac{i}{\sin\theta}\frac{\partial\zeta}{\partial\omega}$$

or $\dfrac{\partial\zeta}{\partial\theta}+\dfrac{i}{\sin\theta}\dfrac{\partial\zeta}{\partial\omega}=0.$

The Lagrange's auxiliary equations to solve this equation are

$$\frac{d\theta}{\sin\theta}=\frac{d\omega}{i}=\frac{\partial\zeta}{0},$$

$\operatorname{cosec}\theta\, d\theta+i\, d\omega=0$ gives $\tan\dfrac{\theta}{2}e^{i\omega}=$ const.

Also $\quad d\zeta=0$ gives $\zeta=$ const., *i.e.*, $\phi+i\psi=$ const.

$$\therefore\ \phi+i\psi=F\left\{e^{i\omega}\tan\frac{\theta}{2}\right\}.$$

This proves the result.

Example 26: *A rigid ellipsoidal envelope, without mass, encloses a perfect incompressible fluid of mass M. The equation of the ellipsoid is*

$$\frac{x^2}{a^2}+\frac{y^2}{b^2}+\frac{z^2}{c^2}-1=0.$$

An impulsive couple in the plane of xy causes the envelope to rotate initially with angular velocity ω. Find the initial velocity potential of the fluid, and prove that the moment of the couple is

$$\frac{1}{5}M\omega\frac{(a^2-b^2)^2}{a^2+b^2}.$$

Solution: Here $\omega_x=0$, $\omega_y=0$, $\omega_z=\omega$. Therefore components of linear velocities of a point (x, y, z) of the ellipsoidal envelope are

$$-y\omega, x\omega, 0.$$

The direction cosines of the normal are proportional to

$$\frac{x}{a^2},\frac{y}{b^2},\frac{z}{c^2}.$$

And if ϕ be the velocity potential, the boundary condition is

$$-\frac{x}{a^2}\frac{\partial\phi}{\partial x}-\frac{y}{b^2}\frac{\partial\phi}{\partial y}-\frac{z}{c^2}\frac{\partial\phi}{\partial z}=\frac{x}{a^2}(-y\omega)+\frac{y}{b^2}(x\omega)+0, \quad ...(1)$$

where $\dfrac{x^2}{a^2}+\dfrac{y^2}{b^2}+\dfrac{z^2}{c^2}=1$. ...(2)

To satisfy this assume a solution of Laplace's equation,

$\phi = Cxy.$

Then (1) gives

$$-\frac{x}{a^2}Cy-\frac{y}{b^2}Cx=xy\omega\left\{\frac{1}{b^2}-\frac{1}{a^2}\right\}.$$

$$\therefore\ C=-\omega\frac{a^2-b^2}{a^2+b^2}.$$

$$\therefore\ \phi=-\omega\frac{a^2-b^2}{a^2+b^2}xy.$$

This is the initial velocity potential.

Now to determine moment of the couple which is equal to the moment of momentum of the fluid about the x-axis (this being zero about the axes of y and z), we have

moment of the couple

$$=\iiint\left[x\left\{-\frac{\partial\phi}{\partial y}\right\}-y\left\{-\frac{\partial\phi}{\partial x}\right\}\right]\rho\,dx\,dy\,dz$$

$$=\rho\omega\frac{a^2-b^2}{a^2+b^2}\iiint(x^2-y^2)dx\,dy\,dz \text{ from (3),}$$

where integration is taken for $\dfrac{x^2}{a^2}+\dfrac{y^2}{b^2}+\dfrac{z^2}{c^2}\le 1$

$$=\rho\omega\frac{a^2-b^2}{a^2+b^2}\iiint[(x^2-z^2)-(y^2-z^2)]dx\,dy\,dz$$

$=\omega\dfrac{a^2-b^2}{a^2+b^2}$ [M.I. of ellipsoid about y-axis-M.I. of ellipsoid about x-axis]

$$=\omega\frac{a^2-b^2}{a^2+b^2}\left[M\frac{a^2-c^2}{5}-M\frac{b^2+c^2}{5}\right]$$

$=\dfrac{1}{5}\omega M\dfrac{(a^2-b^2)^2}{a^2+b^2}$. This proves the result.

Example 27: *An ellipsoidal cavity (semi-axes, a, b, c) in a solid initially at rest is filled with an incompressible frictionless fluid initially at rest. Prove that if the solid be moved with velocities u, v, w parallel to the axes of the cavity, and be rotated with angular velocities p, q. r round the semi-axes, the angular momentum of the fluid round the semi-axis a at any instant is*

$$\frac{4}{15}\pi\rho abc\frac{(b^2-c^2)^2}{b^2+c^2}p.$$

Solution: Here $\omega_x = p$, $\omega_y = q$, $\omega_z = r$,, The velocity potential when the ellipsoid is rotating with these velocities is given by

$$-\frac{b^2-c^2}{b^2+c^2}pyz-\frac{c^2-a^2}{c^2+a^2}qzx-\frac{a^2-b^2}{a^2+b^2}rxy$$

The ellipsoid can be reduced to simple rotation about axes by superimposing velocities - u , = v, -w along the axes and because of this the terms

$$-ux - vy - wz$$

should be added to the velocity potential.

Thus if ϕ be the velocity potential for the motion,

$$\phi = -ux - vy - wz - \frac{b^2-c^2}{b^2+c^2}pyz - \frac{c^2-a^2}{c^2+a^2}qzx\frac{a^2-b^2}{a^2+b^2}rxy.$$

Now let U, V, W be the velocity components at P (x, y, z) ; then

$$U = -\frac{\partial\phi}{\partial x}, V = -\frac{\partial\phi}{\partial y}, W = -\frac{\partial\phi}{\partial z}.$$

The angular momentum of the whole liquid about the axis of

$$x = \iiint(yW - zV)\rho\, dx\, dy\, dz, \text{ where} \frac{x^2}{a^2}+\frac{y^2}{b^2}+\frac{z^2}{c^2}\le 1$$

$$= -\rho\iiint\left(y\frac{\partial\phi}{\partial z} - z\frac{\partial\phi}{\partial y}\right)dx\, dy\, dz$$

$$= \rho\iiint\left[\left(w+\frac{b^2-c^2}{b^2+c^2}py+\frac{c^2-a^2}{c^2+a^2}qx\right)\right.$$

$$\left.-z\left(v+\frac{b^2-c^2}{b^2+c^2}pz+\frac{a^2-b^2}{a^2+b^2}rx\right)\right]dx\, dy\, dz$$

$$= \rho\left(\frac{b^2-c^2}{b^2+c^2}\right)p\iiint(y^2-z^2)dx\, dy\, dz, \text{ other integrals vanish}$$

$$= \rho \frac{b^2 - c^2}{b^2 + c^2} p \int\int\int [(x^2 + y^2) - (x^2 + z^2)] dx\, dy\, dz$$

$$= \frac{b^2 - c^2}{b^2 + c^2} [\text{M.I. about z-axis-M.I. about y-axis}]$$

$$= \frac{b^2 - c^2}{b^2 + c^2} p \left[M \frac{a^2 + b^2}{5} - M \frac{a^2 + c^2}{5} \right], \text{where } M = \frac{4}{3} \pi \rho abc$$

$$= \frac{4}{3} \pi \rho abc \frac{b^2 - c^2}{b^2 + c^2} \cdot \frac{b^2 + c^2}{5}$$

$$= \frac{4}{15} \pi \rho abc \frac{(b^2 - c^2)^2}{b^2 + c^2} p.$$

This proves the result.

5

Boundary Layer Theory

INTRODUCTION

When a real fluid flows past a solid boundary, a layer of fluid which comes in contact with the boundary surface adheres to it on account of viscosity. Since this layer of fluid cannot slip away from the boundary surface it attains the same velocity as that of the boundary. In other words, at the boundary surface there is no relative motion between the fluid and the boundary, This condition is known as no slip condition. If the boundary is moving, the fluid adhering to it will have the same velocity as that of the boundary. However, if the boundary is stationary, the fluid velocity at the boundary surface will be zero. Thus at the boundary surface the layer of fluid undergoes retardation. This retarded layer of fluid further causes retardation for the adjacent layers of the fluid, thereby developing a small region in the immediate vicinity of the boundary surface in which the velocity of flowing fluid increases gradually from zero at the boundary surface to the velocity of the main stream. This region is known as *boundary layer*. In the boundary layer region since there is a larger variation of velocity in a relatively small distance, there exists a fairly large velocity gradient ($\partial v/\partial y$) normal to the boundary surface. As such in this region of boundary layer even if the fluid has small viscosity, the corresponding shear stress $\tau = \mu\ (\partial v/\partial y)$, is of appreciable magnitude. Farther away from the boundary this retardation due to the presence of viscosity is negligible and the velocity there will be equal to that of the main stream. The flow may thus be considered to have two regions, one close to the boundary in the boundary layer zone in which due to larger velocity gradient appreciable viscous forces are produced and hence in this region the effect of viscosity is mostly confined and second outside the boundary layer zone in which the viscous forces are negligible and hence the flow may be treated as non-viscous or inviscid. The concept of boundary layer was first introduced by L. Prandtl in 1904 and since then it has been applied to several fluid flow problems. In the following paragraphs the characteristics of flow in the boundary layer are discussed.

THICKNESS OF BOUNDARY LAYER

The velocity within the boundary layer increases from zero at the boundary surface to the velocity of the main stream asymptotically. Therefore the thickness of the boundary layer represented by d (Greek 'delta') is arbitrarily defined as that distance from the boundary surface in which the velocity reaches 99% of the velocity of the main stream, In other words, the boundary layer thickness δ may be considered equal to the distance y from the boundary surface at which v = 0.99 V. This definition however gives an approximate value of the boundary layer thickness and hence δ is generally termed an *nominal thickness* of the boundary layer. For greater accuracy the boundary layer thickness is defined in terms of certain mathematical expressions which are the measures of the effect of boundary layer on the flow. Three such definitions of the boundary layer thickness which are commonly adopted are the *displacement thickness* δ *the momentum thickness θ and the energy thickness ∂E.

The *displacement thickness* δ^* is defined as the distance by which the boundary surface would have to be displaced outwards so that the total actual discharge would be same as that of an ideal (or frictionless) fluid past the displaced boundary. Thus, it may be expressed by the equation

$$V\,\delta^* = \int_0^\infty (V - v)dy$$

$$\text{or } \delta^* = \int_0^\infty \left(1 - \frac{v}{V}\right)dy \qquad ...(1)$$

Another definition for the displacement thickness may be given which is based on the fact that because of the retardation of the flow within the boundary layer, in order to maintain the continuity of flow the streamlines outside the boundary layer are farther from the boundary than they would be if the fluid was ideal or inviscid upto the boundary. The displacement thickness δ^* may then be defined as the distance by which the external streamlines are shifted or displaced outwards owing to the formation of the boundary layer.

The retardation of flow due to viscosity in the boundary layer would cause the reduction in the momentum flux. Accordingly the *momentum thickness* θ is defined as the distance from the actual boundary surface such that the momentum flux corresponding to the main stream velocity V through this distance θ is equal to the deficiency or loss in momentum due to the boundary layer formation. Thus

$$\rho\; V^2\theta = \rho \int_0^\infty (V - v)v\,dy$$

$$\theta = \int_0^\infty \frac{v}{V}\left(1-\frac{v}{V}\right)dy \qquad ...(2)$$

Further the retardation of flow due to viscosity in the boundary layer would cause the reduction in the flux of energy. At such the energy thickness δ_E is defined as the distance from the actual boundary surface such that the energy flux corresponding to the main stream velocity V through this distance δ_E is equal to the deficiency or loss of energy due to the boundary layer formation. Thus

$$\frac{1}{2}\rho V^3\delta_E = \frac{1}{2}\rho\int_0^\infty (V^2 - v^2)v\,dy$$

or

$$\delta_E = \int_0^\infty \left(1-\frac{v^2}{V^2}\right)dy \qquad ...(3)$$

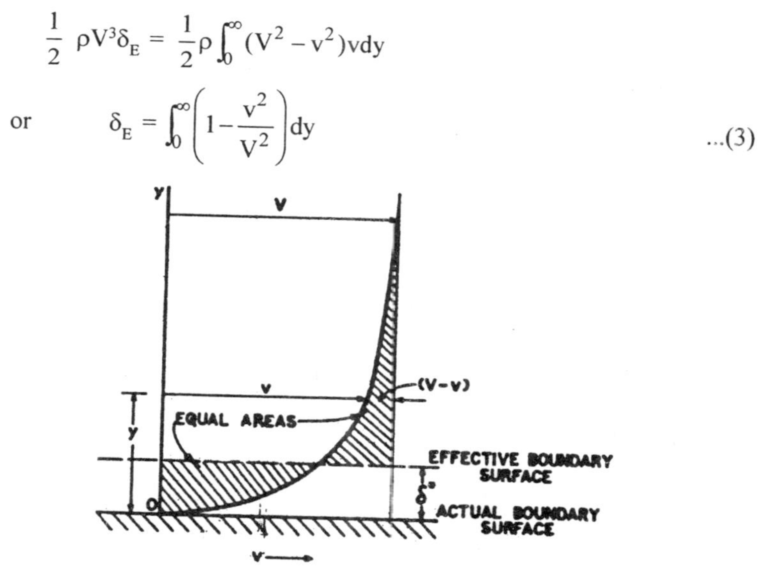

Fig. 5.1 : Displacement thickness of boundary layer.

BOUNDARY LAYER ALONG A LONG THIN PLATE AND ITS CHARACTERISTICS

Consider a long thin plate held stationary in the direction parallel to the flow in a uniform stream of velocity V as shown in Fig. 5.2. The plate is said to be held at zero incidence to the velocity of flow and the velocity of flow is known as 'free stream velocity' or 'ambient velocity' or 'potential velocity.' At the leading edge of the plate the thickness of the boundary layer is zero, but on downstream, for the fluid in contact with the boundary the velocity of flow is reduced to zero and at some distance δ from the boundary the velocity is nearly V. Hence a velocity gradient is set up which develops shear resistance to the flow and retards the motion of the fluid. Near the leading edge of the plate the fluid is retarded in a thin layer. In other words, the boundary layer near the leading edge is relatively thin. As this retarded layer of fluid moves downstream, due to continued action of shear resistance more and more fluid is retarded. Thus the thickness of the boundary layer δ, goes on increasing in the downstream direction as shown in Fig. 2. The various factors which influence the thickness of the boundary layer forming along a flat smooth plate are noted below.

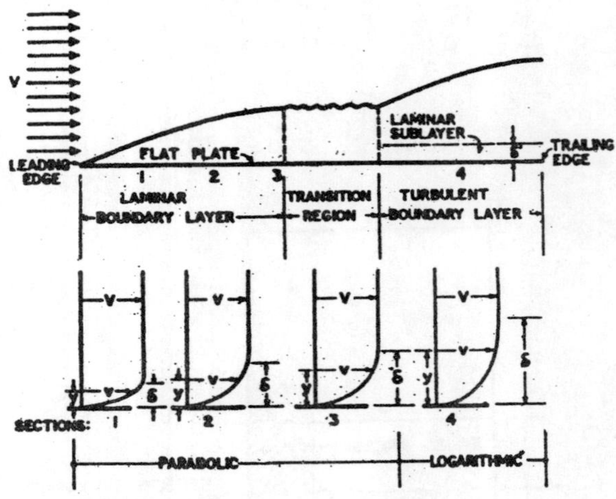

Fig. 5.2 : Boundary layer and velocity distribution at successive points along a flat plate.

1. The boundary layer thickness increases as the distance from the leading increases.
2. The boundary layer thickness decreases with the increase in the velocity of flow of the approaching stream of fluid.
3. Greater is the kinematic viscosity of the fluid greater is the boundary layer thickness.
4. The boundary layer thickness is considerably affected by the pressure gradient $(\partial p/\partial x)$ in the direction of flow. In the case of a flat plate placed in a stream of uniform velocity V, the pressure may also be assumed to be uniform i.e., $(\partial p/\partial x) = 0$. However, if the pressure gradient is negative as in the case of a converging flow, the resulting pressure force acts in the direction of flow and it accelerates the retarded fluid in the boundary layer. As such the boundary layer

growth is retarded in the presence of negative pressure gradient. On the other hand if the pressure gradient is positive as in the case of divergent flow the fluid in the boundary layer is further decelerated and hence assists in thickening of the boundary layer. In the later case back flow and separation may be caused as discussed.

As the boundary layer develops, upto a certain portion of the plate from the leading edge, the flow in the boundary laver exhibits all the characteristics of laminar flow. This is so irrespective of whether the flow of the incoming stream is laminar or turbulent. This is known as *laminar boundary* layer. If the plate is sufficiently long, then beyond some distance from the leading edge the laminar boundary layer becomes unstable and the flow in the boundary layer exhibits the characteristics between those of laminar and turbulent flow. This region of the boundary layer is usually small and is known as transition region. After the transition region the flow in the boundary layer becomes turbulent. In this portion of the boundary layer there is a rapid increase in its thickness and it is known as *turbulent boundary layer* . If the plate is very smooth, even in the region of turbulent boundary layer. If the plate is very smooth, even in the region of turbulent boundary layer, there is a very thin layer just adjacent to the boundary, in which the flow is laminar. This thin is commonly known as *laminar sublayer*, and its thickness is represented by δ.

The velocity distribution in a laminar boundary layer is parabolic $(V - v) \sim (\delta - y)^2$; and for turbulent boundary layer the velocity distribution has been found to follow approximately either the one-seventh power law; $v \sim y^{1/7}$ or it is logarithmic, $v \sim \log y$. For laminar sublayer the velocity distribution is parabolic, but since its thickness δ is usually very small a linear distribution can be assumed.

The change of boundary layer for earned $\delta \times 40.5 \ \frac{g}{n} = \frac{5}{\sqrt{R_e}}$ laminar to turbulent mainly depends on the velocity of flow V, of the approaching stream of fluid, the length x measured along the plate from the leading edge, the mass density ρ of fluid and its dynamic viscosity μ. As such the Reynolds number $Re_x = (\rho Vx/\mu)$ (the suffix x indicating that it is calculated with the distance x as the characteristic length) becomes a significant parameter in indicating the change of boundary layer from laminar to turbulent. The value of Re_x at which the boundary layer may change from laminar to turbulent varies from 3×10^5 to 6×10^5. In exceptional cases, where the approaching flow is free from any disturbance the values of critical Reynolds number can reach 10^6 or even higher. However, change of boundary layer from laminar to turbulent is also affected by several other factors such as roughness of the plate, curvature, pressure gradient and intensity and scale of turbulence.

BOUNDARY LAYER EQUATIONS

The equations of continuity and motion for the steady flow of an incompressible, inviscid fluid in two dimensions without body forces are

$$\frac{\partial u}{\partial x}+\frac{\partial V}{\partial Y}=0 \qquad ...(i)$$

$$u\frac{\partial u}{\partial x}+v\frac{\partial u}{\partial Y}=-\frac{1}{\rho}\frac{\partial p}{\partial x} \qquad ...(ii)$$

$$\text{and } u\frac{\partial u}{\partial x}+v\frac{\partial v}{\partial Y}=-\frac{1}{\rho}\frac{\partial p}{\partial y} \qquad ...(iii)$$

where x and y are rectangular cartesian coordinates and u and v are the corresponding velocity components.

Now if a viscous fluid is considered then the equation of continuity will be unchanged, but in the equations of motion additional terms will be introduced due to viscous stresses as indicated below.

Consider the boundary layer on a flat plate with x measured along the plate and y normal to it as shown if Fig. 5.3. The only viscous stress τ that need be considered is that acting in the direction parallel to the plate. Thus if a small element of fluid of area $\delta x\ \delta y$ in the xy plane and unit length normal to the xy plane is considered, then on its lower surface there is a force, due to shear stress, equal to $-\tau\delta x$ in the x direction. Likewise on the upper surface there is a force in the x direction due to shear stress as

$$(\tau+\partial\tau)\ \delta x=\left[\tau+\frac{\partial\tau}{\partial y}\delta y\right]\delta x$$

neglecting the terms of higher power of δy.

The net force in the x direction acting on the element is therefore $(\partial t/\partial y)\ \delta x,/\delta y$, and hence as δx and δy tend to be zero, the force per unit mass in the x direction due to vicous stresses becomes

$$\frac{1}{\rho}\frac{\partial\tau}{\partial y}$$

Since the left-hand side of equation (ii) noted above represents the acceleration in x direction of a particle of fluid and the right-hand side represents the components of the force per unit mass in that direction due to static pressure, it follows that the above term due to viscous stresses should be added on the right-hand side of the equation to complete the equation of motion in the x direction in the boundary. This equation is then modified as

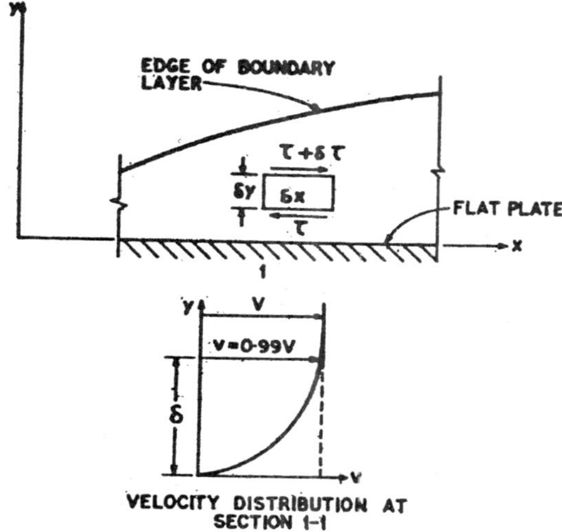

Fig. 5.3. Boundary layer along a flat plate.

$$u\frac{\partial u}{\partial x} + v\frac{\partial u}{\partial y} = -\frac{1}{\rho}\frac{\partial p}{\partial y} + \frac{1}{\rho}\frac{\partial \tau}{\partial y}$$

Further for viscous flow in the laminar boundary layer from Newton's law of viscosity

$$\tau = \mu\frac{\partial \mu}{\partial y}$$

and hence the equation of motion becomes

$$u\frac{\partial u}{\partial x} + v\frac{\partial u}{\partial y} = -\frac{1}{\rho}\frac{\partial p}{\partial x} + v\frac{\partial^2 u}{\partial y^2}$$

where v is kinematic viscosity (μ/ρ).

The second equation of motion i.e., equation (iii) is unchanged by the argument that the shear stress is acting in the x direction only. However, by determining the order of magnitude of each of the terms of equation (iii) and neglecting the terms of smaller order of magnitude, equation (iii) reduces to

$$-\frac{1}{\rho}\frac{\partial p}{\partial y} = 0.$$

Thus the equations that govern the flow in the steady, two dimensional laminar boundary layer on a flat plate are

$$\frac{\partial u}{\partial x}+\frac{\partial v}{\partial y} = 0 \qquad \text{...(4)}$$

$$u\frac{\partial u}{\partial x}+v\frac{\partial u}{\partial y} = -\frac{1}{\rho}\frac{\partial p}{\partial x}+\nu\frac{\partial^2 u}{\partial y^2} \qquad \text{...(5)}$$

and $$-\frac{1}{\rho}\frac{\partial p}{\partial y} = 0 \qquad \text{...(6)}$$

Equations 4, 5 and 6 are generally known as *Prandtl's boundary layer equations* for two-dimensional steady flow of incompressible fluids. In the above analysis, equations 5 and 6 have been derived by adopting a relatively simpler approach. But the same can also be derived directly from the Navier-Stokes equations which are in fact the basic equations of motion for the flow of viscous fluids. The analysis involving the use of Navier-Stokes equations is more accurate and complete, since in this all the viscous stresses are included. However for deriving equations 5 and 6 from the Navier-Stokes equations the order of magnitude of each of the terms of these equations is determined and the terms of smaller order of magnitude are neglected.

By integrating equation 6, we get

$$p = f(x) \qquad \text{for steady flow}$$

and $$p = f(x, t) \qquad \text{for unsteady flow}$$

from which it may be concluded that the pressure p does not vary in the y direction or in other words the pressure remains constant across the boundary layer. Hence the partial derivative $(\partial p/\partial x)$ in equation 5 can be replaced by the total derivative (dp/dx) and equation 5 then becomes

$$u\frac{\partial u}{\partial x}+v\frac{\partial u}{\partial y} = -\frac{1}{\rho}\frac{dp}{dx}+\nu\frac{\partial^2 u}{\partial y^2} \qquad \text{...(7)}$$

Since outside the boundary layer the fluid may be treated as inviscid (or non-viscous), the Euler's equations of motion may be applied according to which

$$u\frac{\partial u}{\partial x}+v\frac{\partial u}{\partial y} = -\frac{1}{\rho}\frac{\partial p}{\partial x}$$

In the region outside the boundary layer since v = 0 and u = V the free stream or ambient velocity of the approaching stream, by substituting these values in the above equation, it becomes

$$V\frac{\partial V}{\partial x} = -\frac{1}{\rho}\frac{\partial p}{\partial x}$$

$$\text{or } V\frac{dV}{dx} = -\frac{1}{\rho}\frac{dp}{dx} \qquad ...(8)$$

as both V and p are the functions of x only. Integration of equation 8 leads to the Bernoulli's equation at any section viz.,

$$\frac{p}{\rho} + \frac{V^2}{2} = \text{constant}$$

Further introducing equation 8 in equation 7 it becomes

$$u\frac{\partial u}{\partial x} + v\frac{\partial u}{\partial y} = V\frac{dV}{dx} + \nu\frac{\partial^2 u}{\partial y^2} \qquad ... (9)$$

MOMENTUM INTEGRAL EQUATION OF THE BOUNDARY LAYER

The boundary layer equations obtained in the preceding sections are non linear partial differential equations and hence difficult to solve. As such approximate method for the solution of these equations have been developed for the case for which either the exact solutions cannot be obtained or the same may involve a lot of computations. One such approximate method developed by therefore von Karman is based on the application of the moment equation to the boundary layer and the corresponding equation derived by him is known as momentum integral equation of the boundary layer or von karman's integral equation. The momentum integral equation of the boundary layer expresses the relation that must exist between the overall rate of flux of momentum across a section of the boundary layer, the shear stress at the boundary surface and the pressure gradient in the direction of flow. For steady, two-dimensional flow in the boundary layer along a flat plate this equation may be derived as indicated below.

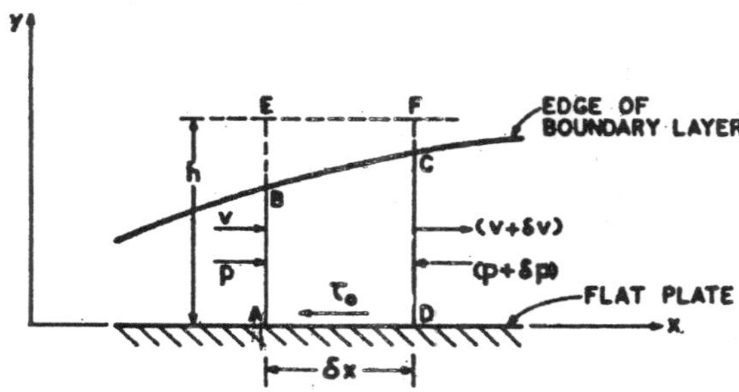

Fig. 5.4 : Definition sketch for the derivation of momentum integral equation.

Consider an element of the boundary layer formed along a flat plate, between sections AB and CD, at a small distance dx apart as shown in Fig. 4. The boundary layer is of thickness δ and its outer edge is represented by BC. Let AB and CD be extended up to E and F where AE = DF = h, such that h is slightly greater than the local boundary layer thickness δ. EEFD may be taken as a suitable control volume for this analysis.

For unit width in the direction perpendicular to the plane of the paper, the rate of mass flow across AE into AEFD

$$= \int_0^h \rho v dy$$

and the corresponding rate of mass flow across DF out of Aefd

$$= \int_0^h \rho v\, dy + \frac{d}{dx}\left[\int_0^h \rho v dy\right]\delta x$$

neglecting the terms of higher power of δx.

Hence the net rate of mass flow across DF and AE, out of aefd

$$= \frac{d}{dx}\left[\int_0^h \rho v dy\right]\delta x \qquad \text{...(i)}$$

If v_h denotes the mean value of velocity in the y direction at height h above the surface over the length EF then the rate of mass flow out of AEFD across EF

$$= \rho v_h\ \delta x$$

But continuity of mass requires that there is no net rate of change of mass inside aefd and hence

$$\rho v_h = -\frac{d}{dx}\left[\int_0^h \rho v\, dy\right] \qquad \text{...(ii)}$$

Now for the fluid in AEFD, the balance of rate of change iof momentum in the x direction and components of applied force in that direction is considered.

The rate of transport of momentum in the x direction across DF minus the rate of transport of momentum in the x direction across AE is

$$\frac{d}{dx}\left[\int_0^h \rho v^{2} dy\right]\delta x$$

The rate of transport of momentum in the x direction in the x direction across EF out of AEFD is

$$\rho v_h V\delta x = -V\frac{d}{dx}\left[\int_0^h \rho v dy\right]\delta x$$

from equation (ii), where V represents the velocity (in the x direction) of the main stream outside the boundary layer at section AB.

The force in the x direction on aefd due to the pressure

$$= -h\delta p = -h\frac{dp}{dx}\delta x$$

and that due to the shear stress at the boundary surface

$$= -t_0\delta x$$

where τ_0 is the mean shear stress over the length AD.

Thus equating the net increase in the rate of transport of momentum to the sum of the forces acting in the x direction, we have

$$\frac{d}{dx}\left[\int_0^h \rho v^2 dy\right]\delta x - V\frac{d}{dx}\left[\int_0^h \rho v dy\right]\delta x$$

$$= h\frac{dp}{dx}\delta x - \tau_0\delta x$$

Dividing both the sides of the above equation by δx and taking the δx → 0, we get

$$\frac{d}{dx}\left[\int_0^h \rho v^2 dy\right]\delta x - V\frac{d}{dx}\left[\int_0^h \rho v dy\right]$$

$$= g\ r\ V\frac{dV}{dx} - \tau_0$$

Further the above equation can be written in the following form

$$\frac{d}{dx}\left[\int_0^h \rho v(v-V)dy\right] + \frac{dV}{dx}\left[\int_0^h \rho v dy\right] = h\rho V\frac{dV}{dx} - \tau_0$$

$$\text{or}\quad \frac{d}{dx}\left[\int_0^h \rho v(V-v)dy\right] + \frac{dV}{dx}\left[\int_0^h \rho(V-v)dy\right] = \tau_0$$

Since (V – v) becomes zero at the edge of the boundary layer the upper limit to both integrals in the above equation may be changed to ∞. Then from the equations 1and 2 defining the displacement thickness δ* and momentum thickness θ the above equation simplifies to

$$\frac{d}{dx}(\rho V^2\theta) + \frac{dV}{dx}(\rho V\delta^*) = \tau_0$$

$$\text{or } \frac{d}{dx}(V^2\theta) + \frac{dV}{dx}(V\delta^*) = \frac{\tau_0}{\rho} \qquad \text{...(11)}$$

Equation 11 is the *monentum integral equation* of the boundary layer which forms the basis for approximate methods of solving boundary layer problems. Since in the derivation of this equation no assumption has been made regarding the nature of flow in the boundary layer, it is applied to both laminar as well as turbulent boundary layers. Further it may be noted that if for a flat plate at zero incidence $(\partial p/\partial x) = 0$ then from equation 8, $(\partial V/\partial x)$ or $(dV/dx) = 0$ and hence equation 11reduces to

$$\frac{\tau_0}{\rho} = \frac{d}{dx}(V^2\theta)$$

$$\text{or } \frac{\tau_0}{\rho V^2} = \frac{d\theta}{dx} \qquad \text{...(12)}$$

The momentum integral equation 11 may also be derived by integration of boundary layer equation 9.

LAMINAR BOUNDARY LAYER

Inside the boundary layer since the viscous forces are predominant, it is reasonable to assume that the inertial and viscous forces are of the same order of magnitude in a laminar boundary layer. The inertial forces per unit volume is given by $(\rho v\partial v/\partial x)$ which is proportional to $(\rho V^2/x)$ for the case of a flat plate. Similarly the viscous force per unit volume is $(\partial\tau/\partial y)$ which for laminar flow becomes $(\partial/\partial y)\,(\mu\partial v/\partial y)$ *i.e.*, $(\mu\partial^2 v/\partial y^2)$. In this boundary layer since $((\partial v/\partial y) \sim (V/\delta)$, hence $(\partial\tau/\partial y) \sim (\mu V/\delta^2)$. Thus if these forces are proportional to each other, then

$$r\frac{V^2}{x} = k'\,\mu\,\frac{V}{\delta^2}$$

$$\text{or } \frac{\delta^2}{x^2} = \frac{k'\mu}{\rho Vx} = \frac{k'}{Re_x}$$

$$\therefore \frac{\delta}{x} = \frac{k}{\sqrt{Re_x}}; \delta = k\sqrt{\frac{xv}{V}} \qquad \text{...(13)}$$

in which k′ and k are constants. By exact analytical solution of the boundary layer equations Blasius has obtained the value of the constant k as 5. Thus

$$\frac{\delta}{x} = \frac{5}{\sqrt{Re_x}}; \text{or } \delta = 5\sqrt{\frac{xv}{V}} \qquad ...(14)$$

Further the expression for the shear stress t0 can also be obtained as indicated below.

Since in the boundary layer

$$\left(\frac{\partial v}{\partial u}\right)_{y=0} \sim x\frac{V}{\delta}; \tau_0 = \mu\left(\frac{\partial v}{\partial y}\right)_{y=0} \sim \mu\frac{V}{\delta}$$

Introducing the value of δ from equation 13 in the above expression it becomes

$$\tau_0 = \text{constant}\sqrt{\frac{V^3\rho\mu}{x}} \qquad ...(15)$$

Equation 15 can also be expressed as

$$\frac{\tau_0}{\frac{\rho V^2}{2}} = c_f = \frac{\text{constant}}{\sqrt{\frac{\rho V_x}{\mu}}} = \frac{\text{constant}}{\sqrt{Re_x}} \qquad ...(16)$$

where the coefficient c_f is known as local drag coefficient. In equation 16 the value of the constant of proportionality has been obtained by Blasius by exact analytical solution of the boundary layer equations as 0.664. Thus equation 16 becomes

$$\frac{\tau_0}{\frac{\rho V^2}{2}} = \frac{0.664}{\sqrt{Re_x}} \qquad ...(17)$$

Further the total horizontal force FD (or skin friction drag) acting on one side of the plate on which laminar boundary layer is developed can be obtained as

$$FD = \int_0^L \tau_0 B dx = (0.664\rho V^{3/2} v^{1/2} L^{1/2} B)$$

in which B is the width of the plate and L is the length of the plate.

The average drag coefficient Cf may be obtained as

$$Cf = \frac{(F_D / BL)}{(\rho V^2 / 2)} = \frac{1.328}{\sqrt{R_{eL}}} \qquad ...(18)$$

in which $ReL = \frac{\rho VL}{\mu}$

Further from the exact analytical solution of the boundary layer equations by Blasius the following expressions for the displacement thickness and the momentum thickness have been obtained.

$$\frac{\delta^*}{x} = \frac{1.729}{\sqrt{Re_x}} \qquad ...(19)$$

and $$\frac{\theta}{x} = \frac{0.664}{\sqrt{Re_x}} \qquad ...(20)$$

The above noted equations for the laminar boundary layer along a flat plate can also be derived by using equation 12. For this it is essential to assume a suitable velocity distribution for the boundary layer. Thus as an example if it is assumed that the velocity distribution across a section of the boundary layer is linear with y up to the edge of the boundary layer, then we have

$$\frac{v}{V} = \frac{y}{\delta} \text{ for } y \le \delta$$

and the boundary conditions are at y = 0, v = 0; and y = d, v = V, where V is the velocity of the approaching stream outside the boundary layer.

Thus $$\tau_0 = \mu\left(\frac{\delta v}{\delta y}\right)_{y=0} = \frac{\mu V}{\delta}$$

$$d^* = \int_0^\delta \left(1 - \frac{v}{V}\right) dy = \int_0^\delta \left(1 - \frac{y}{\delta}\right) dy = \frac{\delta}{2}$$

and $$\theta = \int_0^\delta \frac{v}{V}\left(1 - \frac{v}{V}\right) dy = \int_0^\delta \frac{y}{\delta}\left(1 - \frac{y}{\delta}\right) dy = \frac{\delta}{6}$$

Substituting these values in equation 12 yields

$$\frac{1}{6}\frac{d\delta}{dx} = \frac{\mu}{\rho V \delta} = \frac{\nu}{V\delta}$$

or $$\frac{d}{dx}(\delta^2) = \frac{12\nu}{V}$$

Integrating the above expression, we get

$$\delta^2 = \frac{12\nu x}{V} + C$$

where C is constant of integration.

Since at x = 0, δ = 0, hence C = 0. Thus

$$\delta^2 = \frac{12\nu x}{V}$$

or $\frac{\delta^2}{x^2} = \frac{12\nu}{Vx} = \frac{12}{Re_x}$

$$\therefore \quad \frac{\delta}{x} = \sqrt{\frac{12}{Re_x}} = \frac{3.464}{\sqrt{Re_x}} \qquad ...(21)$$

where $Re_x = \frac{Vx}{\nu}$

Further friction drag coefficient c_f may be obtained as

$$c_f = \frac{\tau_0}{(\rho V^2/2)} = \frac{2\mu}{\rho V \delta} = \frac{0.5774}{\sqrt{Re_x}} \qquad ...(22)$$

Skin friction drag F_D on one side of the plate having laminar boundary layer is obtained as

$$F_D = \int_0^L \tau_0 \, B dx = (0.577\ 4\ \rho V^{3/2}\ \nu^{1/2}\ L^{1/2} B)$$

where B and L are width and the length of the plate respectively.

The average drag coefficient Cf may be obtained as

$$Cf = \frac{(F_D/BL)}{(\rho V^2/2)} = \frac{1.1548}{\sqrt{Re_L}} \qquad ...(23)$$

in which $Re_L = \frac{VL}{\nu}$

The momentum thickness is given by

$$\frac{\theta}{x} = \frac{\delta}{6x} = \frac{0.5774}{\sqrt{Re_x}} \qquad ...(24)$$

and the displacement thickness is given by

$$\frac{\delta^*}{x} = \frac{\delta}{2x} = \frac{1.732}{\sqrt{Re_x}} \qquad ...(25)$$

It may thus be noted that the results obtained by assuming a linear velocity distribution in the boundary layer, are fairly close to the results obtained by Blasius by exact solution of the boundary layer equations. However, by assuming certain other velocity distribution in the boundary layer better agreement with the exact solution can be obtained. Moreover, in laminar boundary layer the flow exhibits the characteristics of laminar flow, and hence the velocity distribution at any section in the laminar boundary layer will follow parabolic law.

TURBULENT BOUNDARY LAYER

Turbulent boundary layers are usually thicker than laminar ones. Further as a result of intermingling of fluid particles between different layers of the fluid in a turbulent boundary layer the velocity distribution in a turbulent boundary layer is much more uniform than that in a laminar boundary layer. However, in a turbulent boundary layer, near the boundary large velocity gradient $(\partial v/\partial y)$ is steeper in a turbulent boundary layer than in a laminar boundary layer. The velocity distribution in a turbulent boundary layer follows a logarithmic law i.e. ~ log y, which can also be represented by a power law of the type

$$\frac{v}{V} = \left(\frac{y}{\delta}\right)^n \qquad ...(26)$$

The value of the exponent n is approximately (1/7) for moderate Reynolds number ($Vx/\nu < 10^7$ for a flat plate) and decreases somewhat with increasing Reynolds number. Although equation 26 satisfactorily describes the velocity distribution for most of the region of turbulent boundary layer, it cannot apply at the boundary itself because $(\partial v/\partial y) = (1/7)\, V\delta^{-1/7}\, y^{-6/7} = \infty$ when $y = 0$. However, immediately adjacent to the boundary there is laminar sublayer, which is so thin that its velocity distribution profile may be taken as linear (instead of parabolic) and tangential to the 'seventh root' profile at the point where the laminar sublayer merges with the turbulent part of the boundary layer.

The expressions for d, cf and Cf for the turbulent boundary layer are as noted below :

$$\frac{\delta}{x} = \frac{0.376}{(Re_x)^{1/5}} \qquad ...(27)$$

$$c_f = \frac{0.059}{(Re_x)^{1/5}} \qquad ...(28)$$

$$C_f = \frac{0.074}{(Re_L)^{1/5}} \qquad ...(29)$$

Equation 29 is applicable only for values of Reynolds number ReL ranging from 5×10^5 to 10^7. For values of Re_L ranging between 10^7 to 10^9, H. Schlichting assumed a logarithmic velocity distribution for the flow in the boundary layer and obtained the semi-empirical relation

$$C_f = \frac{0.455}{(\log_{10} Re_L)^{2.58}} \qquad ...(30)$$

A comparison of equations 13 and 27 indicates that in the case of laminar boundary layer the boundary layer thickness d increases as $x^{1/2}$, but in the case of turbulent boundary layer δ increases as $x^{4/5}$. Hence the thickness of the turbulent boundary layer increases much faster than that of laminar boundary layer.

Equation 18 for the average drag coefficient is applicable only if the boundary layer is laminar for the entire length of the plate, that is when the value of Re_L is less than 5×10^5. Similarly equation 29 (or 30) is applicable if , the boundary layer is turbulent for the entire length of the plate, and the value of Re_L is within the range of 5×10^5 to 10^7 (or 10^7 to 10^9). However, when the plate is of such a length that for some distance from the leading edge of the plate the boundary layer is laminar and it becomes turbulent for the remaining portion of the plate, then combined relations are required. Hence for such cases where the plate is covered with both laminar and turbulent boundary layers, equation 31 as proposed by Prandtl may be used to compute the average drag coefficient C_f

$$C_f = \frac{0.074}{{Re_L}^{1/5}} - \frac{A}{Re_L} \qquad \text{...(31)}$$

The constant A in equation 31 depends on the value of the Reynolds number Re_x at which the laminar boundary layer becomes turbulent. The values of A for various values of critical Reynolds number are as given below :

Critical	3×10^5	5×10^5	10^6	3×10^6
Constant A	1050	1700	3300	8700

It may, however, be mentioned that since for most of the cases the value of critical Re_x may be taken as 5×10^5, equation 31 then becomes

$$C_f = \frac{0.074}{Re_L^{1/5}} - \frac{1700}{Re_L} \qquad \text{...(31a)}$$

Further equation 31 is applicable for values of Re_L up to 10^7. But for higher values of ReL ranging from 107 to 109 the following equation can be used.

$$C_f = \frac{0.455}{(10g_{10}\ Re_L)^{2.58}} - \frac{A}{Re_L} \qquad \text{...(32)}$$

in which value of the constant A again depends on the critical ReL and is the same as given in the above table. Equation 32 is known as *Prandtl-Schilchting Equation.* It may, however, be mentioned that equation 32 is valid in the entire range of Re_L from 5×10^5 upto 10^9, and it agrees with the equatio31 upto $Re_L = 10^7$.

LAMINAR SUBLAYER

If the plate is very smooth, even in the zone of turbulent boundary layer, there exists a very thin layer immediately adjacent to the boundary, in which the flow is laminar. This thin layer is commonly known as *laminar sublayer*, and its thickness is represented by δ'. As stated earlier the exponential as well as the logarithmic velocity distribution cannot apply at the boundary itself, although these satisfactorily describe the velocity distribution for most of the region of turbulent boundary layer. This discrepancy may, however, be explained by the existence of laminar sublayer. Since the flow in the laminar sublayer is laminar, it will follow a parabolic velocity distribution law, which can be approximately taken to be linear as δ' is very small, and hence exponential or logarithmic velocity distribution law cannot be, applied within the laminar sublayer. Thus in this case there will be a gradual change from parabolic velocity distribution very close to the boundary to exponential or logarithmic velocity distribution and the intersection of the parabolic and exponential or logarithmic velocity distribution curves may be arbitrarily taken as the limit of the laminar sublayer. Nikuradse's experimental studies have shown that

$$\delta' = \frac{11.6\nu}{\sqrt{(\tau_0/\rho)}} = \frac{11.6\nu}{V^*} \qquad ...(33)$$

in which $V^* = \sqrt{(\tau_0/\rho)}$ is known as shear or friction velocity. As indicated in Chapter the relative magnitudes of the thickness of laminar sublayer δ' and the average height of roughness projections k facilitates the classification of the boundaries as hydrodynamically smooth and rough.

BOUNDARY LAYER ON ROUGH SURFACES

In the preceding sections the development of boundary layer along smooth plates has been considered. However, in most practical applications connected with the boundary layer development on flat plates such as hips, airplane wings, turbine blades etc., the surfaces cannot be considered smooth. As such the flow past a rough plate is of much practical interest. For a rough plate if k is the average height of roughness projections of the surface of the plate and δ is the thickness of the boundary layer, then the relatie roughness (k/δ) is a significant parameter indicating the behaviour of the boundary surface. For k remaining constant, (k/δ) decreases along the plate because d increases in the downstream direction. As a result the front portion of the plate will behave differently from its year portion as far as the influence of roughness on drag is concerned. If it is assumed that the boundary layer is turbulent from the leading edge of the plate, the front portion of the plate will act as hydrodynamically rough, followed by a transition region and the downstream portion of the plate will be hydrodynamically smooth if the plate is sufficiently long. The limits between these three regimes are determined by the value of a dimensionless roughness parameter $(V_*/k_s/\nu)$ as indicated below :

Hydrodynamically smooth : $\dfrac{V_* k_s}{\nu} < 5$

Transition : $5 < \dfrac{V_* k_s}{\nu} < 70$

Hydrodynamically rough : $\dfrac{V_* k_s}{\nu} > 70$

(or completely rough)

in which k_s is equivalent sand grain roughness defined as that value of the roughness which would offer the same resistance to the flow past the plate as that due to the actual roughness on the surface of the plate. Although the above noted limits have been determined for flow through pipes by Nikuradse on the basis of the results of his experiments on the pipes made artificially rough by gluing sand grains of uniform size on the inner surface of the pipe, but the same may also be applied to the flow past a plate. The introduction of the equivalent sand grain roughness k_s instead of the average height of the roughness projections k is due to the fact that as compared to k, k_s is a more representative parameter of the actual roughness. This is so because usually the actual roughness projections on the surface of the plate are of various heights and they do not have any uniform pattern and hence such roughnesses cannot be truely represented in terms of the average height of the roughness projections. Moreover, the consideration of k_s also facilitates the use of Nikuradse's results, which are in terms of the sand grain roughness, in the analysis of flow past rough plates.

In the completely rough regime the local drag coefficient c_f and the average drag coefficient C_f are given by the following expressions

$$c_f = \left[2.87 + 1.58\log_{10}\left(\frac{x}{k_s}\right)\right]^{-2.5} \qquad ...(34)$$

$$C_f = \left[1.89 + 1.62\log_{10}\left(\frac{L}{k_s}\right)\right]^{-2.5} \qquad ...(35)$$

which are valid for $10^2 < \left(\dfrac{L}{k_s}\right) < 10_{\epsilon}$.

SEPARATION OF BOUNDARY LAYER

As mentioned in section 3 the boundary thickness is considerably affected by the pressure gradient in the direction of flow. If the pressure gradient ($\partial p/\partial x$) is zero, then the boundary layer continues to grow in thickness along a flat plate. With the decreasing pressure in the direction of flow *i.e.* with negative pressure gradient, the boundary layer tends to be reduced in thickens rapidly. The adverse pressure gradient plus the boundary shear decreases the momentum in the boundary layer and if the both act over a sufficient distance they cause the fluid in the boundary lahyer to come to rest i.e., the retarded fluid particles, cannot, in general penetrate too far into the region of increased pressure owing to their small kinetic energy. Thus, the boundary layer is deflected sideways from the boundary, separates from it and moves into the main stream. This phenomenon is called *separation.*

Consider flow of fluid over a curved surface as shown in Fig. 5. As the fluid flows round the surface it is accelerated over the left hand section untill at point C the velocity just outside the boundary layer is a maximum. At this section the pressure is minimum as shown by the graph below the surface. Thus, from A to C the pressure gradient ($\partial p/\partial x$) is negative and the net pressure force on an element of fluid in the boundary layer is in the direction of flow, which counteracts to some extent the slowing down effect of the boundary on the flowing fluid and thus the rate at which the boundary layer thickens is less than that for a flat plate with zero pressure gradient (at a corresponding value of Re_x).

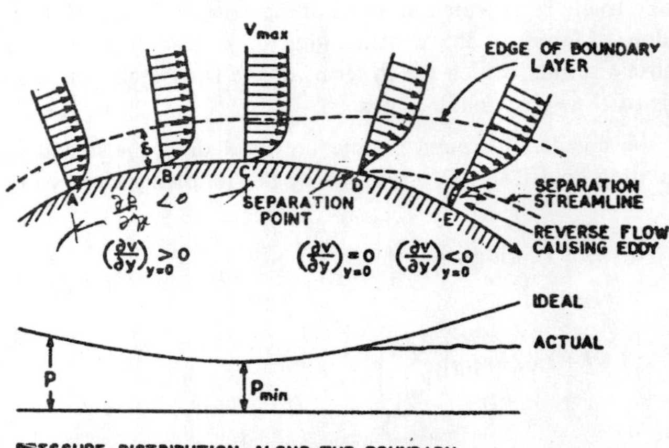

Fig. 5.5 : Separation of boundary layer.

However, beyond C the pressure increases, and hence the net pressure force on an element of fluid in the boundary layer opposes the forward flow. Thus at a certain distance on the downstream of point C the fluid near the boundary surface is soon brought to a standstill. The value of the velocity gradient $(\partial v/\partial y)$ at the boundary surface is then zero as at point D. The fluid is no longer able to follow the contour of the curved surface and it separates from it. The separation of the flowing fluid from the boundary first occurs at a point where $(\partial v/\partial y)$ at the boundary becomes zero and this point is known as *separation point*. On downstream of the separation point D, a further retardation of the fluid close to the boundary can even have a reverse or back flow as at point E shown in Fig. 5. If all the points below which a reverse flow occurs are joined by a smooth curve, a line dividing the forward and reverse flows is obtained which is known as separation streamline. In a region in between the boundary surface and the separation streamline, as a result of the reverse flow, large irregular eddies are formed in which much energy is dissipated as heat. This region of disturbed fluid usually extends for some distance on the downstream. Since the energy of the eddies is dissipated as heat, the pressure downstream remains approximately the same as at the separation point.

Separation occurs with both laminar and turbulent boundary layers, but laminar boundary layer is more susceptible to earlier separation than turbulent boundary layer. This is so because in a laminar boundary layer the increase of velocity with distance from the boundary surface is less rapid, and the adverse pressure gradient can more rapidly halt the slow moving fluid close to the boundary surface. On the other hand in a turbulent boundary layer the velocity distribution is much more uniform than in a laminar boundary layer because of intense lateral mixing. As a result relatively higher velocity prevails within a turbulent boundary layer, which reduces tendency of separation.

Separation of the boundary layer greatly affects the flow as a whole. In particular the formation of a wake zone of disturbed fluid on the downstream, in which the pressure is approximately constant and much less than that on the upstream, gives rise to boundary forces. As indicated in Chapter, the net force in the direction of flow caused by such pressure difference is known as pressure drag or form drage, the magnitude of which depends on the form or shape of the boundary and the resulting flow pattern obtained due to separation. Further the flow in a divergent passage or diffuser is another example in which the separation of the flow may be caused due to adverse pressure gradient prevailing there unless the angle of divergence is very small.

Since the separation of the boundary layer gives rise to additional resistance to flow, attempts should be made to avoid separation by some means. One of these methods may be by developing such boundary shapes for which the separation is as far as possible eliminated. The flow past a streamlined object such as airfoil is one such example. Apart from this the separation may also be avoided by adopting suitable method of controlling the boundary layer which are described in the next section.

METHODS OF CONTROLLING THE BOUNDARY LAYER

Several methods of controlling the boundary layer have been developed which are briefly described below.

(a) Motion of Solid Boundary

The formation of the boundary layer is due to the difference between the velocity of the flowing fluid and that of the solid boundary. As such it is possible to eliminate the formation of a boundary layer by causing the solid boundary to move with the flowing fluid. Such a motion of the boundary may be achieved in the simplest ways by rotating a circular cylinder lying in a stream of fluid [See Fig. 5.6 (a)], so that on the upper side of the cylinder, where the fluid as well as the cylinder move in the same direction, the boundary layer does not form and hence the separation is completely eliminated. However, on the lower side of the cylinder, where the fluid motion is opposite to that of the cylinder, separation would occur.

(b) Acceleration of the Fluid in the Boundary Layer

This method consists of supplying additional energy to the particles of fluid which are being retarded in the boundary layer.

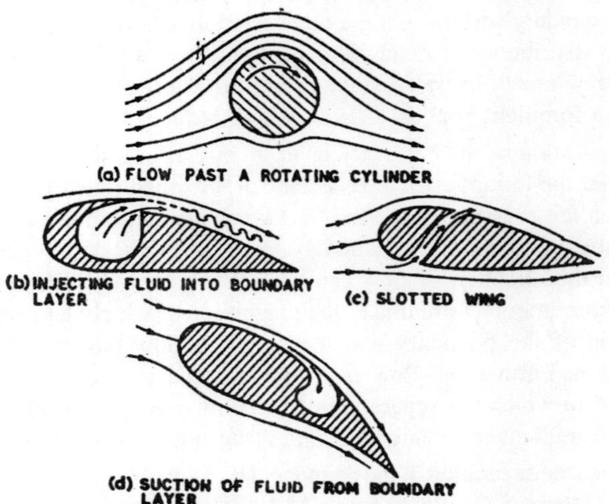

Fig. 5.6 : Methods of controlling the boundary layer.

This may be achieved either by injecting fluid into the region of boundary layer from the interior of the body with the help of some suitable device as shown in Fig. 5.6(b); or by diverting a portion of the fluid of the main stream from the region of high pressure to the retarded region of boundary layer through a slot provided in the body as in the case of the slotted wing shown in Fig. 5.6(c). However, a disadvantage of this method is that if the fluid is injected into a laminar boundary layer, it undergoes a transition to turbulent boundary layer which results in an increased skin friction drag.

(c) Suction of the Fluid from the Boundary Layer

In this method the slow moving fluid in the boundary layer is removed by suction through slots or through a porous surface as shown in Fig. 5.6(d), so that on the downstream of the point of suction a new boundary layer starts developing which is able to withstand an adverse pressure gradient and hence separation is prevented. Moreover, the suction of the fluid from the boundary layer also greatly delays its transition from laminar to turbulent due to which skin friction drag is reduced.

(d) Streamlining of Body Shapes

By the use of suitably shaped bodies the point of transition of the boundary layer from laminar to turbulent can be moved downstream which results in the reduction of the skin friction drag. Furthermore by streamlining of body shapes the separation may be eliminated.

SOLVED EXAMPLES

Example 1: *The velocity distribution in the boundary layer in given as*

$$\frac{v}{V} = \frac{3}{2}\eta - \frac{1}{2}\eta^2$$

in which $\eta = (y/\delta)$. *Compute* (δ^*/δ) *and* (θ/δ).

Solution: $\delta^* = \int_0^\infty \left(1 - \frac{v}{V}\right) dy$

or $\delta^* = \int_0^\delta \left(1 - \frac{v}{V}\right) dy + \int_0^\infty \left(1 - \frac{v}{V}\right) dy$

But outside the boundary layer (v/V) = 1, and hence

$$d^* = \int_0^\delta \left(1 - \frac{v}{V}\right) dy$$

$$\frac{v}{V} = \frac{3}{2}\eta - \frac{1}{2}\eta^2 \text{ and } dy = \delta d\eta$$

Thus $\delta^* = \delta \int_0^1 \left(1 - \frac{3}{2}\eta + \frac{1}{2}\eta^2\right) d\eta = \frac{5}{12}\delta$

$\therefore \frac{\delta^*}{\delta} = \frac{5}{12}$

Similarly $\theta = \int_0^\delta \frac{v}{V}\left(1 - \frac{v}{V}\right) dy$

or $\theta = \delta \int_0^1 \left(\frac{3}{2}\eta - \frac{1}{2}\eta^2\right)\left(1 - \frac{3}{2}\eta + \frac{1}{2}\eta^2\right) d\eta$

$= \frac{19}{120}\delta$

$\frac{\theta}{\delta} = \frac{19}{120}$

Example 2: *Given that a boundary layer at zero pressure gradient over a flat plate is described by the velocity profile*

$$\frac{v}{V} = \sin\left(\frac{\pi}{2}\frac{y}{\delta}\right)$$

determine the momentum correction coefficient (or factor) and the energy correction coefficient (or factor).

Also show that boundary layer thickness δ, wall shear τ_o and coefficient of drag C_D are given by

$$\delta = \frac{4.795x}{\sqrt{Re_x}};\ \tau_o = \frac{0.328\rho V_0^2}{\sqrt{Re_x}};\ C_D = \frac{1.312}{\sqrt{Re_L}}$$

where symbols have their usual meaning.

Solution: Momentum correction coefficient (or factor) is given by

$$\beta = \frac{1}{AV^2}\int_A v^2 dA$$

Mean velocity $V = \frac{Q}{A} = \frac{\int_A v dA}{A}$

Considering unit width of the plate $dA = 1 \times dy$, and $A = (1 \times \delta)$. Thus

$$V = \frac{\int_0^\delta V_0 \sin\left(\frac{\pi}{2}\frac{y}{\delta}\right)(1 \times dy)}{(1 \times \delta)}$$

$$\text{or } V = \frac{V_0}{\delta}\int_0^{\delta}\sin\left(\frac{\pi}{2}\frac{y}{\delta}\right)dy$$

$$\text{or } V = \frac{2V_0}{\pi}$$

$$\text{Thus } \beta = \frac{\pi^2}{4V_0^2}\times\frac{1}{(1\times\delta)}\int_0^{\delta}\left[V_0\sin\left(\frac{\pi}{2}\frac{y}{\delta}\right)\right]^2(1\times dy)$$

$$\text{or } \beta = \frac{\pi^2}{4\delta}\int_0^{\delta}\sin^2\left(\frac{\pi}{2}\frac{y}{\delta}\right)dy$$

$$\text{or } \beta = \frac{\pi^2}{8} = 1.234$$

Energy correction coefficient (or factor) is given as

$$\alpha = \frac{1}{AV^3}\int_A v^3 dA$$

$$\text{or } \alpha = \frac{\pi^3}{8V_0^3}\times\frac{1}{(1\times\delta)}\int_0^{\delta}\left[V_0\sin\left(\frac{\pi}{2}\frac{y}{\delta}\right)\right]^3(1\times dy)$$

$$\text{or } \alpha = \frac{\pi^3}{8\delta}\int_0^{\delta}\sin^3\left(\frac{\pi}{2}\frac{y}{\delta}\right)dy$$

$$\text{or } \alpha = \frac{\pi^2}{6} = 1.645$$

From equation 12 we have

$$\frac{\tau_0}{\rho V_0^2} = \frac{d\theta}{dx} \quad \text{...(i)}$$

At the boundary

$$\tau_0 = \mu\left(\frac{\delta v}{\delta y}\right)_{y=0}$$

$$\left(\frac{\delta v}{\delta y}\right)_{y=0} \frac{\pi}{2}\frac{V_0}{\delta}\left[\cos\left(\frac{\pi}{2}\frac{y}{\delta}\right)\right]_{y=0} = \frac{\pi}{2}\frac{V_0}{\delta}$$

$$\tau_0 = \mu\frac{\pi}{2}\frac{V_0}{\delta} \quad \text{...(ii)}$$

From equation 2

$$\theta = \int_0^{\infty} \frac{v}{V_0}\left(1-\frac{v}{V}\right)dy$$

Outside the boundary layer since $v = V_0$,

$$\theta = \int_0^{\infty} \frac{v}{V_0}\left(1-\frac{v}{V}\right)dy$$

$$\text{or } \theta = \int_0^{\infty} \sin\left(\frac{\pi}{2}\frac{y}{\delta}\right)\left[1-\sin\left(\frac{\pi}{2}\frac{y}{\delta}\right)\right]dy$$

$$\text{or } \theta = \frac{2\delta}{\pi}\left(1-\frac{\pi}{4}\right)$$

By substituting the values of τ_0 and θ in equation (i) we get

$$\frac{1}{\rho V_0^2}\left(\mu\frac{\pi}{2}\frac{V_0}{\delta}\right) = \frac{2}{\pi}\left(1-\frac{\pi}{4}\right)\frac{d\delta}{dx}$$

$$\text{or } \delta\, d\, \delta = \frac{11.4976\mu dx}{\rho V_0} \qquad \text{...(iii)}$$

Since δ is a function of x only, integration of equation (iii) gives

$$\frac{\delta^2}{2} = \frac{11.4976\mu x}{\rho V_0} + \text{const.}$$

$$\text{or } \delta = 4.795\sqrt{\frac{\mu x}{\rho V_0}}$$

$$\text{or } \delta = \frac{4.795x}{\sqrt{Re_x}}$$

$$\text{where } Re_x = \frac{\rho V_0^x}{\mu}$$

Substituting the value of δ in equation (ii) yields

$$\tau_0 = \mu\,\frac{\pi}{2}\frac{V_0}{4.795}\sqrt{\frac{\rho V_0}{\mu x}}$$

$$\text{or } \tau_0 = 0.328\sqrt{\frac{\rho\mu V_0^3}{x}} \qquad \text{...(iv)}$$

$$\text{or } \tau_0 = \frac{0.328\rho V_0^2}{\sqrt{Re_x}}$$

where $Re_x = \left(\frac{\rho V_0^x}{\mu}\right)$

The drag F_D on the side of the plate of unit width is

$$F_D = \int_0^L \tau_0 dx$$

introducing the value of τ_0 from equation (iv) we have

$$\int_0^L 0.328\sqrt{\frac{\rho\mu V_0^3}{x}}dx$$

or $F_D = 0.656\ \sqrt{\rho\mu V_0^3 L}$ ...(v)

the drag can be expressed in terms of a drag coefficient C_D times the stagnation pressure ($\rho V_0^2/2$) and the area of plate L (per unit width). Thus

$$F_D = C_D\ \frac{\rho V_0^2}{2}L$$

Introducing the value of FD from equation (v) we have

$$0.656\sqrt{\rho\mu V_0^3 L}\ = C_D\ \frac{\rho\mu V_0^3}{2}L$$

$$\text{or } C_D = \frac{1.312}{\sqrt{Re_L}}$$

$$\text{where } Re_L = \frac{\rho V_0^L}{\mu}$$

Example 3: *Air flows over a flat plate 1 m long at a velocity of 6m/s. Determine (a) the boundary layer thickness at the end of the plate, (b) shear stress at the middle of the plate, (c) total drag per unit length on the sides of the plate. Take r = 1.226 kg/m³ (0.125 msl/m³) and ν = 0.15 × 10⁻⁴ m²/s (0.15 stokes) for air.*

Solution: $Re_L = \frac{VL}{\nu} = \frac{6\times 1}{0.15\times 10^{-4}} = 4.0 \times 10^5$

Hence the boundary layer is laminar over the entire length of the plate

$$\therefore\ \delta = 5\ \sqrt{\frac{x\nu}{V}}$$

$$= 5\sqrt{\frac{1\times 0.15\times 10^{-4}}{6}} = 7.91 \times 10^{-3}\text{ m} = 7.91\text{ mm}$$

SI units

$$\tau_0 = c_f \frac{\rho V^2}{2} \text{ ; and } c_f = \frac{0.664}{\sqrt{Re_x}}$$

For the middle of the plate x = 0.5 m

$$\text{Thus } Re_x = \frac{6 \times 0.5}{0.15 \times 10^{-4}} = 2.0 \times 10^5$$

$$\text{and } c_f = \frac{0.664}{\sqrt{2.0 \times 10^5}} = 1.485 \times 10^{-3}$$

$$\therefore \tau_0 = \frac{1..485 \times 10^{-3} \times 1.226 \times (6)^2}{2} = 32.77 \times 10^{-3} \text{ N/m}^3$$

$$\text{Further } C_f = \frac{1.328}{\sqrt{Re_L}} = \frac{1.328}{\sqrt{4.0 \times 10^5}} = 2.1 \times 10^{-3}$$

$$\text{and } F_D = 2 \times BL \times C_f \frac{\rho V^2}{2}$$

$$\therefore F_D = 2 \times (1 \times 1) \times 2.1 \times 10^{-3} \frac{1.226 \times (6)^2}{2}$$

$= 92.69 \times 10^{-3}$N

Metric units

$$\tau_0 = c_f \frac{\rho V^2}{2} \text{; and } c_f = \frac{0.664}{\sqrt{Re_x}}$$

For the middle of the plate x = 0.5 m

$$\text{Thus } Re_x = \frac{6 \times 0.5}{0.15 \times 10^{-4}} = 2.0 \times 10^5$$

$$\text{and } c_f = \frac{0.664}{\sqrt{2.0 \times 10^5}} = 1.485 \times 10^{-3}$$

$$\therefore \tau_0 = \frac{1.485 \times 10^{-3} \times 0.125 \times (6)^2}{2}$$

$= 3.34 \times 10^{-3}$ kg (f)/m^2

$$\text{Further } C_f = \frac{1.328}{\sqrt{Re_L}} = \frac{1.328}{\sqrt{4.0 \times 10^5}} = 2.1 \times 10^{-3}$$

$$\text{and } F_D = 2 \times B_L \times C_f \frac{\rho V^2}{2}$$

$$\therefore\ F_D = 2 \times (1 \times 1) \times 2.1 \times 10^{-3} \times \frac{0.125 \times (6)^2}{2}$$

Example 4: *A smooth two-dimensional flat plate is exposed to a wind velocity of 100 km per hour. If laminar boundary layer exists up to a value of Rex equal to 3 × 10⁵, find the maximum distance upto which laminar boundary layer persists, and find its maximum thickness. Assume kinematic viscosity of air as 1.49 × 10⁻⁵ m²/s.*

Solution: $V = \frac{100 \times 10^3}{3600} = 27.78$ m/s

$$Re_x = \frac{Vx}{\nu}$$

$$\text{or } 3 \times 10^5 = \frac{27.78 \times x}{1.49 \times 10^{-5}}$$

$$\therefore\ x = 0.161 \text{ m}$$

$$\delta = 5\sqrt{\frac{x\nu}{V}}$$

$$= 5\sqrt{\frac{0.161 \times 1.49 \times 10^{-5}}{27.78}}$$

$$= 1.47 \times 10^{-3} \text{ m} = 1.47 \text{ mm}$$

Example 5: *Find the power required to tow lengthwise a plate 1.2 m wide and 3 m long at a velocity of 2.4 m/s in water at 23°C. Make allowance for the fact that the boundary layer will change from laminar to turbulent over the plate. ν for water at 23°C is 0.9 × 10⁻⁶ m²/s and ρ = 1000 kg/m³ (102 msl/m³).*

Solution: $Re_L = \frac{VL}{\nu} = \frac{2.4 \times 3}{0.9 \times 10^{-6}} = 8.0 \times 10^6$

$$C_f = \frac{0.074}{(Re_L)^{1/5}} - \frac{1700}{Re_L}$$

$$\text{or } C_f = \left[\frac{0.074}{(8.0 \times 10^6)^{1/5}} - \frac{1700}{8.0 \times 10^6}\right] = 2.87 \times 10^{-3}$$

The drag on the plate is given by

$$F_D = 2C_f \times BL \times \frac{\rho V^2}{2}$$

SI units

$$F_D = \left[2 \times 2.87 \times 10^{-3} \times 1.2 \times 3 \times \frac{1000 \times (2.4)^2}{2}\right]$$

$\therefore$ Power required = 59.51 × 2.4

= 143 W = 0.143 kW

Metric units

$$F_D = \left[2 \times 2.87 \times 10^{-3} \times 1.2 \times 3 \times \frac{102 \times (2.4)^2}{2}\right]$$

= 6.07 kg (f)

$$\therefore \text{ Power required} = \frac{6.07 \times 2.4}{75} = 0.194 \text{ h.p.}$$

Example 6: *Find the ratio of skin friction drag on the front two-third and rear one-third of a flat plate kept in a uniform stream at zero incidence. Assume the boundary layer to be turbulent over the entire plate.*

Solution: $C_f = \dfrac{0.074}{(Re_L)^{1/5}}$

For the front two-third portion of the plate

$$R_{eL} = \frac{V(2/3)L}{\nu}$$

$$\text{Thus } C_f = \frac{0.074}{\left[\dfrac{V(2/3)L}{\nu}\right]^{1/5}}$$

$\therefore$ Drag for the front two-third portion of the plate is

$$FD_1 = C_f \times B_L \times \frac{\rho V^2}{2}$$

$$= \frac{0.074}{\left[\dfrac{V(2/3)L}{\nu}\right]^{1/5}} \times B \times (2/3)\, L \times \frac{\rho V^2}{2}$$

Similarly drag for the entire plate is

$$F_D = \frac{0.074}{\left[\dfrac{VL}{\nu}\right]^{1/5}} \times B \times L \times \frac{\rho V^2}{2}$$

$\therefore$ Drag for the rear one-third portion of the plate is

$$F_{D2} = (F_D - F_{D1})$$

$$= \frac{0.074}{\left[\frac{VL}{\nu}\right]^{1/5}} \times B \times L \times \frac{\rho V^2}{2} \times \left[1 - \frac{2}{3}\left(\frac{3}{2}\right)^{1/5}\right]$$

Example 7: *Calculate the friction drag on a plate 0.15 m wide and 0.45 m long placed longitudinally in a stream of oil flowing with a free stream velocity of 6 m/s. Also find the thickness of the boundary layer and shear stress at the trailing edge. Sp. gr. of oil is 0.925 and its kinematic viscosity is 0.9 × 10–4 m2/s (0.9 stokes).*

Solution:

$$Re_L = \frac{6 \times 0.45}{0.9 \times 10^{-4}} = 3.0 \times 10^4$$

Hence the boundary layer is laminar over the entire length of the plate.

$$\text{Thus } C_f = \frac{1.328}{\sqrt{Re_L}} = \frac{1.328}{\sqrt{3.0 \times 10^4}} = 7.667 \times 10^{-3}$$

SI units

The friction drag for both sides of the plate is

$$F_D = 2 \times BL \times C_f \frac{\rho V^2}{2}$$

$$= 2 \times (0.15 \times 0.45) \times 7.667 \times 10^{-3} \times (0.925 \times 1000) \times \frac{(6)^2}{2}$$

$$= 17.233 \text{ N}$$

Thickness of the boundary layer at the trailing edge is given by equation 12.14 for x = L as

$$\delta = 5\sqrt{\frac{L\nu}{V}}$$

$$= 5\sqrt{\frac{0.45 \times 0.9 \times 10^{-4}}{6}} = 13 \times 10^{-3} \text{ m} = 13 \text{ mm}$$

Shear stress at the trailing edge is given by equation 17 for x = L as

$$\tau_0 = \frac{0.664 \times \rho^{1/2}}{2\sqrt{Re_L}}$$

$$= \frac{0.664 \times (0.925 \times 1000) \times (6)^2}{2 \times \sqrt{3.0 \times 10^4}} = 63.83 \text{ N/m}^2$$

Metric units

The friction drag for both sides of the plate is

$$F_D = 2 \times BL \times C_f \frac{\rho V^2}{2}$$

$$= 2 \times (0.15 \times 0.45) \times 7.667 \times 10^{-3} \times (0.925 \times 102) \times \frac{(6)^2}{2}$$

$$= 1.78 \text{ kg (f)}$$

Thickness of the boundary layer at the traílling edge is given by equation 14 for x = L as

$$\delta = 5\sqrt{\frac{L\nu}{V}}$$

$$= 5\sqrt{\frac{0.45 \times 0.9 \times 10^{-4}}{6}} = 1.3 \times 10^{-2} \text{ m} = 1.3 \text{ cm}$$

Shear stress at the traılling edge is given by equation 17 for x = L, as

$$\tau_0 = \frac{0.664 \times \rho V^2}{2\sqrt{Re_L}}$$

$$= \frac{0.664 \times (0.925 \times 102) \times (6)^2}{2 \times \sqrt{3.0 \times 10^4}} = 6.51 \text{ kg (f)/m}^2$$

Example 8: *Assuming that the velocity distribution in the boundary layer is given by*

$$\frac{v}{V} = \left(\frac{y}{\delta}\right)^{1/7}$$

calculate $\frac{\delta^*}{\delta}, \frac{\theta}{\delta}$ and $\frac{\delta_E}{\delta}$. *If at a certain section, free stream velocity V was observed to be 10 m/s and the thickness of the boundary layer as 25 mm, then calculate the energy loss per unit length due to the formation of the boundary layer. Take* ρ *= 1.226 kg/m³ (0.125 msl/m³).*

Solution: The displacement thinckness is given by equation 1 as

$$\delta^* = \int_0^\delta \left(1 - \frac{v}{V}\right) dy$$

$$= \int_0^\delta \left[1 - \left(\frac{y}{\delta}\right)^{1/7}\right] dy = \frac{\delta}{8}, \quad \frac{\delta^*}{\delta} = \frac{1}{8} = 0.125$$

The momentum thickness is given by equation 2 as

$$\theta = \int_0^{\delta} \frac{v}{V}\left(1-\frac{v}{V}\right)dy$$

$$= \int_0^{\delta}\left(\frac{y}{\delta}\right)^{1/7}\left[1-\left(\frac{y}{\delta}\right)^{1/7}\right]dy = \frac{7}{72}\delta$$

$$\therefore \frac{\theta}{\delta} = \frac{7}{72} = 0.097$$

The energy thickness is given by equation 3 as

$$\delta_E = \int_0^{\delta} \frac{v}{V}\left(1-\frac{v^2}{V^2}\right)dy$$

$$= \int_0^{\delta}\left(\frac{y}{\delta}\right)^{1/7}\left[1-\left(\frac{y}{\delta}\right)^{2/7}\right]dy = \frac{7}{40}\delta$$

$$\therefore \frac{\delta_E}{\delta} = \frac{7}{40} = 0.175$$

The loss of energy per unit length due to formation of the boundary layer is given by

$$E_L = \frac{1}{2}\rho V^3\ \delta_E$$

SI units

$$E_L = \frac{1}{2} \times 1.226 \times (10)^3 \times \frac{7}{40} \times 25 \times 10^{-3} = 2.682 \text{ N.m/s}$$

Metric units

$$EL = \frac{1}{2} \times 0.125 \times (10)^3 \times \frac{7}{40} \times 25 \times 10^{-3} = 0.273 \text{ kg (f)-m/s}$$

Example 9: *It is required to determine the frictional drag of a submarine. The length of the hull is 75 m and its surface area is 3000 m^2. The submarine is travelling at a constant speed of 5 m/s. Critical Reynolds number at which the flow in the boundary layer changes from laminar to turbulent is 5 × 105. Assuming that the boundary layer at the leading edge is laminar obtain the frictional drag and the power required to propel the submarine at 5 m/s. Take $\nu = 1 \times 10^{-6}$ m^2/s (0.01cm^2/s) and $\rho = 1000$ kg/m^3 (102 msl/m^3).*

Solution : $Re_L = \dfrac{VL}{\nu} = \dfrac{5 \times 75}{1 \times 10^{-6}} = 3.75 \times 10^8$

Since at leading edge boundary layer is laminar, it changes from laminar to turbulent on the surface of the submarine.

As critical Reynolds number = 5×10^5

$$C_f = \frac{0.074}{(Re_L)^{1/5}} - \frac{1700}{Re_L}$$

$$= \left[\frac{0.074}{(3.75 \times 10^8)^{1/5}} - \frac{1700}{3.75 \times 10^8}\right]$$

$$= 1.422 \times 10^{-3}$$

SI units

The frictional drag is

$$F_D = C_f \rho A \frac{V^2}{2}$$

$$= 1.422 \times 10^{-3} \times 1000 \times 3000 \times \frac{(5)^2}{2}$$

$$= 53.325 \times 10^3 \text{ N} = 53.325 \text{ kN}$$

Power required = $(53.325 \times 10^3 \times 5)$

Metric units

The frictional drag is

$$F_D = C_f \rho A \frac{V^2}{2}$$

$$= 1.422 \times 10^{-3} \times 102 \times 3000 \times \frac{(5)^2}{2} = 5.44 \times 10^3 \text{ kg (f)}$$

$$\text{Power required} = \frac{5.44 \times 10^3 \times 5}{75} = 362.7 \text{ h.p.}$$

EXERCISES

1. Explain the characteristics of laminar and turbulent boundary layer.
2. What are the factors affecting the boundary layer thickness ?
3. What do you understand by displacement thickness and momentum thickness?
4. For a linear velocity distribution in the boundary layer on a flat plate show that $(\delta^*/\theta) = 3$, and $(\delta_E/\delta) = 0.25$.
5. Discuss the phenomenon of separation in a diverging flow.